C, HULL

MODERN ELECTRICITY/
ELECTRONICS

MODERN ELECTRICITY/ ELECTRONICS

GARY M. MILLER

Dean of Engineering Technology
Monroe Community College

PRENTICE-HALL, INC., *Englewood Cliffs, New Jersey 07632*

Library of Congress Cataloging in Publication Data

Miller, Gary M.
  Modern electricity/electronics.

  Includes index.
  1. Electric engineering.   2.   Electronics.
I. Title
TK145.M6297   621.3   80-20241
ISBN 0-13-593160-6

Editional production supervision
and interior design by: *James M. Chege*
Manufacturing buyer: *Joyce Levatino*

10   9   8   7   6   5   4   3

PRENTICE-HALL INTERNATIONAL, INC., *London*
PRENTICE-HALL OF AUSTRALIA PTY. LIMITED, *Sydney*
PRENTICE-HALL OF CANADA, LTD., *Toronto*
PRENTICE-HALL OF INDIA PRIVATE LIMITED, *New Delhi*
PRENTICE-HALL OF JAPAN, INC., *Tokyo*
PRENTICE-HALL OF SOUTHEAST ASIA PTE. LTD., *Singapore*
WHITEHALL BOOKS LIMITED, *Wellington, New Zealand*

This book is dedicated to parents
throughout the world and especially
*my own—**Dorothy** and **Reginald Miller***

CONTENTS

PREFACE

This book provides a comprehensive study of the Electricity/Electronics field. We live in a society that has technological innovation as a primary force. Electronics has become increasingly important in that regard and to all phases of science and technology. It is the author's view that the currently available texts for this field are either overly simplistic or too detailed to be effective. It is hoped that this volume strikes the appropriate middle ground—a good balance. The mathematics base is limited to an algebraic level (with a minimal use of right angle trigonometry) and yet the topic coverage is rigorous, up-to-date and applications oriented.

Chapters 1–4 introduce the basics of electricity and magnetism. This material may be partially remedial for those who have had previous electricity studies. Chapters 5–7 provide a firm grounding in the area of discreet semi-conductors and their associated circuits with strong coverage of amplifiers and power supplies. Chapters 8 and 9 cover the important fields of linear integrated circuits (with emphasis on operational amplifiers), oscillators and communication systems. Chapters 10–12 introduce the world of digital electronics culminating with a rather complete coverage of the increasingly important microcomputer. The final Chapter provides a survey of control systems and sensors.

The questions and problems at the end of each Chapter are keyed to the appropriate section of the Chapter. For example, problem 1-7-28 indicates Chapter 1, the 28th problem and that Section 1-7 contains the essential material to solve that problem.

I invite your comments and suggestions regarding this textbook. Please address them to my attention, Monroe Community College, Rochester, N.Y. 14623.

GARY M. MILLER

MODERN ELECTRICITY/ ELECTRONICS

1

ELECTRICITY

1-1 ELECTRONICS—ELECTRONS

This book is for anyone—almost. If you are interested in learning about electricity/ electronics out of curiosity, or just for the sake of learning, or out of necessity, you are in the right spot. Perhaps you have gotten interested in the subject because of home computers, radio communications, or studies in the areas of health, science, and technology that utilize electronics. Whatever the reasons, I am prepared to guide you through an introductory, yet comprehensive look at the field. All I ask of you is some of your time and effort.

Atomic Structure

The basis for electronics is the electron. Electrons are extremely small particles that are part of an atom. An atom is basically composed of electrons, protons, and neutrons. The protons and neutrons form the nucleus (center portion) and contain almost all of the atom's mass. The electrons orbit around the nucleus in a fashion similar to that of the planets about the sun. While the electrons have roughly 1/2000 the mass of the protons and neutrons, they occupy roughly 2000 times as much space!

1

An atom having an equal number of electrons and protons is electrically *balanced* and does not possess electrical charge. Any material made up of balanced atoms will not attract or repel other balanced materials. If for some reason an atom acquires more than a normal number of electrons, it is unbalanced and is called a *negative ion*. If, on the other hand, an atom has fewer electrons than protons, it is called a *positive ion* and is also unbalanced.

Electron Displacement

Electrons can be displaced from one material to another by friction. As an example, walking across a carpet can displace electrons from the rug to your body. At this point, the rug has a deficiency of electrons and is therefore said to be positively charged. On the other hand, your body is negatively charged because it has an excess of electrons. If you now get close to a neutral object, such as a light switch, the excess electrons rush from your body with a spark through the air to the switch and a mild electric shock occurs. A *static discharge* has taken place.

An object that has either a positive or a negative charge will attract objects of the opposite charge or neutral materials. Similarly charged objects repel one another. If a comb is run rapidly through hair, the friction causes electron displacement to the comb. If small bits of paper are now brought close to the comb, they are attracted and cling to it. The paper was a neutral material that contained no charge but was, in effect, oppositely charged with respect to the comb. A close examination of this process may also show the static discharge process previously explained. The sparks may be so slight, however, as to be viewable only in a dark room. We have seen that an object with unbalanced atoms will exhibit forces on other materials and that a static discharge can occur to other materials.

Conduction

The very weak static discharge between the comb and paper in the example cited above is predictable. The paper is a poor conductor—that is, it does not easily conduct (carry) electrons. It therefore accepts a low number of electrons from the comb, and the flow of electrons from comb to paper just before they are in contact may or may not be visible as a spark. This is in contrast to the spark from your body to a metallic neutral material after walking across a carpet. In this case, a sizable spark can occur because metallic materials are good conductors (of electrons), and therefore many electrons can "jump the gap" from your body just before touching the neutral metal.

The different ability of electron conduction for different materials is determined by atomic structure. The electrons in the outermost orbit are known as *valence electrons*. If the outer orbit is more than half-filled with electrons, that element is a poor conductor. A poor conductor is known as an *insulator*. Elements with outer orbits less than half-filled make good conductors. Those elements with exactly half-filled outer orbits are known as *semiconductors*. They do not make good conductors or insulators, but do play an extremely important role in electronics—as will be shown in future chapters. Silicon and germanium are examples of semiconductor elements. Helium and argon make good insulators, whereas copper and aluminum are good conductors.

1-2 CURRENT

The preceding discussion has talked of electron movement. Whenever the majority of electrons in a conductor are going in one direction due to some external force, we say that a flow of *current* is occurring. This is to be distinguished from the random movement of electrons that occurs naturally in all conductors. This random action has no net effect in any direction and is therefore *not* a current flow. Current flow occurs most actively in good conductors, where the so-called *free electrons* are loosely held to the nucleus. It occurs in lesser amounts in the spectrum extending from conductors through semiconductors and into insulators.

Current flow, then, is a directed flow of electrons caused by an external force. Current flow is measured in *amperes*. An ampere is the flow of 6.24×10^{18} electrons per second.* That extremely large quantity of those extremely small electrons is defined as 1 *coulomb* (C) of charge.

EXAMPLE 1-1

A flow of 18.72×10^{18} electrons per second occurs in a conductor. Calculate the current flow in amperes.

Solution:

By definition, 1 ampere (A) is the flow of 6.24×10^{18} electrons per second. Therefore,

$$\frac{18.72 \times 10^{18} \text{ electrons per second}}{6.24 \times 10^{18} \text{ electrons per second per ampere}} = \underline{3 \text{ A}}$$

Current flow is usually denoted by the letter I. Thus, the solution could also be given as $I = 3$ A.

---

* Refer to Appendix A if you need an introduction to, or review of, scientific notation.

1-3 VOLTAGE

It has been stated that current is a directed flow of charge caused by some external force. An external source often used for this purpose is a battery. A *battery* is a device that produces electrical energy as a result of chemical reaction. In that respect, a battery is a *transducer*—a device that converts energy from one form to another form.

Basically, a battery is made up of two dissimilar materials that are immersed in an electrolyte. One of the surfaces acquires a surplus of electrons due to the chemical action and thereby becomes negatively charged. The other surface then has a shortage of electrons and is therefore positively charged. These two surfaces are called *electrodes*. The electrodes are brought out of the battery to terminals, where connections can be made to form an *electric circuit*.

The imbalance of electrons between the electrodes can be thought of as creating a pressure or force. This situation is analogous to that of the water pump shown in Fig. 1-1. The pump exerts a force on the water just as the battery does upon the electrons. This force is called an *electromotive force,* usually abbreviated EMF. Because EMF is measured in units of *volts,* the term "voltage" is often used instead of EMF. The term "potential difference" is also used to designate EMF. Thus, the difference

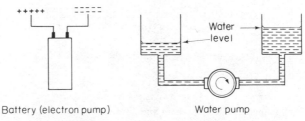

Battery (electron pump)   Water pump

FIGURE 1-1 *Battery/water pump analogy.*

4

in concentration of electrons between the electrodes is called, alternatively, potential difference, voltage, EMF, or electromotive force.

Figure 1-2 shows what is termed a *completed* or *closed circuit*. In this instance, the excess electrons made available at the battery's negative terminal are "pumped" through the wire conductor to the lamp. When the electrons reach the lamp's filament they encounter an increased level of "resistance" to the flow of current (electrons). This does not stop the electrons, but the increased resistance to their flow heats up the lamp filament sufficiently to make it "white" hot, at which point it emits light. The electrons then return to the positive terminal via another wire conductor, where further chemical reactions within the battery "replenish" the strength of the electrons, and so the cycle continues.

The simple lamp circuit shown in Fig. 1-2 illustrates several interesting energy transformations. The battery contains chemical energy which it converts to electrical energy. The electrical energy is then converted to light energy by the lamp. Since the lamp gives off heat in this process, the electrical energy is also converted into heat energy by the lamp—this is usually an undesired energy transformation in the process of creating light.

As you know, eventually the lamp dims and then goes out. The battery's supply of energy is limited—the voltage it creates gradually decreases and thereby does not have as much force. At this point, the battery can be replaced, or for certain types of batteries, *recharged*. Recharging is accomplished by forcing electrons back into the electrode until the original chemical energy is restored.

The symbol for a battery source of EMF is shown in Fig. 1-3(a). The positive terminal is shown with the longer of the two lengths of lines used to represent the battery. The shorter line represents the negative electrode—recall that the electron flow is from negative to positive or counterclockwise in Fig. 1-3(a). If a greater EMF (voltage) is necessary, the connection of two batteries as shown in Fig. 1-3(b) would be possible. An additional battery gives a total of 3 volts (V)—1.5 V from each battery. The increased "pumping pressure" causes the lamp to emit more light—and perhaps soon burn out if not able to handle an EMF of 3 V!

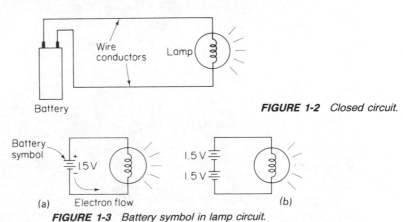

FIGURE 1-2 Closed circuit.

FIGURE 1-3 Battery symbol in lamp circuit.

The battery is a source of EMF. Another common source is the electrical generator—a device that converts the energy of motion (mechanical energy) into electrical
energy. Generators are discussed in Chapter 4.

1-4 OHM'S LAW

The example of the lamp in Section 1-3 introduced (without a definition) the concept
of *resistance*. It was stated that the lamp exhibited resistance to the flow of electrons
and was thereby heated. The most basic law of electricity, *Ohm's Law*, relates voltage,
current, and resistance. It states that voltage is equal to the product of current times
resistance. Stated as an equation,

$$V = I \times R \qquad\qquad (1\text{-}1)$$

where $V =$ voltage, in volts

         $I =$ current, in amperes

         $R =$ resistance, in ohms

The unit of resistance, R, is *ohms*, where 1 ohm is defined as the resistance that
causes 1 ampere of current flow when a voltage of 1 volt exists. The symbol for

---

EXAMPLE 1-3

Two D-cell batteries are used in a flashlight. Calculate the resistance of the
bulb when a current of 0.5 A flows.

Solution:

D-cell batteries are rated to deliver 1.5 V each. In this type of battery configuration, the voltage is additive. Thus, the two batteries would apply 3 V (2 $\times$
1.5 V) to the lamp. Using Ohm's Law,

$$V = I \times R \qquad\qquad (1\text{-}1)$$

we obtain

$$3\,\text{V} = 0.5\,\text{A} \times R$$

Algebraically rearranging yields

$$\frac{3\,\text{V}}{0.5\,\text{A}} = R$$

$$\therefore{}^{*} \quad R = \underline{6\,\Omega}$$

---

* The three dots mean *therefore*.

ohms is the Greek capital letter omega (Ω). Thus, a resistance of 10 ohms is normally written as 10 Ω.

EXAMPLE 1-4

The cigarette lighter in an automobile has a resistance of 1.2 Ω. The car's electrical system operates at a 12-V level. Determine the current "drawn" by the lighter.

Solution:

$$V = I \times R \qquad \text{(1-1)}$$

$$\therefore \quad 12\,V = I \times 1.2\,\Omega$$

$$\therefore \quad I = \frac{12\,V}{1.2\,\Omega}$$

$$= \underline{10\,A}$$

The circuit in Fig. 1-4 shows the symbol for an electrical device called a *resistor*. A resistor is a device manufactured to exhibit a specific amount of resistance. Notice in Fig. 1-4 that a 100-Ω resistor is indicated. Note also that a current flow of 0.1

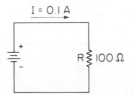

FIGURE 1-4 *Circuit with resistor.*

A is shown and that its direction is opposite to the flow of electrons. Most people assume that current flows from the positive to the negative terminal. Since this convention is prevalent, it will be used for the remainder of this book. This assumption is called *conventional current flow* and assumes current flow in the opposite direction of electron flow.

EXAMPLE 1-5

Calculate the battery voltage for the circuit shown in Fig. 1-4.

Solution:

$$V = I \times R \qquad \text{(1-1)}$$

$$= 0.1\,A \times 100\,\Omega$$

$$= \underline{10\,V}$$

1-5 RESISTANCE

Resistors come in varied sizes and shapes. An example of the most common variety is shown in Fig. 1-5(a). It is a round tubular shape that has at least three color bands as shown. The resistor color code is a means of identifying a resistor's value, as indicated in Fig. 1-5(b). The first and second bands (closest to one end) determine the first and second digits, and the third band indicates the number of zeros to follow the first two digits. The fourth band (if present) is usually either gold or silver and indicates a 5% or 10% tolerance. The absence of a fourth band denotes a 20% tolerance.

EXAMPLE 1-6

Determine the resistance of resistors with the following color bands:

    (a)   Red, red, brown, silver.

    (b)   Green, blue, orange, gold.

    (c)   Brown, black, black.

    (d)   Brown, black, green, gold.

Solution:

From the color-code information shown in Fig. 1-5:

(a)	Red	red	brown	silver
	2	2	1 zero	10% tolerance

$$\therefore \quad R = \underline{220\ \Omega \pm 10\%}$$

(b)	Green	blue	orange	gold
	5	6	3 zeros	5% tolerance

$$\therefore \quad R = \underline{56{,}000\ \Omega} \quad \text{or} \quad \underline{56\ k\Omega^* \pm 5\%}$$

(c)	Brown	black	black	
	1	0	no zeros	20% tolerance

$$\therefore \quad R = \underline{10\ \Omega \pm 20\%}$$

(d)	Brown	black	green	gold
	1	0	5 zeros	5% tolerance

$$\therefore \quad R = \underline{1{,}000{,}000\ \Omega} \quad \text{or} \quad \underline{1\ M\Omega \pm 5\%}$$

---

* Refer to Appendix A if you are not familiar with multiplier values such as kΩ.

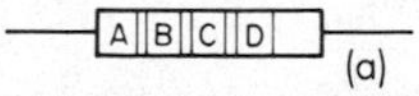

(a)

Color	1st Digit A	2nd Digit B	Multiplier C	Tolerance D
Black	0	0	1	–
Brown	1	1	10	±1%
Red	2	2	100	±2%
Orange	3	3	1,000	±3%
Yellow	4	4	10,000	±4%
Green	5	5	100,000	–
Blue	6	6	1,000,000	–
Violet	7	7	10,000,000	–
Gray	8	8	100,000,000	–
White	9	9	–	–
Gold	–	–	0.1	±5%
Silver	–	–	0.01	±10%
No color	–	–	–	±20%

(b)

FIGURE 1-5   *Resistor color code.*

EXAMPLE 1-7

A resistor is connected in a circuit. The voltage across it is measured at 13.6 V and the current through it measured at 4.3 mA. The resistor's "marked" value is orange, orange, red, gold. Determine the resistor's value and if it is within tolerance.

Solution:

Orange, orange, red, gold means 3, 3 with two zeros and 5% tolerance, so the marked value is 3300 Ω ± 5%. Thus, the permissible range of values is

$$3300 \ \Omega \pm (5\% \times 3300)$$

or

$$3300 \ \Omega \pm 165 \ \Omega$$

or

$$3465 \ \Omega \quad \text{down to} \quad 3135 \ \Omega$$

Now from the measured values:

$$V = I \times R \qquad\qquad \textbf{(1-1)}$$

$$R = \frac{V}{I}$$

$$= \frac{13.6 \text{ V}}{4.3 \text{ mA}}$$

$$= \frac{13.6 \text{ V}}{0.0043 \text{ A}}$$

$$= \underline{3163 \ \Omega}$$

Thus, the resistor is within tolerance because it is between 3135 Ω and 3465 Ω.

1-6 SERIES CIRCUITS

Most circuits consist of more than a single voltage source and a single lamp or one resistor. The circuit shown in Fig. 1-6 contains a battery and two resistors. This is a *series circuit,* since all elements (the battery, R_1, and R_2) must have the same

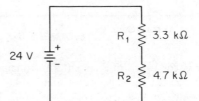

FIGURE 1-6 *Series circuit.*

current through them. There is no other path for the current to flow. It is thus said that the battery, R_1, and R_2 are *in series*. In a series circuit, the total resistance (R_T) is equal to the sum of the individual resistors. Thus, R_T for a series string of *n* resistors is:

$$R_T = R_1 + R_2 + R_3 + \ldots + R_n \qquad (1\text{-}2)$$

Therefore, the circuit in Fig. 1-6 has

$$R_T = R_1 + R_2 = 3.3 \text{ k}\Omega + 4.7 \text{ k}\Omega = 8 \text{ k}\Omega$$

Example 1-8 shows that in a series circuit, the current through each element is the same, but the voltages are (usually) different. A simple way to calculate the voltage across each resistor in a series circuit is to use the *voltage divider rule*. In equation form, the rule is

EXAMPLE 1-8

For the circuit shown in Fig. 1-6, calculate the current through each resistor and the voltage across each resistor.

Solution:

The total resistance equals the sum of R_1 and R_2 or 8 kΩ. The battery "sees" a resistance of 8 kΩ, and thus the current delivered by the battery is

$$I = \frac{V}{R} = \frac{24 \text{ V}}{8 \text{ k}\Omega} = 3 \text{ mA}$$

Since this is a series circuit, the current is the same through each element, and therefore

$$I = I_{R_1} = I_{R_2} = \underline{3 \text{ mA}}$$

Since the current through each resistor is known, the individual resistor voltages can be calculated from Ohm's Law:

$$
\begin{aligned}
V_{R_1} &= I_{R_1} \times R_1 \\
&= 3 \text{ mA} \times 3.3 \text{ k}\Omega \\
&= \underline{9.9 \text{ V}}
\end{aligned}
$$

$$
\begin{aligned}
V_{R_2} &= I_{R_2} \times R_2 \\
&= 3 \text{ mA} \times 4.7 \text{ k}\Omega \\
&= \underline{14.1 \text{ V}}
\end{aligned}
$$

$$V_x = V_s \times \frac{R_x}{R_T} \tag{1-3}$$

where V_x = voltage across R_x

            V_s = source voltage

            R_x = resistor whose voltage is to be determined

            R_T = total resistance in the series circuit

An illustration of the voltage divider rule is provided in Fig. 1-7. Study those results and notice that the voltage across each resistor is in proportion to the resistor's value. That is, the higher the resistor value, the higher the voltage across it. It is

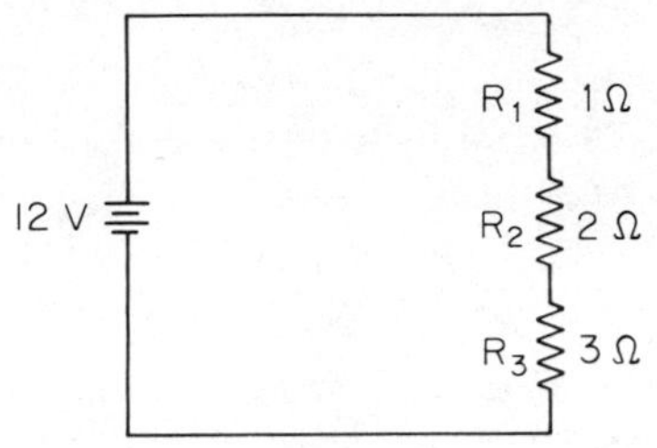

$$V_X = V_s \times \frac{R_X}{R_T}$$

$$R_T = 1\,\Omega + 2\,\Omega + 3\,\Omega = 6\,\Omega$$

$$V_{R_1} = V_s \times \frac{R_1}{R_T}$$

$$= 12\,V \times \frac{1\,\Omega}{6\,\Omega} = 2\,V$$

$$V_{R_2} = 12\,V \times \frac{2\,\Omega}{6\,\Omega} = 4\,V$$

$$V_{R_3} = 12\,V \times \frac{3\,\Omega}{6\,\Omega} = 6\,V$$

FIGURE 1-7 Voltage divider illustration.

also of interest to note that the sum of the individual resistor voltages (2 V + 4 V + 6 V = 12 V) is equal to the supply voltage. That situation will be explored in greater detail in Chapter 2.

EXAMPLE 1-9

Using the voltage divider rule, calculate the voltages across R_1 and R_2 in Fig. 1-6.

Solution:

$$V_x = V_s \times \frac{R_x}{R_T} \tag{1-3}$$

$$V_{R_1} = 24\,V \times \frac{R_1}{R_1 + R_2}$$

$$= 24\,V \times \frac{3.3\,k\Omega}{8\,k\Omega}$$

$$= \underline{9.9\,V}$$

$$V_{R_2} = 24\,V \times \frac{4.7\,k\Omega}{8\,k\Omega}$$

$$= \underline{14.1\,V}$$

The results of Example 1-9 agree with those of Example 1-8, as they should. You should note that the use of the voltage divider rule was a quicker, more direct method of solution.

EXAMPLE 1-10

Determine the current and voltage for each resistor in the circuit shown in Fig. 1-8.

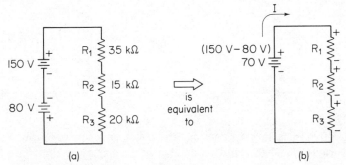

(a)   (b)

FIGURE 1-8 *Circuit for Example 1-10.*

Solution:

Notice that the two batteries in Fig. 1-8(a) are "opposing" one another. That is, the top battery (150 V) causes current flow in a clockwise direction, while the 80-V battery would cause a counterclockwise current flow. The stronger battery (150 V) wins; the resulting current is clockwise with a voltage of 70 V (150 V − 80 V), as shown in Fig. 1-8(b). The total resistance is

$$R_T = R_1 + R_2 + R_3 \qquad (1\text{-}2)$$
$$= 35\ \text{k}\Omega + 15\ \text{k}\Omega + 20\ \text{k}\Omega$$
$$= \underline{70\ \text{k}\Omega}$$

Therefore, the circuit current is

$$I = \frac{V}{R_T}$$

$$= \frac{70\ \text{V}}{70\ \text{k}\Omega}$$

$$= \underline{1\ \text{mA}}$$

The 1-mA current flows through each resistor. Therefore,

$$I_{R_1} = I_{R_2} = I_{R_3} = 1\ \text{mA}$$

The voltage divider rule can be used to determine the resistor voltages:

$$V_x = V_s \times \frac{R_x}{R_T} \qquad (1\text{-}3)$$

$$V_{R_1} = 70 \text{ V} \times \frac{35 \text{ k}\Omega}{70 \text{ k}\Omega}$$

$$= \underline{35 \text{ V}}$$

$$V_{R_2} = 70 \text{ V} \times \frac{15 \text{ k}\Omega}{70 \text{ k}\Omega}$$

$$= \underline{15 \text{ V}}$$

$$V_{R_3} = 70 \text{ V} \times \frac{20 \text{ k}\Omega}{70 \text{ k}\Omega}$$

$$= \underline{20 \text{ V}}$$

It is often necessary to consider the effect of a resistor failure. A failure may occur in various ways. It could be an "open" circuit such that no current can flow. This is the same effect as when a wire connector in a circuit is cut or breaks. Another mode of failure is termed a "short" circuit. A resistor seldom fails this way in practice. A short circuit means that the resistor value has effectively fallen to zero ohms. A third failure mode occurs when a resistor is out of tolerance. Thus, a 1-kΩ 10% resistor that is less than 900 Ω (1000 Ω − 10%) or greater than 1100 Ω (1000 Ω + 10%) can be classified as a failure.

EXAMPLE 1-11

For the circuit shown in Fig. 1-8, consider the circuit effects of R_2 failing in the "open" mode and then the "shorted" mode.

Solution:

If R_2 "opens," the path for current flow is broken and the circuit current is therefore zero. If R_2 "shorts," the total resistance equals the sum of R_1 and R_3 because the resistance of R_2 is now zero. Therefore,

$$V_{R_1} = V_s \times \frac{R_1}{R_T}$$

$$= 70 \text{ V} \times \frac{35 \text{ k}\Omega}{35 \text{ k}\Omega + 20 \text{ k}\Omega}$$

$$= \underline{44.5 \text{ V}}$$

$$V_{R_3} = 70 \text{ V} \times \frac{20 \text{ k}\Omega}{20 \text{ k}\Omega + 35 \text{ k}\Omega}$$

$$= \underline{25.6 \text{ V}}$$

It is seen that the voltages across R_1 and R_3 go up when R_2 is shorted. The circuit current also goes up, as follows:

$$I = \frac{V}{R_T}$$
$$= \frac{70 \text{ V}}{55 \text{ k}\Omega}$$
$$= \underline{1.27 \text{ mA}}$$

1-7 PARALLEL CIRCUITS

The circuit shown in Fig. 1-9 is a simple *parallel circuit*. The name is derived by the "parallel" alignment of elements as they are normally drawn in a circuit schematic.

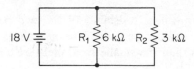

FIGURE 1-9 *Simple parallel circuit.*

Circuit elements are said to be *"in parallel"* when both ends of both elements are connected directly together. Thus, in Fig. 1-9, the battery, R_1, and R_2 are all in parallel because the top and bottom ends of all three are connected together. A commonly used shorthand designation for "in parallel with" is the use of two vertical parallel lines. Thus, saying R_1 in parallel with R_2 is represented by

$$R_1 \parallel R_2$$

You will recall that to get the total circuit resistance in a series circuit requires the simple addition of the individual resistors. It is a bit more complicated with parallel circuits, as the total equivalent resistance on n parallel resistors is

$$\frac{1}{R_{eq}} = \frac{1}{R_1} + \frac{1}{R_2} + \ldots + \frac{1}{R_n} \qquad \textbf{(1-4)}$$

where R_{eq} = total equivalent resistance of n parallel resistors.

EXAMPLE 1-12

Determine the equivalent resistance of the circuit shown in Fig. 1-9.

Solution:

$$\frac{1}{R_{eq}} = \frac{1}{R_1} + \frac{1}{R_2}$$

$$= \frac{1}{6\ k\Omega} + \frac{1}{3\ k\Omega}$$

$$= 0.167 \times 10^{-3} + 0.333 \times 10^{-3}$$

$$= 0.5 \times 10^{-3}$$

$$R_{eq} = \frac{1}{0.5 \times 10^{-3}} = 2 \times 10^3\ \Omega$$

$$= 2\ k\Omega$$

Notice that the total equivalent resistance of the two parallel resistors in Example 1-12 was *less* than the smallest resistor. This is true of parallel circuits—the total resistance is always less than the smallest parallel resistor.

EXAMPLE 1-13

Calculate the resistance of five 10-Ω resistors in parallel. Repeat for four and then two 10-Ω resistors in parallel.

Solution:

For five parallel 10-Ω resistors:

$$\frac{1}{R_{eq}} = \frac{1}{R_1} + \frac{1}{R_2} + \frac{1}{R_3} + \frac{1}{R_4} + \frac{1}{R_5} \tag{1-4}$$

$$= \frac{1}{10\ \Omega} + \frac{1}{10\ \Omega} + \frac{1}{10\ \Omega} + \frac{1}{10\ \Omega} + \frac{1}{10\ \Omega}$$

$$= 0.1 + 0.1 + 0.1 + 0.1 + 0.1$$

$$= 0.5$$

$$R_{eq} = \frac{1}{0.5} = \underline{2\ \Omega}$$

For four parallel 10-Ω resistors:

$$\frac{1}{R_{eq}} = \frac{1}{10\ \Omega} + \frac{1}{10\ \Omega} + \frac{1}{10\ \Omega} + \frac{1}{10\ \Omega}$$

$$= 0.1 + 0.1 + 0.1 + 0.1$$
$$= 0.4$$

$$R_{eq} = \frac{1}{0.4} = \underline{2.5 \; \Omega}$$

For two parallel 10-Ω resistors:

$$\frac{1}{R_{eq}} = \frac{1}{10 \; \Omega} + \frac{1}{10 \; \Omega}$$
$$= 0.1 + 0.1$$
$$= 0.2$$

$$R_{eq} = \frac{1}{0.2} = \underline{5 \; \Omega}$$

Example 1-13 illustrates the validity of the following equation:

$$R_{eq} = \frac{R}{n} \tag{1-5}$$

where　　R_{eq} = equivalent resistance of n equal-valued resistors with a resistance of R ohms

Thus, five 10-Ω resistors in parallel have an equivalent resistance of 10 Ω/5, or 2 Ω, as shown in Example 1-13. Similarly, the resistance of the four parallel 10-Ω resistors is 10 Ω/4 = 2.5 Ω, and that of the two parallel 10-Ω resistors is 10 Ω/2 = 5 Ω. These results are verified in Example 1-13.

The case of just two parallel resistors leads to the *product-over-sum formula:*

$$R_{eq} = \frac{R_1 \times R_2}{R_1 + R_2} \tag{1-6}$$

where R_{eq} is the equivalent resistance of R_1 and R_2 in parallel.

Thus, the two 10-Ω resistors in parallel are equal to

$$\frac{10 \times 10}{10 + 10} \quad \text{or} \quad \frac{100}{20} = 5 \; \Omega$$

as previously calculated.

The voltage across all parallel resistors in a circuit is the same. For instance, in Fig. 1-9 the two resistors are in parallel and both have 18 V across them. The

EXAMPLE 1-14

Calculate the resistance of a 20-Ω and a 50-Ω resistor in parallel.

Solution:

$$R_{eq} = \frac{R_1 \times R_2}{R_1 + R_2} \qquad (1\text{-}6)$$

$$= \frac{20\Omega \times 50\Omega}{20\Omega + 50\Omega}$$

$$= \frac{1000}{70}$$

$$= 14.3\Omega$$

This result can be verified by Example 1-4:

$$\frac{1}{R_{eq}} = \frac{1}{R_1} + \frac{1}{R_2} \qquad (1\text{-}4)$$

$$= \frac{1}{20\ \Omega} + \frac{1}{20\ \Omega} \cdot 50 \wedge$$

$$= 0.05 + 0.02 = 0.07$$

$$R_{eq} = \frac{1}{0.07}$$

$$= 14.3\ \Omega$$

current through each resistor can then be calculated by using Ohm's Law since the voltage and resistance are both known.

There are occasions when the fact that resistors are in parallel is not immediately obvious from a circuit diagram. In Fig. 1-10(a), the three resistors are actually in parallel, even though a first glance may not indicate that to you. The circuit has been redrawn in Fig. 1-10(b) in a more recognizable fashion. You should verify in your own mind that the two are identical. If a circuit diagram is confusing to you, it is often helpful to try and redraw it in different form.

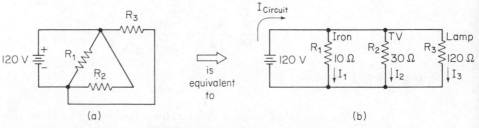

FIGURE 1-10 *Parallel circuit redrawn.*

The electrical wiring of most homes is in the form of parallel circuits. The supply voltage is typically 120 V, as shown in Fig. 1-10(b). Also shown are three connected loads, a TV set that "looks" like a 30-Ω resistance, a lamp at 120 Ω, and an iron at 10 Ω. The parallel wiring configuration allows a voltage of 120 V, regardless of the number of items "plugged" (connected) into the circuit.

EXAMPLE 1-15

For the current shown in Fig. 1-10(b), calculate the equivalent resistance, the circuit current, I_1, I_2, and I_3.

Solution:

$$\frac{1}{R_{eq}} = \frac{1}{R_1} + \frac{1}{R_2} + \frac{1}{R_3} \qquad (1\text{-}5)$$

$$= \frac{1}{10\ \Omega} + \frac{1}{30\ \Omega} + \frac{1}{120\ \Omega}$$

$$= 0.1 + 0.0333 + 0.00833$$

$$= 0.142$$

$$R_{eq} = \underline{7.06\ \Omega}$$

The circuit current can now be calculated:

$$V = IR \qquad (1\text{-}1)$$

$$I = \frac{V}{R}$$

$$= \frac{120\ V}{7.06}$$

$$I_{circuit} = \underline{17\ A}$$

$$I_1 = \frac{V}{R_1}$$

$$= \frac{120\ V}{10\ \Omega}$$

$$= \underline{12\ A}$$

$$I_2 = \frac{V}{R_2}$$

$$= \frac{120\ V}{30\ \Omega}$$

$$= \underline{4\ A}$$

$$I_3 = \frac{V}{R_3}$$

$$= \frac{120 \text{ V}}{120 \text{ }\Omega}$$

$$= \underline{1 \text{ A}}$$

It is noteworthy that $I_1 + I_2 + I_3 = I_{\text{circuit}}$ (12 A + 4 A + 1 A = 17 A)—a result that makes sense and will be explored further in Chapter 2.

QUESTIONS AND PROBLEMS

1-1-1. Describe in some detail the structure of an atom.

1-1-2. Explain the process whereby an electric shock might occur when a metallic light switch is touched.

1-1-3. Define the meaning of the word "conductor." What makes a material a good or bad conductor? What is another name for a nonconductor?

1-1-4. Describe the basic properties of a semiconductor. List two semiconductor materials.

1-2-5. Define the meaning of current flow. Explain why the random motion of electrons is *not* a current flow.

1-2-6. What is the unit of current flow, and how is it defined? Describe the relationship between current flow and a coulomb.

1-2-7. Calculate the current flow (in amperes) when 5×10^{17} electrons per second is flowing.

1-2-8. A charge of 50 C passes a point in 1 hour. Calculate the current flow.

1-3-9. In basic terms, describe what a battery is and what it does.

1-3-10. List four terms used to describe the "force" exerted by a battery.

1-3-11. Explain why both a battery and a lamp can be thought of as transducers.

1-3-12. A flashlight uses two D-cell batteries to energize its lamp. What is the voltage applied to the lamp? If only one battery were used to light the same lamp, what do you suspect would be the result?

1-4-13. In words and with an equation, state Ohm's Law.

1-4-14. Calculate the voltage for a circuit that has a resistance of 10 Ω and a current of 2 A.

1-4-15. Calculate the current drawn by an automobile's headlamps ($R = 3$ Ω) and clock ($R = 900$ Ω).

1-4-16. Explain the difference between electron flow and conventional current flow.

1–4–17. Calculate the resistance of a bulb that draws 23 mA from an EMF source of 28 V.

1–5–18. Determine the nominal resistance and allowable resistance range for resistors with the following color bands:
(a) Green, brown, brown, silver.
(b) Yellow, violet, orange, gold.
(c) Orange, white, red.
(d) Brown, black, black, silver.

1–5–19. Provide the color code for the following resistors:
(a) 1.5 kΩ   10%
(b) 27 Ω   5%
(c) 1.5 MΩ   20%
(d) 470 kΩ   10%
(e) 10 kΩ   10%

1–6–20. Define "series circuit."

1–6–21. A series circuit contains a 36-V battery and resistors of 10 kΩ, 18 kΩ, and 33 kΩ. Calculate the total resistance, and the current through and the voltage across each resistor.

1–6–22. Calculate the voltage across each resistor in Problem 1–6–21 using a different method than you used previously, and compare the results.

1–6–23. A series string of Christmas tree bulbs containing 15 lamps with 20 Ω resistance each is plugged into a 120-V outlet. Calculate the current flow and voltage for each bulb.

1–6–24. If one bulb fails in the string in Problem 1–6–23, and does so by shorting, calculate the new current and voltage for the remaining bulbs. What do you think will happen to the expected life of the remaining bulbs?

1–6–25. If a bulb from Problem 1–6–24 fails in the "open" mode, what is the result?

1–7–26. Define "parallel circuit."

1–7–27. Determine the resistance of a 10-Ω and a 15-Ω resistor in parallel.

1–7–28. A 27-kΩ, a 33-kΩ, and a 56-kΩ resistor are connected in parallel. Calculate the equivalent resistance.

1–7–29. A 24-V battery and the following resistors are all in parallel: 50 Ω, 200 Ω, 350 Ω, and 75 Ω. Calculate the equivalent resistance, the current delivered by the battery, and the individual resistor currents and voltages.

1–7–30. A string of 20 Christmas tree bulbs are connected in parallel. Each bulb has 320 Ω resistance. Calculate the equivalent resistance "seen" by the 120-V source and the current drawn from it. Calculate the current and voltage for each bulb.

1–7–31. If one of the bulbs in Problem 1–7–30 burns out ("opens"), what is the effect on the circuit? Explain the relative merits of series versus parallel strung lamps.

2

CIRCUIT ANALYSIS
AND ELEMENTS

2-1 SERIES/PARALLEL CIRCUITS

The circuits examined in Chapter 1 were either simple series or simple parallel circuits. In actual practice, however, most electronic circuits are combinations of series and parallel connections. For example, in Fig. 2-1(a) resistors R_2 and R_3 are in parallel with each other. They $(R_2 \| R_3)$ are in series with R_1. If you were to find the equivalent resistance at the x-y terminals in Fig. 2-1(a), the appropriate expression would be

$$
\begin{aligned}
R_{xy} &= (R_2 \| R_3) + R_1 \\
&= (2 \text{ k}\Omega \| 2 \text{ k}\Omega) + 1 \text{ k}\Omega \\
&= 1 \text{ k}\Omega + 1 \text{ k}\Omega \\
&= 2 \text{ k}\Omega
\end{aligned}
$$

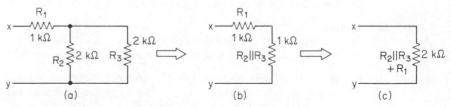

FIGURE 2-1 Circuit simplification.

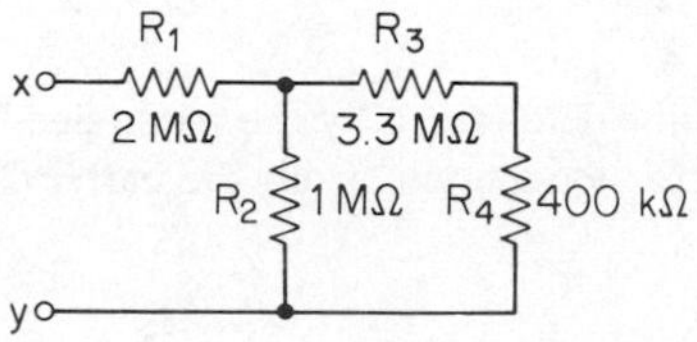

R_1 R_3

x○—WW—•—WW—

2 MΩ   3.3 MΩ

R_2 ≷1 MΩ R_4≷400 kΩ

y○————————•————

FIGURE 2-2 *Circuit for Example 2-1.*

The circuit simplification process is shown in Fig. 2-1(b) and (c). In terms of the x-y terminals, the circuits in (a), (b), and (c) are all equivalent. As a general procedure when simplifying a series/parallel circuit, start at the opposite end of the circuit and work back toward the open terminals. For the circuit shown in Fig. 2-2, that means to start simplifying at R_4 and work back to the x-y terminals.

EXAMPLE 2-1

Determine the equivalent resistance "seen" at the x-y terminals of Fig. 2-2.

Solution:

To simplify, start at the opposite end of the circuit from the x-y terminals and work back. Initially, it is seen that R_3 and R_4 are in series and they are in parallel with R_2. That entire combination is then in series with R_1. Stated as an equation:

$$R_{xy} = (R_4 + R_3)\|R_2 + R_1$$
$$= (3.3 \text{ MΩ} + 400 \text{ kΩ})\|1 \text{ MΩ} + 2 \text{ MΩ}$$
$$= (3.7 \text{ MΩ}\|1 \text{ MΩ}) + 2 \text{ MΩ}$$
$$= 0.79 \text{ MΩ} + 2 \text{ MΩ}$$
$$= \underline{2.79 \text{ MΩ}}$$

EXAMPLE 2-2

Calculate the battery current for the circuit shown in Fig. 2-3.

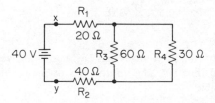

FIGURE 2-3 *Circuit for Example 2-2.*

Solution:

The battery current will be determined by the resistance "seen" by the battery. The resistance seen is that looking out from the battery's terminals, or R_{xy} in Fig. 2-3.

$$
\begin{aligned}
R_{xy} &= (R_4 \| R_3) + R_2 + R_1 \\
&= (30\ \Omega \| 60\ \Omega) + 40\ \Omega + 20\ \Omega \\
&= 20\ \Omega + 40\ \Omega + 20\ \Omega \\
&= 80\ \Omega
\end{aligned}
$$

Therefore, the circuit current can now be calculated using Ohm's Law:

$$V = I \times R \tag{1-1}$$

$$
\begin{aligned}
I &= \frac{V}{R} \\
&= \frac{40\ \text{V}}{80\ \Omega} \\
&= \underline{0.5\ \text{A}}
\end{aligned}
$$

EXAMPLE 2-3

Determine the voltage and current for each resistor in Fig. 2-3.

Solution:

From Example 2-2, it is known that the source (battery) current is 0.5 A. Thus, the currents I_{R_1} and I_{R_2} must also be 0.5 A, because they are in series with the battery. The voltages across R_1 and R_2 can then be calculated using Ohm's Law:

$$
\begin{aligned}
V_{R_1} &= I_{R_1} \times R_1 \\
&= 0.5\ \text{A} \times 20\ \Omega \\
&= \underline{10\ \text{V}}
\end{aligned}
$$

$$
\begin{aligned}
V_{R_2} &= I_{R_2} \times R_2 \\
&= 0.5\ \text{A} \times 40\ \Omega \\
&= \underline{20\ \text{V}}
\end{aligned}
$$

A calculation of the R_3 and R_4 voltages and currents is not yet possible from the theory introduced thus far. The information presented in the next section will make that possible.

2-2 KIRCHHOFF'S LAWS

The two laws to be discussed in this section allow circuit analysis of the vast majority of circuits normally encountered. There are a number of advanced techniques that will not be developed in this book, but they are rarely needed anyway. Textbooks devoted exclusively to circuit analysis should be consulted in cases where advanced techniques are required.

The simple series circuit shown in Fig. 2-4 is easily analyzed. The total resistance seen by the battery is $R_1 + R_2 + R_3$ or 20 kΩ. The circuit current is therefore 20

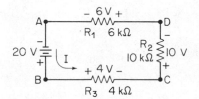

FIGURE 2-4 *Circuit for KVL analysis.*

V/20 kΩ or 1 mA. Since this is a series circuit, the 1-mA current flows through all three resistors. The voltage across each resistor is determined by Ohm's Law.

$$V_{R_1} = 1 \text{ mA} \times 6 \text{ k}\Omega = 6 \text{ V}$$
$$V_{R_2} = 1 \text{ mA} \times 10 \text{ k}\Omega = 10 \text{ V}$$
$$V_{R_3} = 1 \text{ mA} \times 4 \text{ k}\Omega = 4 \text{ V}$$

These voltages and their polarity are shown in Fig. 2-4. Notice that the conventional current flow enters the side of the resistor that has positive polarity. The voltage V_{AB} in Fig. 2-4 is -20 V, since V_{AB} means the voltage at A with respect to B. Since A is negative with respect to B, V_{AB} is negative. On the other hand, V_{BA} is $+20$ V, since B is positive with respect to A.

The simple circuit in Fig. 2-4 can now be used to illustrate a very useful circuit analysis tool. *Kirchhoff's Voltage Law* (KVL) states that the sum of voltages (including the voltage source) in any closed-circuit path must equal zero. This means that if you start at any point in a circuit and sum the voltages in either direction until you are back where you started, the sum should equal zero. In Fig. 2-4 there is only one complete circuit path (usually termed a *loop*), formed by the junctions A, B, C, and D. These junctions are usually termed *nodes*.

EXAMPLE 2-4

Verify Kirchhoff's Voltage Law (KVL) for the circuit shown in Fig. 2-4.

Solution:

KVL states that the sum of voltages in any loop must equal zero. The circuit in Fig. 2-4 has only one loop, so starting at node A and summing counterclockwise (CCW):

25

Does $V_{AB} + V_{BC} + V_{CD} + V_{DA} = 0$?

$$-20 \text{ V} + 4 \text{ V} + 10 \text{ V} + 6 \text{ V} = 0 \qquad \underline{\text{yes}}$$

Thus, KVL is verified. Starting at node A and summing clockwise (CW) should also provide verification:

Does $V_{AD} + V_{DC} + V_{CB} + V_{BA} = 0$?

$$-6 \text{ V} + (-10 \text{ V}) + (-4 \text{ V}) + 20 \text{ V} = 0$$

$$-6 \text{ V} - 10 \text{ V} - 4 \text{ V} + 20 \text{ V} = 0 \qquad \underline{\text{yes}}$$

Starting at any other node should also provide verification. Thus, starting at node C and summing CCW:

Does $V_{CD} + V_{DA} + V_{AB} + V_{BC} = 0$?

$$10 \text{ V} + 6 \text{ V} + (-20 \text{ V}) + 4 \text{ V} = 0 \qquad \underline{\text{yes}}$$

Kirchhoff's Current Law

Kirchhoff's Current Law (KCL) states that the sum of currents into a node must equal the current sum leaving it. The simple series circuit in Fig. 2-4 provides an elementary proof of KCL. The current into any one of the circuit's four nodes obviously equals the current out, since they are the same current. The simple parallel circuit in Fig. 2-5 is a bit more interesting with respect to KCL analysis.

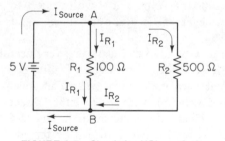

FIGURE 2-5 Circuit for KCL analysis.

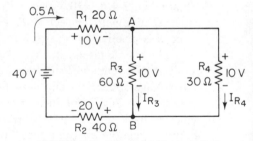

FIGURE 2-6 Circuit for Example 2–6.

EXAMPLE 2-5

Verify Kirchhoff's Current Law for nodes A and B in Fig. 2-5.

Solution:

The first step is to calculate the three currents in the circuit.

$$I_{R_1} = \frac{V_{R_1}}{R_1}$$

$$= \frac{5 \text{ V}}{100 \ \Omega}$$

$$= \underline{50 \text{ mA}}$$

$$I_{R_2} = \frac{V_{R_2}}{R_2}$$

$$= \frac{5 \text{ V}}{500 \ \Omega}$$

$$= \underline{10 \text{ mA}}$$

The resistance seen by the battery source is 100 Ω‖500 Ω, or 83.3 Ω. Therefore,

$$I_{\text{source}} = \frac{5 \text{ V}}{83.3 \ \Omega}$$

$$= 60 \text{ mA}$$

We are now in a position to verify KCL. At node *A*, the source current flows in while I_{R_1} and I_{R_2} flow out, as shown in Fig. 2-5. In equation form, KCL states:

$$\text{Does } I_{\text{source}} = I_{R_1} + I_{R_2}?$$

$$60 \text{ mA} = 50 \text{ mA} + 10 \text{ mA} \qquad \underline{\text{yes}}$$

The KCL proof for node *B* is essentially the same except that now I_{R_1} and I_{R_2} flow in and I_{source} flows out. Therefore,

$$\text{Does } I_{R_1} + I_{R_2} = I_{\text{source}}?$$

$$50 \text{ mA} + 10 \text{ mA} = 60 \text{ mA} \qquad \underline{\text{yes}}$$

Recall Example 2-3, where we were unable to calculate V_{R_3} and V_{R_4} before introduction of KVL and KCL. Let us take another look. The circuit and the known information from Example 2-3 are shown in Fig. 2-6.

EXAMPLE 2-6

Using KVL and KCL, determine V_{R_3}, V_{R_4}, I_{R_3}, and I_{R_4} for the circuit shown in Fig. 2-6.

Solution:

Notice in Fig. 2-6 that the voltages across R_1 and R_2 have been given a polarity. Recall that conventional current flows into the positive side of a resistor.

The voltage across R_3 can be determined by applying KVL to the loop that includes the 40-V battery, R_2, R_3, and R_1. Starting at the node between the battery and R_1 and proceeding CCW, we have

$$-40\text{ V} + 20\text{ V} + V_{R_3} + 10\text{ V} = 0$$
$$V_{R_3} = 40\text{ V} - 20\text{ V} - 10\text{ V}$$
$$= \underline{10\text{ V}}$$

Note that R_3 and R_4 are in parallel, so the voltage across them is the same.

$$V_{R_4} = V_{R_3} = 10\text{ V}$$

The current I_{R_3} is

$$I_{R_3} = \frac{V_{R_3}}{R_3}$$
$$= \frac{10\text{ V}}{60\ \Omega}$$
$$= \underline{0.167\text{ A}}$$

The current I_{R_4} can be determined using KCL at node A.

$$I_{\text{source}} = I_{R_3} + I_{R_4}$$
$$0.5\text{ A} = 0.167\text{ A} + I_{R_4}$$
$$I_{R_4} = 0.5\text{ A} - 0.167\text{ A}$$
$$= \underline{0.333\text{ A}}$$

This result can be verified by

$$I_{R_4} = \frac{V_{R_4}}{R_4}$$
$$= \frac{10\text{ V}}{30\ \Omega}$$
$$= \underline{0.333\text{ A}}$$

2-3   POTENTIOMETERS

A *potentiometer* is a specially constructed resistor. The schematic symbol for a potentiometer is shown in Fig. 2-7(a). Notice that the "pot" is a three-terminal device. It has the terminals of a regular resistor [*a* and *c* in Fig. 2-7(a)] and a third terminal at *b* in Fig. 2-7(a). The third terminal, with the arrowhead on it, is called the "wiper." The wiper can be physically moved from one end of the pot's resistance element to the other. The physical construction of some typical potentiometers is shown in Fig. 2-7(b).

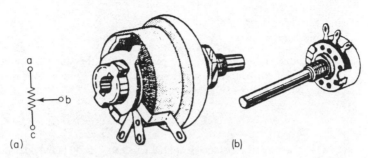

(a)   (b)

FIGURE 2-7   *Potentiometer.*

When connected in a circuit, a pot can be used to vary the voltage to a load. A common application is shown in Fig. 2-8. It represents the way that the dashboard lights in a car can be made to vary in brilliance from an "off" condition all the way to a maximum value. When the wiper *(b)* in Fig. 2-8 is at *a,* the lamps will have the full battery voltage (12 V) across them. When the wiper is at *c,* the lamp voltage is zero and the lights are therefore "off." The circuit is a series/parallel connection.

Notice the strange-looking symbol at the bottom of the circuit in Fig. 2-8. It represents an electrical *ground,* the point that serves as the reference that all voltages in the circuit are taken with respect to. The negative battery terminal is at "ground potential," and this circuit is therefore referred to as a negative ground system. Most automobiles are wired with a negative ground connection. This means that the entire conducting (metal) portion of the car is grounded. The term "ground" is sometimes referred to as "earth," since the earth is the common ground for most utilities' electrical distribution systems.

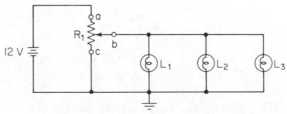

FIGURE 2-8   *Potentiometer circuit.*

EXAMPLE 2-7

The three dashboard lamps in Fig. 2-8 each have a resistance of 100 Ω. When the pot is set to a specific setting, its a-to-b resistance is 50 Ω. The pot's full resistance (a to c) is 150 Ω. Calculate the lamp voltages when the a-to-b resistance is 50 Ω.

Solution:

The b-to-c pot resistance is in parallel with the three lamps. Since this is a 150-Ω pot, the b-to-c resistance is 150 Ω minus the a-to-b resistance of 50 Ω. The b-to-c resistance is therefore

$$R_{b\,to\,c} = 150\ \Omega - 50\ \Omega = 100\ \Omega$$

The total circuit resistance from the wiper to ground is the pot's b-to-c resistance in parallel with the three lamps, or

$$R_{\text{tot}} = R_{bc} \| L_1 \| L_2 \| L_3$$
$$= 100\ \Omega \| 100\ \Omega \| 100\ \Omega \| 100\ \Omega$$

Now

$$R_{\text{tot}} = \frac{R_n}{n} \qquad \text{(1-5)}$$

$$= \frac{100\ \Omega}{4}$$

$$= 25\ \Omega$$

Since the lamps are all in parallel, their voltage and therefore their brilliance will be identical. The circuit resistance from a to b was given as 50 Ω, and the total resistance from b to c was just calculated as 25 Ω. The lamp voltages can now be calculated using the voltage divider equation from Chapter 1.

$$V_x = V_s \times \frac{R_x}{R_T} \qquad \text{(1-3)}$$

$$= 12\ \text{V} \times \frac{25\ \Omega}{25\ \Omega + 50\ \Omega}$$

$$= 12\ \text{V} \times \frac{25\ \Omega}{75\ \Omega}$$

$$= \underline{4\ \text{V}}$$

The 4-V lamp voltage would make the dashboard lighting very dim.

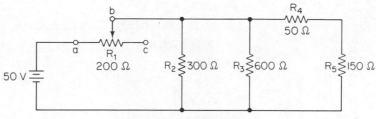

FIGURE 2-9 *Rheostat circuit.*

A potentiometer is often used with one of the two end terminals unconnected. This situation is shown in Fig. 2-9. In this situation, the potentiometer acts like a simple variable resistor and is called a *rheostat*. A rheostat can be thought of as a potentiometer in which one of the end terminals is not used. In Fig. 2-9, the *c* terminal of the rheostat R_1 is not connected in this series/parallel circuit.

EXAMPLE 2-8

Calculate the voltage range for R_5 in Fig. 2-9 as the rheostat is varied over its full range.

Solution:

(a) Condition 1—wiper at terminal *a:* The rheostat is effectively not in the circuit when the wiper is adjusted to terminal *a*. Therefore, the full battery voltage is across R_2, R_3, and the R_4, R_5 series combinations. The R_5 voltage can be calculated using the voltage divider equation.

$$V_x = V_s \times \frac{R_x}{R_T} \tag{1-3}$$

$$= 50 \text{ V} \times \frac{R_5}{R_4 + R_5}$$

$$= 50 \text{ V} \times \frac{150 \ \Omega}{50 \ \Omega + 150 \ \Omega}$$

$$= 50 \text{ V} \times \frac{150 \ \Omega}{200 \ \Omega}$$

$$= \underline{37.5 \text{ V}}$$

(b) Condition 2—wiper at terminal *c:* In this case, the full rheostat resistance of 200 Ω is in the circuit. The total resistance seen by the battery is

$$[(R_4 + R_5)\|R_3\|R_2] + R_1$$
$$= [(50\ \Omega + 150\ \Omega)\|600\ \Omega\|300\ \Omega] + 200\ \Omega$$
$$= 100\ \Omega + 200\ \Omega$$
$$= 300\ \Omega$$

$$I_{\text{battery}} = \frac{50\ \text{V}}{300\ \Omega}$$
$$= 0.167\ \text{A}$$

$$V_{R_1} = 0.167\ \text{A} \times 200\ \Omega$$
$$= 33.3\ \text{V}$$

The voltage across R_2 can now be determined using KVL.

$$50\ \text{V} = V_{R_1} + V_{R_2}$$
$$= 33.3\ \text{V} + V_{R_2}$$
$$V_{R_2} = 50\ \text{V} - 33.3\ \text{V}$$
$$= 16.7\ \text{V}$$

The voltage across R_3 and the R_4, R_5 series combination will also equal V_{R_2}, or 16.7 V. The R_5 voltage can now be calculated by the voltage divider equation:

$$V_{R_5} = 16.7\ \text{V} \times \frac{R_5}{R_4 + R_5}$$
$$= 16.7\ \text{V} \times \frac{150\ \Omega}{50\ \Omega + 150\ \Omega}$$
$$= \underline{12.5\ \text{V}}$$

Thus, the R_5 voltage varies from 12.5 V to 37.5 V as the rheostat is varied from one extreme to the other.

2-4   POWER

Power is the rate of doing work. *Work* is a force operating at some distance. Lifting 20 pounds (lb) a distance of 5 feet (ft) in 1 second (s) represents 100 foot-pounds (ft-lb) of work and a power of 100 ft-lb/s. A unit of power, the *horsepower* (hp), is defined as 550 ft-lb/s.

EXAMPLE 2-9

A hoist is able to lift 300 lb a distance of 20 ft in 4 s. Determine the work performed and the power involved.

Solution:

The work performed is

$$300 \text{ lb} \times 20 \text{ ft} = 6000 \text{ ft-lb}$$

The power involved is

$$6000 \text{ ft-lb/4 s} \quad \text{or} \quad \underline{1500 \text{ ft-lb/s}}$$

Expressed in terms of horsepower, the power is

$$\frac{1500 \text{ ft-lb/s}}{550 \text{ ft-lb/s per horsepower}} = \underline{2.73 \text{ hp}}$$

Power of an electrical nature results from current and voltage operating together. Expressed as an equation, we have

$$P = V \times I \tag{2-1}$$

where the power P is expressed in watts. There are 746 watts (W) per horsepower.

The power equation (2-1) can be algebraically manipulated by applying Ohm's Law to yield two alternative equations for electrical power computation.

$$P = I^2 R \tag{2-2}$$

and

$$P = \frac{V^2}{R} \tag{2-3}$$

EXAMPLE 2-10

Calculate the individual resistor and total power when five 100-Ω resistors are connected in parallel across a 20-V battery.

Solution:

Each 100-Ω resistor will have 20 V across it. Therefore,

$$P = \frac{V^2}{R} \tag{2-3}$$

$$= \frac{20^2}{100 \ \Omega}$$

$$= \underline{4 \text{ W}}$$

Since each resistor is dissipating (consuming) 4 W of power, the total power is 5 × 4 W, or 20 W. This total power can also be determined by using the equivalent resistance of five 100-Ω resistors in parallel. Their equivalent resistance is 100 Ω/5, or 20 Ω. Thus, the total power is

$$P = \frac{V^2}{R} \tag{2-3}$$

$$= \frac{20^2}{20\ \Omega}$$

$$= \underline{20\ W}$$

EXAMPLE 2-11

Determine the following for a household electrical system that operates on 120 V.

    (a)   The current drawn by a 100-W light bulb.

    (b)   The wattage rating of a toaster that draws 10 A.

    (c)   The current drawn by a clock with a 5-W rating.

    (d)   The resistance of and current drawn for a 60-W light bulb.

Solution:

(a)
$$P = V \times I \tag{2-1}$$
$$100\ W = 120\ V \times I$$
$$I = \frac{100}{120}$$
$$= \underline{0.833\ A}$$

(b)
$$P = V \times I$$
$$= 120\ V \times 10\ A$$
$$= \underline{1200\ W}$$

(c)
$$P = V \times I$$
$$5\ W = 120\ V \times I$$
$$I = \frac{5}{120}$$
$$= 0.0417\ A$$
$$= \underline{41.7\ mA}$$

(d)

$$P = \frac{V^2}{R} \qquad\qquad (2\text{-}3)$$

$$60 \text{ W} = \frac{120 \text{ V}^2}{R}$$

$$R = \frac{120^2}{60}$$

$$= \underline{240\ \Omega}$$

$$P = I^2 R \qquad\qquad (2\text{-}2)$$

$$60 \text{ W} = I^2 \times 240\ \Omega$$

$$I^2 = \frac{60}{240}$$

$$I = \underline{0.5 \text{ A}}$$

The electrical power consumed (dissipated) by the devices in Example 2-11 is converted into heat and light or, in the case of the clock, heat and mechanical motion. The desired output for the light bulb is light, for the toaster it is heat, and for the clock it is mechanical motion. The amount of energy involved is *different* from the powers we have been talking about. *Energy* is power for a certain period of time.

$$W = P \cdot t \qquad\qquad (2\text{-}4)$$

where $W =$ energy, in joules

            $P =$ power, in watts

            $t =$ time, in seconds

The user of electricity is charged by utility companies on the basis of energy usage. It is common practice to be charged on the basis of kilowatt-hours (kWh). Thus, a 1.5-kW heater operated for 2 hours (h) has consumed 3 kWh (1.5 kW $\times$ 2 h) of energy.

EXAMPLE 2-12

Calculate the electricity cost for the following cases in an area where the cost of electricity is 6.2 cents per kWh.

    (a)   A 1200-W iron for 3 h.

    (b)   A 5-W clock for 30 days.

(c) A month of TV usage; the set consumes 200 W and is used an average of 4 h/day.

Solution:

(a)
$$1200 \text{ W} \times 3 \text{ h} = 3600 \text{ Wh}$$
$$= 3.6 \text{ kWh}$$

$$3.6 \text{ kWh} \times 6.2 \text{ cents/kWh} = \underline{22.3 \text{ cents}}$$

(b)
$$5 \text{ W} \times 30 \text{ days} \times 24 \text{ h/day} = 3600 \text{ Wh}$$
$$= 3.6 \text{ kWh}$$

$$3.6 \text{ kWh} \times 6.2 \text{ cents/kWh} = \underline{22.3 \text{ cents}}$$

(c)
$$200 \text{ W} \times 30 \text{ days} \times 4 \text{ h/day} = 24{,}000 \text{ Wh}$$
$$= 24 \text{ kWh}$$

$$24 \text{ kWh} \times 6.2 \text{ cents/kWh} = \underline{149 \text{ cents}} \quad \text{or} \quad \$1.49$$

The *efficiency* of a device is important in many applications. By definition, efficiency is

$$\% = \frac{P_o}{P_i} \times 100\% \tag{2-5}$$

where $\%$ = efficiency, in percent

P_o = power output

P_i = power input

EXAMPLE 2-13

Calculate the efficiency of a ½-hp motor that draws 3.5 A at 120 V. Determine the power lost by the motor.

Solution:

$$\% = \frac{P_o}{P_i} \times 100\% \tag{2-5}$$

$$= \frac{746 \text{ W/hp} \times 0.5 \text{ hp}}{3.5 \text{ A} \times 120 \text{ V}}$$

$$= \frac{373 \text{ W}}{420 \text{ W}} \times 100\%$$

$$= \underline{83.8\%}$$

The power lost is 420 W − 373 W, or 47 W. It is dissipated in the form of heat.

2-5 AC CIRCUITS

Until now we have been concerned only with power sources that were constant in nature. The output of a 12-V battery is illustrated as a function of time in Fig. 2-10(a). Its output is shown to be a constant unchanging value and is termed *pure dc*. The dc terminology comes from the expression *direct current*. The voltage shown in Fig. 2-10(b) is also dc, but it is not pure dc. By definition any voltage (or current) that does not change polarity is dc. The voltage in Fig. 2-10(b) is varying slightly around a 12-V level. It is therefore not pure dc but *pulsating DC*. The voltage waveform shown in Fig. 2-10(c) extends from a peak value of 20 V down to 0 V on a repetitive basis. It is also pulsating dc, because it never changes polarity.

The waveforms shown in Fig. 2-11, on the other hand, are termed *ac signals*.

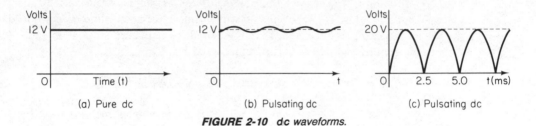

| (a) Pure dc | (b) Pulsating dc | (c) Pulsating dc |

FIGURE 2-10 *dc waveforms.*

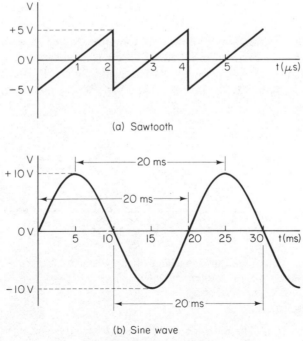

(a) Sawtooth

(b) Sine wave

FIGURE 2-11 *ac waveforms.*

37

The term "ac voltage or current" is derived from the expression *alternating current.* The voltages in Fig. 2-11 do, in fact, alternate between positive and negative polarities, which is the criterion for ac. In Fig. 2-11(a) a *sawtooth* voltage waveform is shown. Its polarity is alternating between a +5-V and −5-V level. Notice that it gradually builds up from the −5-V level to +5 V and then rapidly shifts to −5 V and starts building up to +5 V again.

The ac waveform shown in Fig. 2-11(b) is called a *sine wave.* Its shape is determined by the sine function in trigonometry. This is the most commonly encountered ac waveform and therefore merits closer study. The voltage delivered by utilities for home use is a sine-wave ac signal. The sine wave in Fig. 2-11(b) varies from a maximum positive value of +10 V to a maximum negative value of −10 V. This waveform is often described as a 20-v peak-to-peak sine wave, usually abbreviated as 20 v p-p. Notice that a lowercase v is used to designate ac voltage. Capitals are used for dc values; lowercase letters are used for ac signals. This is true for both voltages and currents. All repetitive ac waveforms have some time or period before they start repeating themselves. This is defined as one *cycle* or a complete cycle. This time is called the *period* and is represented by *T.*

$$T = \text{period} = \text{time it takes for one complete cycle} \qquad \text{(2-6)}$$

The period, *T,* for the sine wave in Fig. 2-11(b) is shown to be 20 ms in three different locations. It does not matter where *T* is measured; just be sure that a complete cycle is included.

The rate at which an ac signal repeats itself is termed its *frequency.* Thus, a signal that repeats itself 60 times each second is said to have a frequency of 60 cycles per second. This is more commonly designated as 60 hertz (Hz). The standard line frequency in the United States is 60 Hz, whereas in many other parts of the world it is 50 Hz (50 complete variations per second). The frequency of a waveform can be calculated as the inverse of the period, *T.* Thus, frequency, *f,* is

$$f = \frac{1}{T} \qquad \text{(2-7)}$$

For the waveform in Fig. 2-11(b), $f = \frac{1}{20}$ ms = 50 Hz.

EXAMPLE 2-14

(a) Determine the period and (b) calculate the frequency for the waveforms shown in Figs. 2-10(c) and 2-11(a).

Solution:

(a) The pulsating dc voltage in Fig. 2-10(c) completes one cycle in 2.5 ms. Thus, $T = 2.5$ ms and

$$f = \frac{1}{T} \tag{2-7}$$

$$= \frac{1}{2.5 \text{ ms}}$$

$$= \frac{1}{2.5 \times 10^{-3}}$$

$$= 0.4 \times 10^3$$

$$= \underline{400 \text{ Hz}}$$

(b) The sawtooth waveform in Fig. 2-11(a) repeats itself every 2 μs. Thus, $T = 2$ μs and

$$f = \frac{1}{T} \tag{2-7}$$

$$= \frac{1}{2 \text{ } \mu\text{s}}$$

$$= \frac{1}{2 \times 10^{-6}}$$

$$= 0.5 \times 10^6$$

$$= \underline{500 \text{ kHz}}$$

EXAMPLE 2-15

A television circuit causes lines on the picture to be scanned at a rate of 15,750 times per second. Determine the time for each line.

Solution:

In this case, the frequency is 15,750 Hz, or 15.75 kHz. The time for each line is the period. Therefore,

$$f = \frac{1}{T} \tag{2-7}$$

and rearranging the equation yields

$$T = \frac{1}{f}$$

$$= \frac{1}{15.75 \text{ kHz}}$$

$$= 0.0635 \text{ ms}$$

$$= \underline{63.5 \text{ } \mu\text{s}}$$

If the 20-v p-p sine wave in Fig. 2-11(b) were applied to a 20-Ω resistor, the current would be 20 v p-p/20 Ω, or 1 a p-p. The current waveform through the resistor is identical to the voltage waveform—a sine wave in this case. Do you think that the power being dissipated in the resistor would equal that of a pure 20-V dc voltage applied to the 20-Ω resistor? The answer is no—rather a value called the *root-mean-square* (rms) voltage and current must be used to calculate power. The rms values for a sine wave can be obtained from the p-p values by dividing by $2\sqrt{2}$. Thus,

$$\text{rms (volts or amps)} = \frac{\text{p-p value}}{2\sqrt{2}} \qquad \text{(2-8)}$$

The rms values of sine-wave voltage and current are often left undesignated. Thus, a 120-v sine wave is assumed to be an rms value.

EXAMPLE 2-16

The voltage waveform shown in Fig. 2-11(b) is applied to a 20-Ω resistor. Calculate:

    (a)   The p-p current.

    (b)   The rms voltage.

    (c)   The rms current.

    (d)   The power developed in the resistor.

    (e)   The dc voltage that would develop the same power.

Solution:

    (a) $i\,\text{p-p} = \dfrac{v\,\text{p-p}}{R}$

            $= \dfrac{20\ \text{v p-p}}{20\ \Omega}$

            $= 1\ \text{a p-p}$

    (b) $\text{rms} = \dfrac{\text{p-p value}}{2\sqrt{2}}$ (2-8)

              $= \dfrac{20\ \text{v p-p}}{2\sqrt{2}}$

              $= 7.07\ \text{v rms} \qquad \text{or} \qquad 7.07\ \text{v}$

(c) $i\,\text{rms} = \dfrac{v\,\text{rms}}{R} = \dfrac{7.07\,\text{v}}{20\,\Omega}$

   $= \underline{0.35\ \text{a rms}}$ or $\underline{0.35\ \text{a}}$

(d) Recall that to calculate power, rms values must be used.

$$P = v \times i \tag{2-1}$$
$$= 7.07\ \text{v} \times 0.35\ \text{a}$$
$$= \underline{2.5\ \text{W}}$$

   Alternative solution:

$$P = \frac{V^2}{R} \tag{2-3}$$
$$= \frac{(7.07\ \text{V})^2}{20\ \Omega}$$
$$= \underline{2.5\ \text{W}}$$

(e) If 2.5 W were developed by a dc signal, the voltage can be calculated by

$$P = \frac{V^2}{R} \tag{2-3}$$
$$2.5\ \text{W} = \frac{V^2}{20}$$
$$V^2 = 2.5\ \text{W} \times 20$$
$$V = \underline{7.07\ \text{V dc}}$$

EXAMPLE 2-17

The household line voltage is 120 v ac, 60 Hz. Determine the rms and peak-to-peak values.

Solution:

Since the 120 v ac was given with no other designators, you must assume it to be the rms value:

$$\text{rms} = \frac{\text{p-p value}}{2\sqrt{2}} \tag{2-8}$$

$$\text{p-p value} = \text{rms} \times 2\sqrt{2}$$
$$= 120\ \text{v} \times 2\sqrt{2}$$
$$= \underline{339\ \text{v p-p}}$$

2-6 METER MEASUREMENTS

The most basic and widely used test instrument in electronics is the *volt-ohm-meter,* or VOM. A typical example of a VOM is shown in Fig. 2-12. It can be used to measure ohms, dc volts, dc current, and ac volts. These measurements can be made on a variety of scales selected by adjusting the large function control switch to the desired setting.

FIGURE 2-12 *Typical VOM (Courtesy, Triplett Electric Co.).*

Ohms Measurement

The value of a resistor (or resistor combination) can be measured with the VOM. This is accomplished by connecting the meter (via its two leads) in parallel with the resistor. Just prior to the measurement, the meter leads must be shorted together and then the meter is "zero-adjusted." This is done by varying the "ohms adjust" control shown in Fig. 2-12 until the meter reads zero on the ohms scale. The meter is then calibrated and ready for resistor measurement. Notice that the ohms scale (the top one) for the meter scale shown in Fig. 2-13 is the only reverse-reading scale on the meter.

Zero is read at the right and increasing resistance is read from right to left. Notice also that a number of scale factors can be chosen by the function selector switch when reading resistance. As an example, a reading of 5 on the R × 1 scale means 5 Ω, whereas the same reading on the R × 1000 scale means 5 kΩ.

When making resistance readings, maximum accuracy is achieved by selecting

FIGURE 2-13   *Scale for VOM (Courtesy, Triplett Electric Co.).*

the multiplying factor that gives a reading roughly in the middle one-third of the meter scale. When changing scales, be sure to rezero the meter before making the measurement. Resistance measurements *must* be made with the resistor(s) disconnected from any other source of power. Damage to the VOM might result otherwise and provide an incorrect reading. An internal battery is utilized when measuring resistance with a VOM. The other measurements (voltage and current) rely on external circuit power and do not use the battery.

Meter Reading

A meter scale such as the one shown in Fig. 2-13 requires some explanation before it can be used. Like the scale on a good ruler, all the readings are not exactly specified; the user is required to "read between the lines." The function selector switch determines *which* scale to read. If the switch is set to 12 V dc, that means that the scale to be used has a 12 at full deflection (on the right) and the reading should be read along that scale. All scales, except ohms, are read from left to right with zero on the left.

In addition, for best accuracy, all voltage and current measurements should be made on the lowest range that does not "peg the meter." To "peg the meter" means to drive the needle past full-scale deflection, a practice to be avoided because it may damage the meter. Recall that resistance readings are made on the middle one-third of the meter for maximum accuracy, in contrast with good practice on voltage and current measurements. The latter readings should be made with the highest-possible

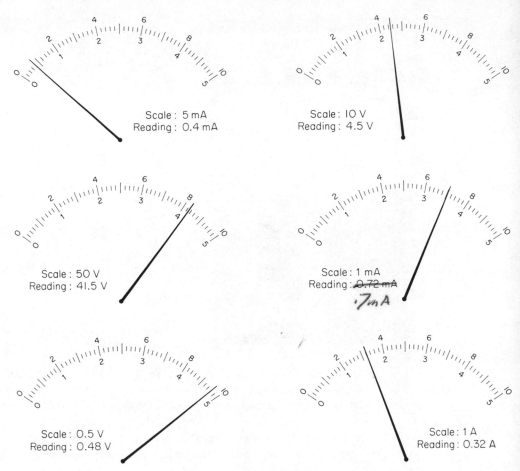

Scale: 5 mA
Reading: 0.4 mA

Scale: 10 V
Reading: 4.5 V

Scale: 50 V
Reading: 41.5 V

Scale: 1 mA
Reading: ~~0.72 mA~~
.7 m A

Scale: 0.5 V
Reading: 0.48 V

Scale: 1 A
Reading: 0.32 A

FIGURE 2-14   Meter scale readings.

scale deflections. A number of meter readings and the proper values are shown in Fig. 2-14. Be sure to study them until you understand the results.

The meter scale on most good-quality VOMs includes a mirror surface. When taking a reading, you should view the meter needle such that no image of the needle can be seen in the mirror. This ensures that you are viewing and reading the meter "straight on" and minimizes reading error. The error created by viewing at an angle is called *parallax* error.

Voltage Readings

The characteristic of voltage occurs across a circuit element. For example, a resistor voltage occurs across, or from one end to the other, of the resistor. It seems logical, then, that a meter in parallel with the device for which voltage is to be measured. This connection is shown in Fig. 2-15. The meter is measuring the voltage across

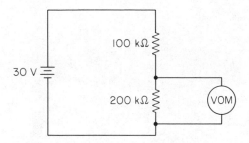

FIGURE 2-15 *VOM voltage measurement.*

the 200-kΩ resistor. It is not uncommon for a voltage to be less than its desired result when measuring across high-valued resistors. This effect is called *meter loading*. The voltmeter's resistance, although normally quite high, when in parallel with the resistor whose voltage is being read actually causes the circuit voltage to be less than it would be without the meter in the circuit. This effect is explored in Example 2-18.

EXAMPLE 2-18

The VOM in Fig. 2-15 is set to the 30-V scale and has a rated resistance of 20,000 Ω/V. Calculate the actual resistor voltage and the "loaded-down" value caused by meter loading.

Solution:

The actual circuit voltage is easily calculated using the voltage divider rule.

$$V_x = V_s \times \frac{R_x}{R_T} \qquad \text{(1-3)}$$

$$V_{200\text{k}\Omega} = 30 \text{ V} \times \frac{200 \text{ k}\Omega}{200 \text{ k}\Omega + 100 \text{ k}\Omega}$$

$$= 20 \text{ V}$$

With the meter in the circuit a change in circuit resistance occurs. The meter is rated at 20,000 Ω/V, so its resistance on the 30-V scale is

$$30 \text{ V} \times 20,000 \text{ } \Omega/\text{V} = 600 \text{ k}\Omega$$

Thus, the circuit voltage ("loaded" value) with the meter included will be

$$V_{200\text{k}\Omega} = 30 \text{ V} \times \frac{200 \text{ k}\Omega \| 600 \text{ k}\Omega}{(200 \text{ k}\Omega \| 600 \text{ k}\Omega) + 100 \text{ k}\Omega}$$

$$= 30 \text{ V} \times \frac{150 \text{ k}\Omega}{150 \text{ k}\Omega + 100 \text{ k}\Omega}$$

$$= \underline{18 \text{ V}}$$

FIGURE 2-16   Digital VOM (DVOM) (Courtesy, Simpson Electric Co.).

The loading effect demonstrated in Example 2-18 would be minimized by using a VOM with a higher resistance, although this usually involves use of a more costly meter. Another solution is to use a *digital voltmeter* (DVM), such as the one illustrated in Fig. 2-16. A DVM offers a very high resistance when used for voltage measurement, typically 10 MΩ, and usually offers better accuracy than its VOM counterpart. It is also free of parallax error.

EXAMPLE 2-19

Calculate the loading effect in the circuit shown in Fig. 2-15 when a DVM with 10 MΩ resistance is used.

Solution:

$$V_{200\text{k}\Omega} = 30 \text{ V} \times \frac{200 \text{ k}\Omega \| 10 \text{ M}\Omega}{(200 \text{ k}\Omega \| 10 \text{ M}\Omega) + 100 \text{ k}\Omega}$$

$$= \underline{19.9 \text{ V}}$$

Current Readings

Current flows through circuit elements, and to measure it the meter must be in series. This process is shown in Fig. 2-17. This is in contrast to voltage and resistance measurements, in which the meter is in parallel with a circuit element. To measure current, the circuit must be "broken" and the meter put in series with the element for which current is being measured. In Fig. 2-17, the VOM is measuring the current through the 10-Ω resistor.

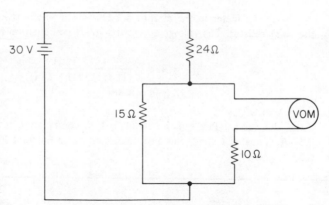

FIGURE 2-17 *VOM current measurement.*

It was previously shown that the meter resistance should ideally be very high when measuring voltage. The opposite is true for current measurements. The ideal case would be zero resistance. The situation for current meter loading is illustrated in the following example.

EXAMPLE 2-20

Calculate the current through the 10-Ω resistor shown in Fig. 2-17, with and without the meter in the circuit. The VOM has a 1-Ω resistance on the current scale being used.

Solution:

Without the meter, the voltage across the 10-Ω resistor can be calculated using the voltage divider equation.

$$V_x = V_s \times \frac{R_x}{R_T} \qquad \text{(1-3)}$$

$$V_{10\Omega} = 30 \text{ V} \times \frac{10\ \Omega \| 15\ \Omega}{(10\ \Omega \| 15\ \Omega) + 24\ \Omega}$$

$$= 30 \text{ V} \times \frac{6\ \Omega}{6\ \Omega + 24\ \Omega}$$

$$= 6 \text{ V}$$

$$I_{10\Omega} = \frac{6 \text{ V}}{10\ \Omega}$$

$$= \underline{0.6 \text{ A}}$$

With the meter in the circuit, a 1-Ω resistor is effectively in series with the 10-Ω resistor. The voltage across this 11-Ω combination is

$$V_{11\Omega} = 30 \text{ V} \times \frac{11 \text{ } \Omega \| 15 \text{ } \Omega}{(11 \text{ } \Omega \| 15 \text{ } \Omega) + 24 \text{ } \Omega}$$

$$= \underline{6.27 \text{ V}}$$

Thus, the current through the 10 Ω + 1 Ω combination is 6.27 V/11 Ω = 0.57 A. Thus, the meter has introduced an error (current loading) of 0.6 − 0.57 = 0.03 A.

2-7 THE OSCILLOSCOPE

Section 2-6 dealt with one (the VOM or DVM) of the two indispensable instruments to those working with electronics. The other required instrument is the *oscilloscope*, often referred to simply as the scope. It is able to show very rapidly changing voltages as a function of time. In effect, it draws a graph of voltage versus time. Thus, the scope display shows voltage on the vertical scale and time on the horizontal scale.

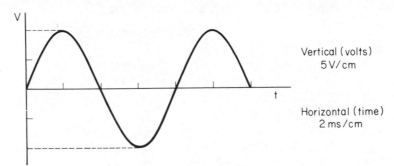

FIGURE 2-18 *Oscilloscope display.*

The display is provided by a cathode-ray tube (CRT) very similar to the tube used in a TV set.

The two major controls on a scope are the vertical and horizontal adjustments. They allow the volts per centimeter (V/cm) in the vertical direction and seconds per centimeter (s/cm) in the horizontal direction to be variable. For instance, the sine wave shown on the scope display in Fig. 2-18 occurs with the horizontal control set to 2 ms/cm while the vertical control is at 5 V/cm. In the vertical direction, the sine wave is 4 cm from peak to peak. Thus, the amplitude is 5 V/cm × 4 cm, or 20 v p-p. In the horizontal direction, the sine wave is seen to have a period *(T)* of 4 cm. Thus, the period is 2 ms/cm × 4 cm = 8 ms. Thus, the frequency can be calculated as

$$f = \frac{1}{T}$$ (2-7)

$$= \frac{1}{8 \text{ ms}}$$

$$= 125 \text{ Hz}$$

EXAMPLE 2-21

Determine (a) the amplitude and (b) the period and frequency for the square wave shown in Fig. 2-19.

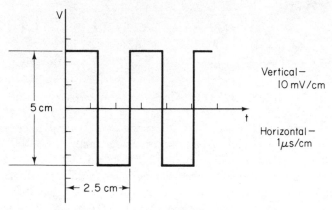

FIGURE 2-19   *Oscilloscope display for Example 2-21.*

Solution:

(a) The control setting is 10 mv/cm and the amplitude is 5 cm p-p. Thus,

$$V = 10 \text{ mv/cm} \times 5 \text{ cm p-p}$$
$$= \underline{50 \text{ mv p-p}}$$

(b) The control setting is 1 μs/cm and the displacement for one cycle is 2½ cm. Thus,

$$T = 1 \text{ μs/cm} \times 2\frac{1}{2} \text{ cm}$$
$$= 2.5 \text{ μs}$$

$$f = \frac{1}{T}$$ (2-7)

$$= \frac{1}{2.5 \text{ μs}}$$

$$= \underline{400 \text{ kHz}}$$

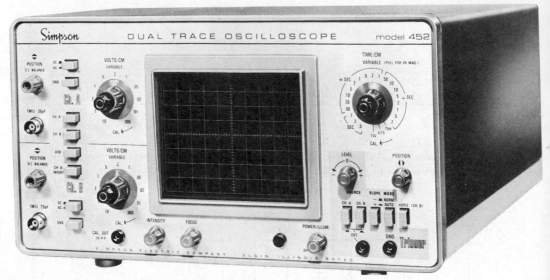

FIGURE 2-20   *Oscilloscope (Courtesy*, Simpson Electric Co.).

Example 2-21 shows that the scope can be used to measure events with very
high frequencies and low amplitudes. Some scopes have vertical settings (and thus
sensitivity) as low as 1 mV/cm and horizontal capability to 1 ns (10^{-9} s)/cm. Figure
2-20 shows a good-quality oscilloscope. As can be seen, there are a large number of
controls over and above the two basic ones that we have discussed. Do not let that
frighten you. If you have understood the basics, an afternoon's experience with a
scope and its manual should make you a reasonably proficient user of this valuable
tool.

QUESTIONS AND PROBLEMS

2–1–1. Explain what is meant by the terms "series," "parallel," and "series/parallel" circuits.

2–1–2. Describe the general process used when simplifying a series/parallel resistor network
down to one equivalent resistance.

2–1–3. Calculate the equivalent resistance for the circuit in Fig. 2-1 if the resistor values
were changed so that $R_1 = 4.7$ kΩ, $R_2 = 10$ kΩ, and $R_3 = 27$ kΩ.

2–1–4. If R_2 in Fig. 2-3 were changed to 60 Ω, calculate the value for R_{xy} and the battery
current.

2–2–5. State Kirchhoff's Voltage Law (KVL) in words.

2–2–6. Use KVL to solve for the resistor voltages in the circuit described in Problem
2–1–4.

2-2-7. Solve for R_{xy} and use KVL to solve for the resistor voltages for the circuit shown in Fig. 2-21.

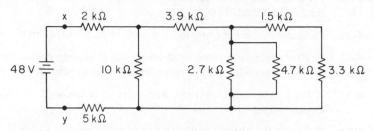

FIGURE 2-21 *Circuit for Problem 2-2-7.*

2-2-8. State Kirchhoff's Current Law (KCL) in words. Explain what a node is.

2-2-9. Determine the unknown currents for the situations depicted in Fig. 2-22.

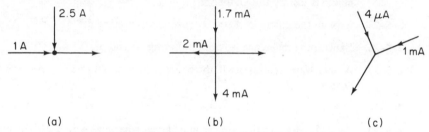

FIGURE 2-22 *KCL illustration for Problem 2-2-9.*

2-2-10. An automobile engine is running. The lighting system is drawing 14 A, the radio 0.5 A, and the coil 1 A. The alternator is delivering 12 A. Draw a circuit diagram representing this situation and determine the battery current. Is it supplying current or being recharged?

2-3-11. What is a potentiometer? Briefly explain its construction.

2-3-12. The circuit described in Example (2-7) is to be adjusted such that the lamp voltage is 6 V. Calculate the *a*-to-*b* and *b*-to-*c* potentiometer resistance that is required.

2-3-13. Explain the difference between a potentiometer and a rheostat. Use simple circuit diagrams to show the difference.

2-3-14. Design a circuit that will vary a lamp voltage from 0 to 6 V given a 12-V battery, a lamp resistance of 50 Ω, and a 100-Ω potentiometer. (*Hint:* Consider the use of a resistor in series with the battery.)

2-4-15. Define what is meant by the terms "work," "power," and "horsepower."

2-4-16. A 150-lb person runs up a flight of stairs in 3 s. The vertical height increase is 20 ft. Calculate the work performed and the horsepower expended.

2–4–17. Write down three equations that can be used to calculate power in an electrical circuit.

2–4–18. Show how two of the electrical power equations can be derived from the third with the help of Ohm's Law.

2–4–19. Calculate the power dissipated for each resistor in the circuit shown in Fig. 2-6.

2–4–20. A 100-Ω resistor is rated to handle 2 W. It is connected across a 15-V battery. Is it being operated within its rated limits? If not, what do you think might happen?

2–4–21. A lamp is operated on 120-v line voltage. It has a 72-Ω resistance. Calculate its wattage rating.

2–4–22. The lamp in Problem 2–4–21 is used 55 h/month. At 5.5 cents/kWh, determine the cost for electricity per month.

2–4–23. Calculate the cost of operating a 7-W night light continuously for a year at 6 cents/ kWh.

2–4–24. Determine the efficiency of a 2-hp motor that draws 15 A at 120 V. How much power is lost by the motor?

2–5–25. Explain the difference between pure and pulsating dc.

2–5–26. Explain the difference between pulsating dc and ac.

2–5–27. A sine wave repeats itself three times in 1 s. What is its period? How is period defined?

2–5–28. Define frequency. What is the relationship between period and frequency?

2–5–29. An FM radio station operates on a carrier sine wave with a 0.01-μs period. Calculate the frequency of its carrier sine wave.

2–5–30. Describe the relationship between resistor voltage and current waveforms.

2–5–31. A 100-V p-p sine wave is applied to a 50-Ω resistor. Calculate the rms voltage and current, the peak-to-peak current, and the power dissipation. What value of dc voltage would provide the same power in the resistor?

2–6–32. Describe the procedure for measuring resistance using a VOM such as the one shown in Fig. 2-12.

2–6–33. Explain parallax error and how it is minimized in a high-quality VOM.

2–6–34. Explain the procedure for measuring voltage and current using a VOM. Be sure to state the method of proper scale selection.

2–6–35. How does meter loading occur when using the VOM to measure voltage? Describe what should be done to minimize loading.

2–6–36. The voltage across a 150-kΩ resistor in a simple series circuit is to be measured. The voltage source is 70 V and the other circuit resistance is 300 kΩ. Calculate the actual resistor voltage and the measured value when using a 10-kΩ/V VOM on the 30-V scale.

2–6–37. How does meter loading occur when using the VOM to measure current? What characteristic of the meter would be ideal and eliminate this form of loading?

2–6–38. A current range on a VOM has 10 Ω resistance and is used to measure the current through R_4 in Fig. 2-6. Determine the actual and "loaded" current values.

2–7–39. In basic terms, explain what an oscilloscope can be used for.

2–7–40. An oscilloscope displays a sine wave with amplitude of 4 cm p-p and 5 cm for a complete cycle. The vertical scale is set at 10 V/cm and the horizontal scale at 4 ms/cm. Determine the peak-to-peak and rms voltage and the frequency.

3

CAPACITANCE AND INDUCTANCE

3-1 CAPACITANCE

Consider the situation shown in Fig. 3-1(a), two metal plates separated by a nonconductor and connected via a battery and switch. If the switch were closed at time $t = 0$, the following events would occur:

    1. The "electrical force" of the battery would cause electrons from plate A in Fig. 3-1(a) to be "pumped" counterclockwise (CCW) to be *stored* on plate B. The

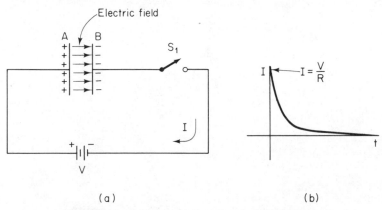

(a)   (b)

FIGURE 3-1 *Capacitance and electric field illustration.*

metal plates are said to have a *capacity* to store electrons. The CCW flow of electrons results in a clockwise (CW) conventional flow, as shown in Fig. 3-1(a).

2. The current flow will be maximum at time $t = 0$ and gradually decline to 0, as shown in Fig. 3-1(b). The amount of initial current flow depends upon the amount of voltage and amount of resistance *(R)* in the circuit. Even if there is no resistor in such a circuit, a small amount of wire resistance and internal battery resistance exists to limit the initial amount of current flow. If that circuit resistance is represented by R, the initial current flow at $t = 0$ is simply V/R, as shown in Fig. 3-1(b).

3. The metal plates in Fig. 3-1(a) have a voltage potential across them equal to the battery voltage, V. An *electric field, E,* exists from plate A to plate B, as shown by the lines in Fig. 3-1(a) with arrowheads pointing at plate B. The electric field intensity depends upon the amount of voltage and the distance between the plates. The electric field contains the energy given up by the battery in "charging" the metal plates. They "store" that energy until it is released. If, for instance, the battery in Fig. 3-1(a) were shorted out, the excess electrons on plate B would flow back to plate A and the electric field collapses. The energy that had been stored in the electric field is returned to the circuit in the form of heat generated by I^2R power losses (recall that $P = I^2R$) in the wire resistance.

A device made up of two metal plates and separated by a nonconductor (dielectric) is called a *capacitor*. The *capacitance, C,* provided by a capacitor is predicted by the following equation:

$$C = \epsilon_0 \cdot \epsilon_r \cdot \frac{A}{d} \tag{3-1}$$

where C = capacitance in farads (F)

  ϵ_0 = permittivity of free space (8.85×10^{-12})

  ϵ_r = relative permittivity of the capacitor's dielectric

  A = plate area in square meters (m²)

  d = distance between plates, in meters (m)

It should be noted from Eq. (3-1) that the capacitance of a capacitor is directly proportional to plate area and inversely proportional to plate spacing. Thus, doubling the plate area doubles the capacitance, and doubling the plate spacing would halve the capacitance. The relative permittivities of some common materials used for capacitor dielectrics are shown in Table 3-1. The relative permittivity of mica is shown to be about five times that of an air dielectric and results in five times the capacitance, given equal plate areas and spacings.

The physical appearance of capacitors is varied; some common configurations are shown in Fig. 3-2. Their size also varies, depending upon the amount of capacitance desired, the type of dielectric used, and the voltage rating. The voltage rating of a capacitor should not be exceeded because the capacitor would be damaged or de-

TABLE 3-1 *Relative permittivities.*

Dielectric	ϵ_r
Vacuum	1.0
Air	1.0006
Teflon	2.0
Paper, paraffined	2.5
Rubber	3.0
Transformer oil	4.0
Mica	5.0
Porcelain	6.0
Bakelite	7.0
Glass	7.5
Water	80.0

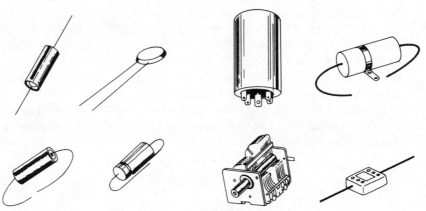

FIGURE 3-2 *Typical capacitors.*

stroyed. This rating is determined by the plate spacing and dielectric material. The closer the plates, the sooner an applied voltage will "break down" the dielectric and cause electrons to flow between the plates. Thus, for a capacitor to be able to withstand a higher voltage requires greater spacing, and thus a physically larger capacitor results. To maintain the same amount of capacitance when doubling the spacing, the plate area would have to be doubled.

EXAMPLE 3-1

A capacitor using a mica dielectric has an effective plate area of 1×10^{-3} m² and plate spacing of 0.1×10^{-6} m. Calculate the amount of capacitance.

Solution:

$$C = \epsilon_0 \epsilon_r \frac{A}{d} \qquad (3\text{-}1)$$

From Table 3-1, ϵ_r for mica is 5. Therefore,

$$C = 8.85 \times 10^{-12} \times 5 \times \frac{1 \times 10^{-3} \text{ m}^2}{0.1 \times 10^{-6} \text{ m}}$$

$$= 44.25 \times 10^{-9} \text{ F}$$
$$= \underline{44.25 \text{ nF}} \text{ (nanofarads)}$$

44.25 nF can also be written as 0.044 μF.

The capacitor values typically encountered range from roughly 1×10^{-12} F up to 1×10^{-2} F (1 pF up to 10,000 μF). Consider the effect of two equal capacitors in parallel. A little thought leads to the realization that the new effective plate area is doubled and hence the capacitance is doubled. Thus, capacitors in parallel are treated like resistors in series when determining total capacitance.

$$C_T = C_1 + C_2 + C_3 + \ldots + C_n \tag{3-2}$$

where C_T = total capacitance of n parallel capacitors

Capacitors in series behave like resistors in parallel. Thus,

$$\frac{1}{C_T} = \frac{1}{C_1} + \frac{1}{C_2} + \frac{1}{C_3} + \ldots + \frac{1}{C_n} \tag{3-3}$$

where C_T = total capacitance of n series capacitors

The case of two series capacitors allows use of the formula

$$C_T = \frac{C_1 \times C_2}{C_1 + C_2} \tag{3-4}$$

where C_T = total capacitance of two series capacitors

The schematic symbol used to represent capacitance and some capacitor networks (circuits) is shown in Fig. 3-3. Notice the symbol used in Fig. 3-3(f) for a variable 3- to 15-pF capacitor.

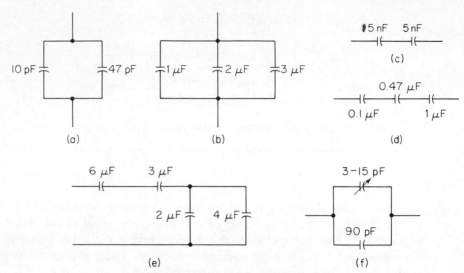

FIGURE 3-3 *Capacitor networks.*

EXAMPLE 3-2

Calculate the equivalent total capacitance for the circuits in Fig. 3-3(a) through (f).

Solution:

(a) $C_T = C_1 + C_2$ **(3-2)**

   $= 10 \text{ pF} + 47 \text{ pF} = \underline{57 \text{ pF}}$

(b) $C_T = C_1 + C_2 + C_3$ **(3-2)**

   $= 1 \ \mu\text{F} + 2 \ \mu\text{F} + 3 \ \mu\text{F}$

   $= \underline{6 \ \mu\text{F}}$

(c) $C_T = \dfrac{C_1 \times C_2}{C_1 + C_2}$ **(3-4)**

   $= \dfrac{5 \text{ nF} \times 15 \text{ nF}}{5 \text{ nF} + 15 \text{ nF}}$

   $= \underline{3.75 \text{ nF}}$

(d) $\dfrac{1}{C_T} = \dfrac{1}{C_1} + \dfrac{1}{C_2} + \dfrac{1}{C_3}$ **(3-3)**

   $= \dfrac{1}{0.1 \ \mu\text{F}} + \dfrac{1}{0.47 \ \mu\text{F}} + \dfrac{1}{1 \ \mu\text{F}}$

$$= 13.12$$
$$C_T = \underline{0.075 \ \mu F}$$

(e) $\dfrac{1}{C_T} = \dfrac{1}{2 \ \mu F + 4 \ \mu F} + \dfrac{1}{3 \ \mu F} + \dfrac{1}{6 \ \mu F}$

$$= 0.67$$
$$C_T = \underline{1.5 \ \mu F}$$

(f) $C_T = (90 \ pF + 3 \ pF)$ up to $(90 \ pF + 15 \ pF)$

$$= \underline{93 \ pF \ to \ 105 \ pF}$$

Capacitance exhibits a form of "resistance" to the flow of alternating current. The proper term here is *reactance*, however, rather than resistance. The reactance exhibited by a capacitor is termed *capacitive reactance* and symbolized as X_c. It is frequency-dependent, has ohms as its units, and is equal to

$$X_c = \frac{1}{2\pi fc} \tag{3-5}$$

where X_c = capacitive reactance, in ohms

           f = frequency, in hertz

          C = capacitance, in farads

The current flow resulting from an applied sine-wave voltage can be calculated using Ohm's Law $(i = v/X_c)$. It is of interest to consider the reactance (sometimes called *impedance*) of a capacitor to a pure dc signal. The frequency of dc is effectively 0 Hz and thus [from Eq. (3-5)] the reactance is infinite and the current flow (by Ohm's Law) is therefore zero. This result is shown in Fig. 3-1(b), where after an initial surge of current to redistribute the plate electrons, the current flow was zero. In this case, the *steady-state* (final) value of current flow is zero. The flow of current from an applied sine wave *does* exist, however, since the capacitor plates are being first charged up in one direction and then discharged to the opposite plate as the applied voltage changes polarity. This process continues on as the applied sine wave continues to change polarity.

     Theoretically, there is no power dissipation in a capacitor. The energy delivered by a source is "stored" by the electric field developed between the plates [see Fig. 3-1(a)]. The energy stored is available or redelivered back to the circuit when the electric field is allowed to collapse (e.g., remove the voltage source from the capacitor). This property of energy storage is of practical value, as will be seen later.

     In actual practice, a very small amount of power is dissipated in a capacitor. This is caused by a so-called "leakage" current flow from one plate to the other, through the dielectric. Even though the dielectric between the plates is a "nonconductor," it is not perfect and a very small leakage current flow results. A typical leakage current is in the nanoamp (10^{-9} A) range or even less.

EXAMPLE 3-3

A 30-v rms sine wave at a frequency of 1000 Hz is applied to a 0.01-μF capacitor. Calculate the reactance exhibited by the capacitor and the resulting current flow, and the power dissipation in the capacitor.

Solution:

The capacitive reactance is calculated as

$$X_c = \frac{1}{2\pi fC} \tag{3-5}$$

$$= \frac{1}{2\pi \times 1 \text{ kHz} \times 0.01 \text{ } \mu\text{F}}$$

$$= 15.9 \times 10^3 \text{ } \Omega$$

$$= \underline{15.9 \text{ k}\Omega}$$

The current flow can be calculated using Ohm's Law, so that $i = v/X_c$, or

$$i = \frac{30 \text{ v}}{15.9 \text{ k}\Omega}$$

$$= \underline{1.88 \text{ ma rms}}$$

The power dissipated is theoretically zero. A very small amount is dissipated due to the leakage current.

3-2 RESISTIVE/CAPACITIVE CIRCUITS

A simple resistive/capacitive *(RC)* circuit is shown in Fig. 3-4(a). If the switch is closed at $t = 0$, the voltage across the capacitor, V_c, as a function of time is graphed at Fig. 3-4(b). Notice that initially the voltage is increasing rapidly but that it slows down and gradually approaches the final steady-state value of 10 V. Also shown is the resistor voltage, V_R, which at $t = 0$ equals 10 V and then decreases to zero. If you study Fig. 3-4(b) you will see that V_c and V_R add up to 10 V at any instant of time—a result that is predicted by KVL. Thus, at an instant when $V_R = 3.6$ V, V_c must equal 10 V $-$ 3.6 V, or 6.4 V.

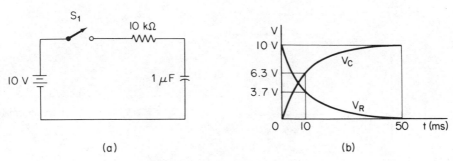

(a)   (b)

FIGURE 3-4 *Charging capacitor.*

The voltage across a capacitor *cannot* instantaneously change. At the instant the switch is closed in Fig. 3-4(a), the resistor voltage instantaneously goes from zero to 10 V and the capacitor voltage remains at 0 V. Then the resistor voltage decays to zero and the capacitor voltage increases from 0 to 10 V. The rate at which these voltage changes occur in *RC* circuits is determined by the value of resistance and capacitance in the circuit. The time taken for final steady-state conditions to be reached is approximately 5τ, where τ (tau), the time constant, is

$$\tau = RC \qquad \text{(seconds)} \tag{3-6}$$

Thus, for the circuit in Fig. 3-4(a), the time constant, τ, is RC or 10 kΩ $\times$ 1 μF = 10 ms. Thus, in 5τ or 50 ms, the final values from V_R and V_c are reached as shown in Fig. 3-4(b). The rate of voltage change in *RC* circuits is fast when the *RC* product is small. Thus, increasing either R or C (or both) slows down the voltage change. For instance, doubling the resistor value would double the time required to reach final steady-state conditions. In one time constant, the voltage changes in an *RC* circuit are 63% of the way toward their final value. Thus, for the circuit in Fig. 3-4, $\tau = 10$ ms, and at $t = 10$ ms, $V_c = 63\%$ of 10 V, or 6.3 V. The resistor voltage has changed 63% of the way from 10 V to zero (10 V $-$ 63% $\times$ 10 V) = 3.7 V. As previously indicated, $V_R + V_c$ must equal the supply voltage at all times and agreement exists at $t = 10$ ms, since $6.3 + 3.7 = 10$.

The curves shown in Fig. 3-4(b) are termed *exponential* and can be mathematically solved using the e^{-x} function. Consult a detailed circuit analysis book should you need to make such calculations.

EXAMPLE 3-4

Sketch the capacitor and resistor voltage waveforms and the circuit current waveform starting at time $t = 0$ when the switch is closed for the circuit shown in Fig. 3-5. Label the time and voltage values after 1 and 5 time constants.

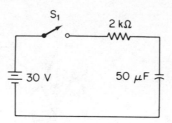

FIGURE 3-5 *Circuit for Example 3-4.*

Solution:

The capacitor voltage will exponentially increase to 30 V starting from zero when the switch is closed at $t = 0$. The resistor voltage starts at 30 V and exponentially decays to zero, as shown in Fig. 3-6.

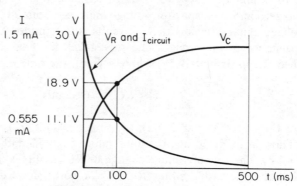

FIGURE 3-6 *Solution for Example 3-4.*

$$\tau = RC \qquad\qquad\qquad\qquad (3\text{-}6)$$
$$= 20 \text{ k}\Omega \times 50 \text{ }\mu\text{F}$$
$$= 100 \text{ ms}$$

Thus, in 1 time constant, τ, of 100 ms, 63% of the final values are reached.

$$30 \text{ V} \times 63\% = 18.9 \text{ V}$$

The capacitor voltage is shown at 18.9 V in Fig. 3-6 when $t = 100$ ms and reached 30 V in 5τ or 500 ms. At 1τ, $V_R = 30 \text{ V} - 18.9 \text{ V} = 11.1 \text{ V}$, as shown, and reaches zero at 5τ. The graph of circuit current i in Fig. 3-6 coincides with the resistor voltage. When $V_R = 30 \text{ V}$ at $t = 0$, $i_{\text{circuit}} = 30 \text{ V}/20 \text{ k}\Omega = 1.5 \text{ mA}$, as shown in Fig. 3-6. When $t = \tau = 100$ ms, the current is 11.1 V/20 kΩ = 0.555 mA, as shown. The circuit current decays to zero in 5τ when final steady-state conditions have been reached.

EXAMPLE 3-5

Determine when the capacitor voltage reaches its final steady-state value for the circuit shown in Fig. 3-7. The switch is closed at $t = 0$.

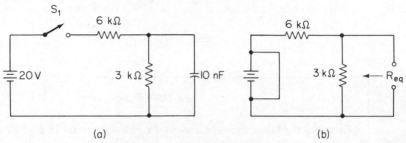

FIGURE 3-7 *Circuit for Example 3-5.*

Solution:

The final steady-state value is attained in 5τ. The only problem here is to determine what value of R to use in the calculation of τ. The proper value is the equivalent resistance "seen" by the capacitor with the battery shorted out. This process is shown in Fig. 3-7(b).

$$R_{eq} = 3\ \text{k}\Omega \parallel 6\ \text{k}\Omega$$
$$= \frac{3\ \text{k}\Omega \times 6\ \text{k}\Omega}{3\ \text{k}\Omega + 6\ \text{k}\Omega}$$
$$= 2\ \text{k}\Omega$$

$$\tau = R_{eq}C \tag{3-6}$$
$$= 2\ \text{k}\Omega \times 10\ \text{nF}$$
$$= 2 \times 10^3 \times 10 \times 10^{-9}$$
$$= 20\ \mu\text{s}$$

Since one time constant is 20 μs, the steady voltage (20 V) is reached in 5τ or $\underline{100\ \mu\text{s}}$.

3-3   *RC* CIRCUITS—AC ANALYSIS

The circuit in Fig. 3-8 shows a series *RC* circuit driven by a sinusoidal ac source. Notice the symbol used for the ac source. The total "resistance" to ac current flow of an *RC* circuit is termed *impedance, Z,* with units of ohms, and is given by the following formula:

$$Z = \sqrt{R^2 + X_c^2} \tag{3-7}$$

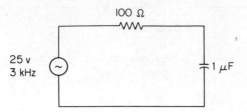

FIGURE 3-8 *RC circuit with ac voltage source.*

EXAMPLE 3-6

Calculate the impedance and current flow for the circuit in Fig. 3-8.

Solution:

$$X_c = \frac{1}{2\pi f C} \tag{3-5}$$

$$= \frac{1}{2\pi \times 3 \times 10^3 \times 1 \times 10^{-6}}$$

$$= 0.053 \times 10^3$$

$$= 53 \ \Omega$$

$$Z = \sqrt{R^2 + X_c^2} \tag{3-7}$$

$$= \sqrt{100^2 + 53^2}$$

$$= \underline{113 \ \Omega}$$

Now, i can be found using Ohm's Law, $i = v/Z$.

$$i = \frac{25 \ v}{113 \ \Omega}$$

$$= \underline{221 \ ma}$$

In a pure resistive circuit, the ac voltage and current are in *phase*. That is, when the voltage reaches a peak value, so will the current. The in-phase condition is shown in Fig. 3-9(a). In a circuit with capacitance only, a phase shift between voltage and current as shown in Fig. 3-9(b) occurs. The current "leads" the voltage by ¼ cycle. Since a full cycle is 360°, the ¼ cycle phase shift is ¼ × 360° = 90°, as shown. Notice that when the voltage is maximum, the current is minimum, and vice versa.

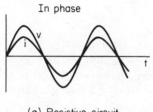

In phase

(a) Resistive circuit

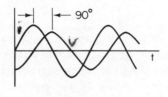

90°

(b) Capacitive circuit

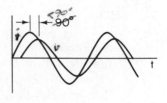

90°

(c) RC circuit

FIGURE 3-9 *Phase shift.*

The case for an *RC* circuit is shown in Fig. 3-9(c). The current still leads the voltage but by less than 90° (< 90°). The amount of phase shift θ (theta) is

$$\theta = \tan^{-1} \frac{X_c}{R} \qquad \text{(3-8)}$$

This equation means that the tangent (tan) of θ equals X_c/R.

EXAMPLE 3-7

Calculate the phase shift between the voltage and current for the circuit used in Example 3-7.

Solution:

$$\theta = \tan^{-1} \frac{X_c}{R} \qquad \text{(3-8)}$$

$$= \tan^{-1} \frac{53 \ \Omega}{113 \ \Omega}$$

or

$$\tan \theta = \frac{53 \ \Omega}{113 \ \Omega}$$

$$= 0.469$$

Now from a table of trigonometry functions or using a scientific calculator, the angle whose tangent is 0.469 is 25.1°. Therefore,

$$\theta = 25.1° \qquad \text{the phase-shift angle between } v \text{ and } i$$

3-4 *RC* FILTERS

An interesting effect occurs if a graph of v_{out} over v_{in} versus frequency is taken for the circuit in Fig. 3-10(a). At zero frequency (dc) $v_{out} = v_{in}$ and hence $v_{out}/v_{in} = 1$, as shown in Fig. 3-10(b). As the frequency is increased, however, the output starts dropping and then falls to zero at higher frequencies. Thus, the low frequencies are passed from input to output while the high frequencies are blocked. For this reason,

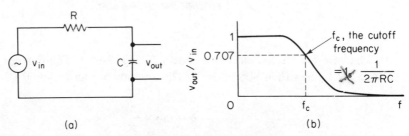

FIGURE 3-10 *Low-pass filter.*

the circuit is called a *low-pass filter* circuit—it "filters" out the high frequencies and passes the lows.

The frequency at which the output voltage has fallen to 0.707 times the input voltage is called the *cutoff* frequency, f_c. It occurs when the resistance of the resistor equals the reactance of the capacitor. Thus, f_c occurs when

$$R = X_c$$

but

$$X_c = \frac{1}{2\pi f_c C} \tag{3-5}$$

and therefore when $R = X_c$

$$R = \frac{1}{2\pi f_c C}$$

solving for f_c yields

$$f_c = \frac{1}{2\pi RC} \tag{3-9}$$

The cutoff frequency, f_c, is seen to depend upon the values of R and C in the filter circuit. The tone control in audio equipment contains a filter circuit with the resistor being a potentiometer so as to allow a variable cutoff frequency. Changing the resistor value changes the frequency response of the filter and therefore changes the "tone" heard from the speaker.

66

EXAMPLE 3-8

(a) Calculate f_c for the circuit shown in Fig. 3-10 if $R = 5 \text{ k}\Omega$ and $C = 0.47 \ \mu\text{F}$.

(b) Calculate v_{out}/v_{in} when $f = 100$ Hz.

Solution:

(a)
$$f_c = \frac{1}{2\pi RC} \qquad (3\text{-}9)$$

$$= \frac{1}{2\pi \times 5 \text{ k}\Omega \times 0.47 \ \mu\text{F}}$$

$$= \frac{1}{2\pi \times 5 \times 10^3 \times 0.47 \times 10^{-6}}$$

$$= 0.0677 \times 10^3 \text{ Hz}$$

$$= \underline{67.7 \text{ Hz}}$$

(b) By the voltage divider equation,

$$v_{out} = v_{in} \times \frac{X_c}{Z}$$

where $Z = \sqrt{R^2 + X_c^2}$ (3-7)

$$X_c = \frac{1}{2\pi f_c} \qquad X_c = \frac{1}{2\pi f C} \qquad (3\text{-}5)$$

$$= \frac{1}{2\pi \times 100 \text{ Hz} \times 0.47 \ \mu\text{F}}$$

$$= \frac{1}{295 \times 10^{-6}} = 3.39 \times 10^3 = 3.39 \text{ k}\Omega$$

$$Z = \sqrt{(5 \text{ k}\Omega)^2 + (3.39 \text{ k}\Omega)^2}$$

$$= 6.04 \text{ k}\Omega$$

$$v_{out} = v_{in} \times \frac{X_c}{Z}$$

$$\frac{v_{out}}{v_{in}} = \frac{X_c}{Z}$$

$$= \frac{3.39 \text{ k}\Omega}{6.04 \text{ k}\Omega}$$

$$= \underline{0.56}$$

This value of v_{out}/v_{in} is reasonable since 100 Hz is above f_c (67.7 Hz), and therefore v_{out}/v_{in} should be less than 0.707 [see Fig. 3-10(b)].

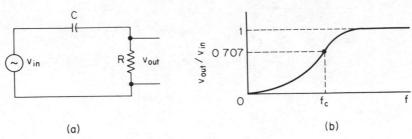

FIGURE 3-11 *High-pass filter.*

Interchanging the position of R and C in the low-pass filter, as shown in Fig. 3-11(a), results in an v_{out}/v_{in} response as shown in (b). The low frequencies are now blocked and the high frequencies passed—hence the name *high-pass filter.* The frequency where $v_{out}/v_{in} = 0.707$ is once again called the *cutoff frequency, f_c.* It is calculated using the same formula as for the low-pass filter, Eq. (3-9). The reason that RC circuits such as these are frequency-selective can be logically deduced. For instance, for the circuit in Fig. 3-10(a), the capacitor's reactance (X_c) is dependent on frequency. For low frequencies, X_c is high and thus v_{out} will be high at low frequencies.

This is seen since $v_{out}/v_{in} = X_c/Z$, as shown in Example 3-8. On the other hand, at high frequencies X_c gets very small and thus v_{out}/v_{in} is very small. The same logic can be applied to the high-pass filter in Fig. 3-11(a) to understand its frequency characteristics.

EXAMPLE 3-9

Select the proper resistor value such that the cutoff frequency is 15 kHz for a high-pass filter. The capacitor used is to be 0.01 μF. The circuit configuration is shown in Fig. 3-11(a).

Solution:

$$f_c = \frac{1}{2\pi RC} \qquad\qquad (3\text{-}9)$$

$$15\text{ kHz} = \frac{1}{2\pi R \times 0.01 \times 10^{-6}}$$

$$R = \frac{1}{2\pi \times 15 \times 10^3 \times 0.01 \times 10^{-6}}$$

$$= 1.06 \times 10^3$$

$$= \underline{1.06 \text{ k}\Omega}$$

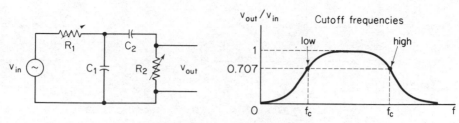

FIGURE 3-12 *Bandpass filter.*

When a low-pass and a high-pass filter are combined as shown in Fig. 3-12(a), the circuit is known as a *bandpass filter*. The v_{out}/v_{in} versus frequency characteristic is shown in Fig. 3-12(b). The resistors are shown as variable to illustrate an audio-tone-control circuit that offers both bass (low-frequency) and treble (high-frequency) controls. The R_1 control affects the high-frequency cutoff f_c, shown in Fig. 3-12(b), and R_2 affects the low-frequency cutoff, f_c.

3-5   INDUCTANCE

An *inductor* is typically a coil or coils of wire that look like the devices shown in Fig. 3-13. An inductor sometimes has an iron core and at other times is wound on a nonconductor or with an air core. An inductor is sometimes termed a *coil* or a

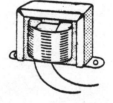

FIGURE 3-13 *Inductors.*

choke. It has characteristics that are similar and yet in many ways reversed from those of a capacitor. An inductor possesses a certain amount of inductance, L, measured in units called *henrys*. The amount of inductance exhibited by an inductor is determined by the number of wire turns and the core characteristics. The use of certain core materials, such as steel or iron, can increase the amount of inductance by 1000 or more as compared to using an air core.

The schematic symbol for inductors is shown in Fig. 3-14. Inductors in series behave like resistors in series. As shown in Fig. 3-14(a),

$$L_T = L_1 + L_2 + L_3 + \ldots + L_n \tag{3-10}$$

where L_T = total inductance of n series inductors

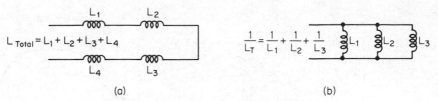

FIGURE 3-14 *Inductors in series and parallel.*

The total value of n parallel inductors can be obtained from

$$\frac{1}{L_T} = \frac{1}{L_1} + \frac{1}{L_2} + \frac{1}{L_3} + \ldots + \frac{1}{L_n} \qquad \text{(3-11)}$$

This situation is shown in Fig. 3-14(b). For two parallel inductors,

$$L_T = \frac{L_1 \times L_2}{L_1 + L_2} \qquad \text{(3-12)}$$

You will recall that a capacitor creates an electric field between its plates. An inductor, on the other hand, creates a magnetic field linking (passing) through the inductor, as shown in Fig. 3-15. This field exhibits the force that a natural magnet displays when it aligns iron filings from one end to the other. Actually, a small amount of

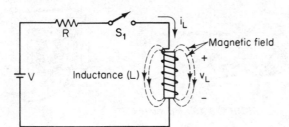

FIGURE 3-15 *Series inductive network.*

inductance and magnetic field exist in any current-carrying wire. The effect is greatly increased by coiling the wire, especially when using a magnetic-core material.

Whereas the voltage across a capacitor cannot instantaneously change, it is the current through an inductor that cannot change instantaneously. If the switch in Fig. 3-15 is closed at time $t = 0$, the resulting current flow and inductor voltage are as shown in Fig. 3-16. Notice that the current rises exponentially from zero and reaches its final steady-state value in about 5 time constants, 5τ. The τ of an *LR* circuit is

$$\tau = \frac{L}{R} \qquad \text{(seconds)} \qquad \text{(3-13)}$$

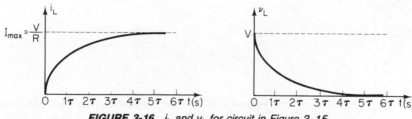

FIGURE 3-16 i_L and v_L for circuit in Figure 3–15.

The current waveform has the same shape as the resistor voltage waveform. The inductor voltage starts off at the battery voltage and decays to zero in 5τ [Fig. 3-16(b)], whereas the resistor voltage starts at zero and rises to the battery voltage in 5τ. As with *RC* circuits, 63% of the final change will occur in 1 time constant.

EXAMPLE 3-10

Calculate the total inductance of the circuit shown in Fig. 3-17(a), the circuit's time constant, and provide a sketch of current versus time if the switch is closed at $t = 0$.

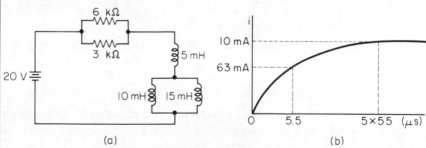

          (a)   (b)

FIGURE 3-17   Example 3-10.

Solution:

The total inductance is 5 mH in series with the 10 mH, 15 mH parallel combination. Thus,

$$L_T = 5\ \text{mH} + (10\ \text{mH} \| 15\ \text{mH})$$

$$= 5\ \text{mH} + \frac{10\ \text{mH} \times 15\ \text{mH}}{10\ \text{mH} + 15\ \text{mH}}$$

$$= 5\ \text{mH} + 6\ \text{mH}$$

$$= \underline{11\ \text{mH}}$$

The resistance seen by the inductance is 6 kΩ‖3 kΩ, or 6 kΩ × 3 kΩ ÷ (6 kΩ + 3 kΩ), or 2 kΩ. Thus,

$$\tau = \frac{L}{R} \tag{3-13}$$

$$= \frac{11 \text{ mH}}{2 \text{ K}\Omega}$$

$$= 5.5 \ \mu\text{s}$$

The current flow in an LR circuit starts at zero and rises exponentially to its final steady-state value. That value will be 20 V/2 kΩ, or 10 ma, and this result is shown at Fig. 3-17(b). Notice that the current reaches 63% of its final value in 1τ and 10 ma in about 5τ.

3-6   *RL* CIRCUITS—AC ANALYSIS

An inductor looks like a short circuit to a dc level in the final steady-state condition. This contrasts to a capacitor, which looks like an open circuit under the same condition. An inductor resists instantaneous current changes, while a capacitor resists instantaneous voltage changes. An inductor is sometimes called a *choke* because of its ability to "choke" off current variations in a circuit. The impedance, Z, of an inductor in an ac sine-wave circuit is termed inductive reactance, X_L, and is predicted by

$$X_L = 2\pi f L \tag{3-14}$$

Notice that X is directly proportional to the frequency, f, as opposed to X_c that is inversely proportional. In a series RL circuit, the total impedance, Z, is

$$Z = \sqrt{R^2 + X_L^2} \tag{3-15}$$

In a circuit with just a pure inductance, the phase angle (θ) between voltage and current is 90° with the current lagging the voltage. Recall that in an RC circuit, the opposite is true—the current leads the voltage by 90°. In an RL circuit, the current will lag voltage by something less than 90°, as predicted by

$$\theta = \tan^{-1} \frac{X_L}{R} \tag{3-16}$$

You will recall a similar situation for an RC circuit [Eq. (3-8)] except that the current leads the voltage by θ in that case.

EXAMPLE 3-11

For the circuits shown in Fig. 3-18, determine the phase angle between v and i and circuit current flow.

Solution:

   (a)   The total inductance is

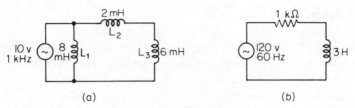

(a)   (b)

FIGURE 3-18 *Example 3-11.*

$$L_T = L_1 \| (L_2 + L_3)$$
$$= 8 \text{ mH} \| (2 \text{ mH} + 6 \text{ mH})$$
$$= 8 \text{ mH} \| 8 \text{ mH}$$
$$= 4 \text{ mH}$$

$$X_L = 2\pi fL \qquad\qquad\qquad\qquad (3\text{-}14)$$
$$= 2\pi \times 1 \text{ kHz} \times 4 \text{ mH}$$
$$= 25.1 \ \Omega$$

$$Z = \sqrt{R^2 + X_L^2} \qquad\qquad\qquad (3\text{-}15)$$
$$= 25.1 \ \Omega \qquad \text{since } R = 0$$

$$i = \frac{v}{Z} = \frac{10 \text{ v}}{25.1 \ \Omega}$$
$$= \underline{0.398 \text{ a}}$$

Since there is no resistance in this circuit, the phase angle θ is 90° with current lagging the voltage.

(b) $$X_L = 2\pi fL \qquad\qquad\qquad\qquad (3\text{-}14)$$
$$= 2\pi \times 60 \text{ Hz} \times 3 \text{ H}$$
$$= 1.13 \text{ k}\Omega$$

$$Z = \sqrt{R^2 + X_L^2} \qquad\qquad\qquad (3\text{-}15)$$
$$= \sqrt{(1 \text{ k}\Omega)^2 + (1.13 \text{ k}\Omega)^2}$$
$$= 1.51 \text{ k}\Omega$$

$$i = \frac{v}{Z} = \frac{120 \text{ v}}{1.51 \text{ k}\Omega}$$
$$= \underline{79.5 \text{ ma}}$$

$$\theta = \tan^{-1} \frac{X_L}{R} \qquad\qquad\qquad (3\text{-}16)$$

$$= \tan^{-1} \frac{1.13 \text{ k}\Omega}{1 \text{ k}\Omega}$$

$$= \tan^{-1} 1.13$$
$$= \underline{48.5°} \qquad i \text{ lagging } e$$

3-7 RESONANCE

In this section the major effects of circuits with resistors, capacitors, and inductors will be considered. As has previously been indicated, to a dc signal (0 Hz), a capacitor "looks" like an open circuit and an inductor "looks" like a short circuit. The *RLC* circuit in Fig. 3-19(a) is therefore equivalent to the circuit in (b) with respect to final steady-state conditions. The inductor is replaced with a short circuit and the capacitor is replaced by an open circuit. The final steady-state voltages can simply be calculated as 5 V across the 5-kΩ resistor and 15 V across the 15-kΩ resistor as shown in Fig. 3-19(b).

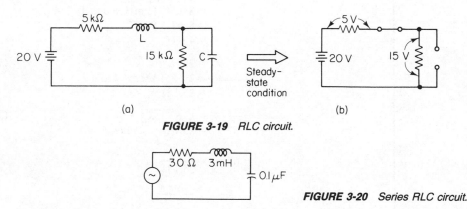

(a)

FIGURE 3-19 RLC circuit.

(b)

FIGURE 3-20 Series RLC circuit.

The effects in *RLC* circuits as a response to ac sine-wave voltages is of greater significance than the dc response. Consider the series *RLC* circuit shown in Fig. 3-20. In this case, the total impedance, *Z*, is predicted by the following formula:

$$Z = \sqrt{R^2 + (X_L - X_c)^2} \qquad \text{(3-17)}$$

An interesting effect occurs at the frequency when X_L is equal to X_c. That frequency is termed the *resonant frequency, f_r*. At f_r the circuit impedance is equal to the resistor value (which may just be the series winding resistance of the inductor). That result can be shown from Eq. (3-17), in that when $X_L = X_c$, $X_L - X_c$ equals zero, so that $Z = \sqrt{R^2 + 0^2} = \sqrt{R^2} = R$. The resonant frequency can be determined by finding the frequency where $X_L = X_c$.

$$X_L = X_c$$

$$2\pi f_r L = \frac{1}{2\pi f_r C}$$

$$f_r{}^2 = \frac{1}{4\pi^2 LC}$$

$$f_r = \frac{1}{2\pi \sqrt{LC}} \qquad \text{(Hz)} \qquad \text{(3-18)}$$

EXAMPLE 3-12

Determine the resonant frequency for the circuit shown in Fig. 3-20. Calculate its impedance when $f = 12$ kHz.

Solution:

$$f_r = \frac{1}{2\pi\sqrt{LC}} \tag{3-18}$$

$$= \frac{1}{2\pi\sqrt{3 \text{ mH} \times 0.1 \text{ } \mu\text{F}}}$$

$$= \frac{1}{2\pi\sqrt{3 \times 10^{-3} \times 0.1 \times 10^{-6}}}$$

$$= \frac{1}{2\pi \times \sqrt{3 \times 10^{-10}}}$$

$$= \frac{1}{10.9 \times 10^{-5}}$$

$$= 0.0919 \times 10^5$$

$$= \underline{9.19 \text{ kHz}}$$

At 12 kHz,

$$X_L = 2\pi fL \tag{3-14}$$
$$= 2\pi \times 12 \text{ kHz} \times 3 \text{ mH}$$
$$= 226 \text{ } \Omega$$

$$X_c = \frac{1}{2\pi fC} \tag{3-9}$$

$$= \frac{1}{2\pi \times 12 \text{ kHz} \times 0.1 \text{ } \mu\text{F}}$$

$$= \frac{1}{7.54 \times 10^{-3}}$$

$$= 133 \text{ } \Omega$$

$$Z = \sqrt{R^2 + (X_L - X_c)^2}$$
$$= \sqrt{30^2 + (226 - 133)^2}$$
$$= \sqrt{900 + 8,755}$$
$$= \underline{98.3 \text{ } \Omega}$$

The circuit contains more inductive than capacitive reactance at 12 kHz and is therefore said to look inductive.

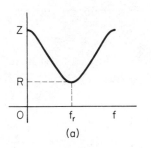

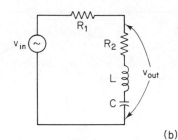

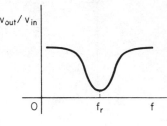

(a)   (b)

FIGURE 3-21 *Series RLC circuit effects.*

The impedance of the series *RLC* circuit is minimum at its resonant frequency and equal to the value of *R*. A graph of its impedance, *Z*, versus frequency has the shape of the curve shown in Fig. 3-21(a). At low frequencies the circuit's impedance is very high, since X_c is high. At high frequencies X_L is very high and thus *Z* is high. At *resonance*, when $f = f_r$, the circuit's $Z = R$ and is at its minimum value. This impedance characteristic can provide a filter effect, as shown in Fig. 3-21(b). At f_r, $X_L = X_c$ and thus

$$v_{out} = v_{in} \times \frac{R_2}{R_1 + R_2}$$

by the voltage divider equation. At all other frequencies, the *RLC* circuit's impedance is greater than R_2 and therefore v_{out} goes up. The response for the circuit in Fig. 3-21(b) is termed a *band-reject* or *notch filter*. A "band" of frequencies are being "rejected" and a "notch" is cut into the output at the resonant frequency, f_r.

EXAMPLE 3-13

Determine f_r for the circuit shown in Fig. 3-21(b) when $R_1 = 20$ Ω, $R_2 = 1$ Ω, $L = 1$ mH, $C = 0.4$ μF, and $v_{in} = 50$ mv. Calculate v_{out} at f_r, 12 kHz, and 24 kHz.

Solution:

The resonant frequency is

$$f_r = \frac{1}{2\pi\sqrt{LC}} \tag{3-18}$$

$$= 7.95 \text{ kHz}$$

At resonance

$$v_{out} = v_{in} \times \frac{R_2}{R_2 + R_1}$$

$$= 50 \text{ mv} \times \frac{1 \ \Omega}{1 \ \Omega + 20 \ \Omega}$$

$$= \underline{2.38 \text{ mv}}$$

At $f = 12$ kHz,

$$\begin{aligned} X_L &= 2\pi f L \\ &= 2\pi \times 12 \text{ kHz} \times 1 \text{ mH} \\ &= 75.4 \ \Omega \end{aligned} \qquad \text{(3-14)}$$

and

$$X_c = \frac{1}{2\pi f C}$$

$$= \frac{1}{2\pi \times 12 \text{ kHz} \times 0.4 \ \mu\text{F}} \qquad \text{(3-5)}$$

$$= 33.2 \ \Omega$$

and the impedance at 12 kHz is

$$\begin{aligned} Z &= \sqrt{(R_1 + R_2)^2 + (X_L - X_c)^2} \\ &= \sqrt{(20 \ \Omega + 1 \ \Omega)^2 + (75.4 \ \Omega - 33.2 \ \Omega)^2} \qquad \text{(3-17)} \\ &= 47.1 \ \Omega \end{aligned}$$

Thus, at 12 kHz, v_{out} is

$$V_{out} = V_{in} \times \frac{\sqrt{R_2^2 + (X_L - X_c)^2}}{Z}$$

$$v_{out} = v_{in} \times \frac{Z}{Z + R_1}$$

$$= 50 \text{ mv} \times \frac{47.1 \ \Omega}{47.1 \ \Omega + 20 \ \Omega}$$

$$= 35.1 \text{ mv} \qquad 44.8 \text{ mV}$$

Now repeating this process at 24 kHz, we obtain

$$X_L = 2\pi \times 24 \text{ kHz} \times 1 \text{ mH} = 151 \ \Omega$$

$$X_c = \frac{1}{2\pi \times 24 \text{ kHz} \times 0.4 \ \mu\text{F}} = 16.6 \ \Omega$$

$$Z = \sqrt{(20 + 1)^2 + (151 \ \Omega - 16.6 \ \Omega)^2}$$

$$= 136 \ \Omega$$

Thus, at 24 kHz, v_{out} is

$$v_{out} = 50 \text{ mv} \times \frac{136 \ \Omega}{136 \ \Omega + 20 \ \Omega}$$

$$= 43.6 \text{ mv} \qquad 49.4 \text{ mV}$$

Example 3-13 shows that the filter's output will approach 50 mv as the frequency is increased. Calculation of the circuit's output for frequencies below resonance would show the same result and is left as an exercise at the end of the chapter. The band-reject or notch filter is sometimes called a *trap*, since it can "trap" or get rid of a specific range of frequencies near f_r. A trap is commonly used in a television receiver, where rejection of some specific frequencies is necessary for good picture quality.

3-8 *LC* BANDPASS FILTER

If the filter's configuration is changed to that shown in Fig. 3-22(a), it is called a *bandpass filter*. The bandpass filter has a response as shown at Fig. 3-22(b). You will recall from the earlier discussion of bandpass filters using just *RC* elements

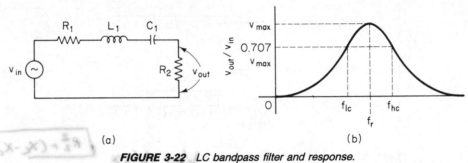

(a)   (b)

FIGURE 3-22 *LC bandpass filter and response.*

that f_{lc} is the low-frequency cutoff where the output has fallen to 0.707 times its maximum value. Similarly, f_{hc} is the high-frequency cutoff. The frequency range between f_{lc} and f_{hc} is called the filter's *bandwidth*, usually abbreviated BW. The BW is equal to $f_{hc} - f_{lc}$ and it can mathematically be shown that

$$BW = \frac{R}{2\pi L} \tag{3-19}$$

where   BW = bandwidth, in hertz

R = total circuit resistance

L = circuit inductance

The *quality factor, Q,* of the filter is a measure of how selective or narrow is its *passband,* as compared to its center frequency, f_r. Thus,

$$Q = \frac{f_r}{BW} \tag{3-20}$$

The quality factor, Q, can also be determined as

$$Q = \frac{X_L}{R}$$

(3-21)

where X_L = inductive reactance at resonance

    R = total circuit resistance

As Q increases, the filter becomes more *selective;* that is, a smaller passband or narrower bandwidth is allowed. A major limiting factor in the highest attainable Q is the resistance factor shown in Eq. (3-21). In order to obtain a high Q, the circuit resistance must be low. Quite often, the limiting factor becomes the *winding resistance* of the inductor itself. The turns of wire (and its associated resistance) used to make an inductor provide this limiting factor.

To obtain the highest Q possible, larger wire (with less resistance) could be used, but then greater cost and physical size results to end up with the same number of turns and thus the same amount of inductance. The typical Q's attainable approach 1000 with very high "quality" inductors.

EXAMPLE 3-14

A filter circuit of the form shown at Fig. 3-22(a) has a response as shown in Fig. 3-23. Determine the bandwidth, the filter's Q, the value of inductance if $C = 0.001$ μF, and the total circuit resistance.

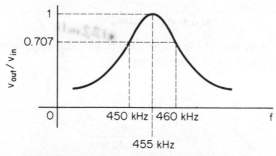

FIGURE 3-23 *Response curve for Example 3-14.*

Solution:

(a) From Fig. 3-23, the BW is simply the frequency range between f_{hc} and f_{lc} or 460 kHz − 450 kHz = <u>10 kHz</u>.

(b) The filter's peak output occurs at 455 kHz, which is f_r.

$$Q = \frac{f_r}{BW} \qquad (3\text{-}20)$$

$$= \frac{455 \text{ kHz}}{10 \text{ kHz}}$$

$$= \underline{45.5}$$

(c) Equation (3-18) can be used to solve for L since f_r and C are known.

$$f_r = \frac{1}{2\pi \sqrt{LC}} \qquad (3\text{-}18)$$

$$455 \text{ kHz} = \frac{1}{2\pi \sqrt{L \times 0.001 \text{ } \mu\text{F}}}$$

Squaring both sides of the equation yields

$$(455 \text{ kHz})^2 = \left(\frac{1}{2\pi \sqrt{L \times 10 \times 10^{-10}}} \right)^2$$

$$(0.455 \times 10^6)^2 = \frac{1}{4\pi^2 \times L \times 10 \times 10^{-10}}$$

$$0.207 \times 10^{12} = \frac{1}{4\pi^2 \times 10 \times 10^{-10} \times L}$$

$$L = \frac{1}{4\pi^2 \times 10 \times 10^{-10} \times 0.207 \times 10^{12}}$$

$$= \frac{1}{0.817 \times 10^4}$$

$$= 1.2 \times 10^{-4}$$

$$= \underline{0.12 \text{ mH}}$$

$\cdot$ *122 mH*

(d) Equation (3-19) can be used to solve for total circuit resistance because the BW and L are known.

$$BW = \frac{R}{2\pi L} \qquad (3\text{-}19)$$

$$10 \text{ kHz} = \frac{R}{2\pi \times 0.12 \text{ mH}}$$

$$R = 10 \times 10^3 \text{ Hz} \times 2\pi \times 0.12 \times 10^{-3} \text{ mH}$$

$$= 7.52 \times 10^0$$

$$= \underline{7.52 \text{ } \Omega} \quad \text{7.67} \Lambda$$

The filter analyzed in Example 3-14 is typical of the response necessary in an AM radio receiver. Further information on this subject is provided in Chapter 9.

3-9 PARALLEL *LC* CIRCUITS

A parallel *LC* circuit and its impedance versus frequency characteristic is shown in Fig. 3-24. The only resistance shown for this circuit is the coil's winding resistance and it is effectively in series with the inductor (coil) as shown. Notice that the impe-

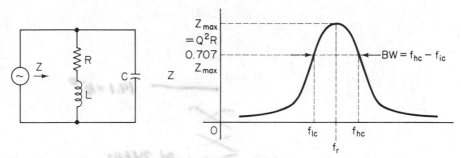

FIGURE 3-24 *Parallel LC circuit and response.*

dance of the parallel *LC* circuit reaches a maximum value at the resonant frequency, f_r, and falls to a low value on either side of resonance. As shown in Fig. 3-24, the maximum impedance is

$$Z_{max} = Q^2 \times R \qquad (3\text{-}22)$$

Equations (3-18) through (3-21) for series *LC* circuits also apply to parallel *LC* circuits as long as Q is greater than 10 ($Q > 10$), which is normally the case.

The parallel *LC* circuit is sometimes called a *tank* circuit. Energy is stored in each reactive element (*L* and *C*), first in one and then the other. The transfer of energy between the two elements will occur at a natural rate equal to the resonant frequency and is sinusoidal in form.

EXAMPLE 3-15

A parallel *LC* tank circuit is made up of an inductor of 3 mH and a winding resistance of 2 Ω. The capacitor is 0.47 μF. Determine

(a) f_r

(b) Q

(c) Z_{max}

(d)   BW

(e)   The frequencies where $Z = 0.707\ Z_{max}$

Solution:

(a)
$$f_r = \frac{1}{2\pi\sqrt{LC}} \tag{3-18}$$

$$= \frac{1}{2\pi\sqrt{3\ \text{mH} \times 0.47\ \mu\text{F}}}$$

$$= \frac{1}{2\pi\sqrt{3 \times 10^{-3} \times 0.47 \times 10^{-6}}}$$

$$= \frac{1}{2\pi\sqrt{\cancel{12.1 \times 10^{-10}}}} \quad 14.1 \times 10^{-10}$$

$$= \frac{1}{21.9 \times 10^{-5}}$$

$$= 0.0458 \times 10^{5}$$

$$= \cancel{4.58\ \text{kHz}} \quad 4.24\ kHz$$

(b)
$$Q = \frac{X_L}{R} \tag{3-21}$$

where $X_L = 2\pi f L$ (3-14)

$$= 2\pi \times 4.58\ \text{kHz} \times 3\ \text{mH}$$

$$= \cancel{86.2\ \Omega} \quad 79.9\ \Lambda$$

$$Q = \frac{86.2\ \Omega}{2\ \Omega}$$

$$= \cancel{43.1} \quad \cancel{79.9\ \Lambda} \quad 39.9$$

(c)
$$Z_{max} = Q^2 \times R$$
$$= (43.1)^2 \times 2\ \Omega \tag{3-22}$$
$$= \cancel{3.72\ \text{k}\Omega} \quad 3.19\ k\Lambda$$

(d)
$$BW = \frac{R}{2\pi L} \tag{3-19}$$

$$= \frac{2\ \Omega}{2\pi \times 3\ \text{mH}}$$

$$= 106\ \text{Hz}$$

(e)   The frequencies where Z is $0.707 \times Z_{max}$ are f_{lc} and f_{hc}, respectively. They are 106 Hz (the BW) apart. Therefore,

$$f_{lc} = 4.58 \text{ kHz} - \frac{106 \text{ Hz}}{2} = \cancel{4527 \text{ Hz}}$$

4187Hz

and

$$f_{hc} = 4.58 \text{ kHz} + \frac{106 \text{ Hz}}{2} = \cancel{4633 \text{ Hz}}$$ *4293 Hz*

The parallel *LC* tank circuit finds much application in circuits used to generate sine waves. These circuits will be studied in Chapter 9 and are called *oscillators*. If a dc voltage is momentarily applied to a tank circuit as shown in Fig. 3-25, an exchange of energy between the inductor and capacitor is set up that occurs with a

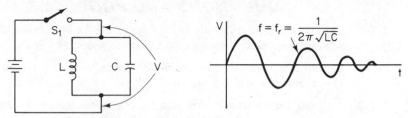

FIGURE 3-25 *Generation of a damped sinewave.*

natural sinusoidal shape. The energy gradually decreases due mainly to the dissipation in the inductor's winding resistance. The resulting sine wave generated (shown in Fig. 3-25) gradually is reduced to zero and is called a damped sine wave. In an oscillator, a means of maintaining a constant amplitude output is provided, as will be shown in chapter 9. The frequency of sine wave generated will be equal to $1/2\pi \sqrt{LC}$, as shown in Fig. 3-25.

EXAMPLE 3-16

Determine the frequency of the damped sine wave shown in Fig. 3-25 if $L = 1$ nH and $C = 47$ pF.

Solution:

The frequency of oscillation will be equal to the resonant frequency, f_r, of the tank circuit.

$$f_r = \frac{1}{2\pi \sqrt{LC}} \qquad\qquad (3\text{-}18)$$

$$= \frac{1}{2\pi \sqrt{1 \text{ nH} \times 47 \text{ pF}}}$$

$$= \frac{1}{2\pi \sqrt{1 \times 10^{-9} \times 47 \times 10^{-12}}}$$

$$= \frac{1}{2\pi \sqrt{4.7 \times 10^{-20}}}$$

$$= \frac{1}{2\pi \times 2.17 \times 10^{-10}}$$

$$= 0.0734 \times 10^{10}$$

$$= 734 \times 10^{6}$$

$$= \underline{734 \text{ MHz}}$$

QUESTIONS AND PROBLEMS

3–1–1. Describe the physical characteristics of a capacitor. In your own words, briefly explain what happens when it is connected to a battery as shown in Fig. 3-1.

3–1–2. Explain how a capacitor can be used to store energy for release at a later time. How is energy stored in a capacitor?

3–1–3. Calculate the capacitance exhibited by a capacitor with plate area of 0.3×10^{-3} m^2 and spacing of 0.035×10^{-6} m. The dielectric is air. Repeat for a Teflon, a mica, and a glass dielectric.

3–1–4. Determine the plate area required for an air dielectric capacitor with plate spacing 1×10^{-4} m if 500 μF is the necessary capacitance.

3–1–5. Explain the interrelationships among a capacitor's voltage rating, physical size, and the amount of capacitance it exhibits.

3–1–6. A 10-μF, a 15-μF, and a 5-μF capacitor are placed in parallel. Together they are in series with a 60-μF capacitor. What is the total equivalent capacitance?

3–1–7. A 25-nF and 45-nF capacitor are in series. Calculate the capacitive reactance and current flow for a 20-v 100-MHz signal.

3–1–8. What is the theoretical amount of power dissipation in a capacitor? Why is the value not quite realized in actual practice? Explain the meaning of leakage resistance and current.

3–2–9. In a series RC circuit, explain what happens when a dc voltage is suddenly applied. Include the circuit current, resistor voltage, and capacitor voltage in your explanation. What determines the initial and final values of current flow? Sketch the voltage and current waveforms as a function of time.

3–2–10. For a series RC circuit, explain what determines the rate at which charging takes place. How long does it take to reach 63% of a final charging value?

3–2–11. A series RC circuit is connected to a 12-V battery at time $t = 0$. The values are $C = 0.47 \mu$F and $R = 3.3$ kΩ. Sketch the capacitor and resistor voltage waveforms and the circuit current waveform. Label the time *and* values after 1 and 5 time constants.

3–2–12. A steady-state charging condition is to be reached in 1 s. Calculate the capacitor value required if a 15-kΩ resistor is to be used.

3–3–13. A 10-kΩ resistor in series with a 0.68-μF capacitor is driven by a 20-V, 1-kHz voltage. Calculate X_c, Z, and the resulting current flow.

3–3–14. Explain the effect of phase shift between voltage and current in an *RC* circuit driven by an ac source. Calculate the phase shift for the circuit described in Problem 3–3–13.

3–3–15. A series *RC* circuit has an impedance of 1.25 kΩ. Calculate X_c, C, and θ if $R = 1$ kΩ and the ac signal has a frequency of 10 kHz.

3–4–16. Describe the major characteristics of a filter circuit. How is the cutoff frequency, f_c, defined?

3–4–17. Show mathematically how a voltage reduction factor of 0.707 results in a half-power condition to a fixed value of load resistance. (*Hint:* Recall that power is equal to V^2/R.)

3–4–18. Provide the schematic of a low-pass *RC* filter that offers a cutoff frequency of 1 kHz using a 10-kΩ resistor. Indicate the required capacitor value on your schematic.

3–4–19. Explain how an *RC* circuit can be used to provide a "tone control" to a radio receiver.

3–4–20. Provide the schematic of a high-pass *RC* filter that offers a cutoff frequency of 1 kHz using a 10-kΩ resistor. Indicate the required capacitor value on your schematic.

3–4–21. A high-pass filter has $R = 75$ Ω and $C = 1$ μF. Calculate the cutoff frequency. What could be done to lower f_c?

3–4–22. For the filter described in Problem 3–4–21, calculate v_{out}/v_{in} at f_c and at 1 kHz and 3 kHz.

3–4–23. Explain in a logical fashion how an *RC* circuit provides a frequency-selective characteristic.

3–4–24. Provide a schematic of a bandpass *RC* circuit and show a sketch of its v_{out}/v_{in} versus frequency characteristic.

3–5–25. Describe the physical construction of an inductor. What factors determine the amount of inductance that an inductor exhibits?

3–5–26. Draw a schematic diagram of the following inductors and determine the total inductance:
(a) 1 mH in series with 5.5 mH.
(b) 3 mH in series with a 7 mH and 4 mH parallel combination.
(c) 8 H in parallel with 12 H.
(d) A parallel combination of 27 μH, 100 μH, and 320 μH.

3–5–27. What is a magnetic field? Describe the shape of a magnetic field around an inductor.

3–5–28. A 5-kΩ resistor is in series with a 27-mH inductor. At time $t = 0$ an 8 V dc source is applied to the circuit. How long will it take for the final steady-state condition to be reached?

3–5–29. For the circuit described in Problem 3–5–28, sketch circuit current versus time. Show the current value after 1 and 5 time constants.

3–6–30. Calculate the inductive reactance of a 5-mH inductor for a 5-v 60-kHz source of voltage. Determine the current flow.

3–6–31. A 2.7-kΩ resistor is added in series with the inductor in Problem 3–6–30. Calculate the impedance, i, and the phase angle (θ) between v and i.

3–6–32. A doorbell buzzer operates from a line voltage of 120 v, 60 Hz. It draws 100 ma of current and has a winding resistance of 800 Ω. Calculate X_L and the value of inductance the buzzer winding has.

3–7–33. Describe a procedure for analyzing an RLC circuit with respect to final steady-state conditions as a response to a dc signal.

3–7–34. Calculate the impedance of a series RLC circuit to a 10-kHz signal when $R = 100$ Ω, $L = 1.5$ mH, and $C = 0.25$ μF.

3–7–35. Calculate the resonant frequency and impedance at resonance for the circuit described in Problem 3–7–34. Provide a sketch of impedance versus frequency for this circuit.

3–7–36. The filter circuit shown in Fig. ~~3-22(a)~~ 3-21(b) has $R_1 = 15$ Ω, $R_2 = 0.5$ Ω, $L = 0.5$ mH, $C = 0.068$ μF, and $v_{in} = 300$ mv. Calculate v_{out} at f_r and at ($f_r + 1$ kHz) and ($f_r - 1$ kHz).

3–7–37. For the filter described in Example 3-13, calculate v_{out} at 4 kHz and 2 kHz. Based on these results and those provided in Example 3-13, provide a graph of v_{out}/v_{in} for this band-reject filter.

3–8–38. Sketch the v_{out}/v_{in} versus frequency characteristic for a LC bandpass filter. Show f_{lc} and f_{hc} on the sketch and explain how they are defined. On this sketch, show the bandwidth (BW) of the filter and explain how it is defined.

3–8–39. Define the quality factor (Q) of an LC bandpass filter. Explain how it relates to the "selectivity" of the filter. Describe the major limiting value on the Q of a filter.

3–8–40. An FM radio receiver uses an LC bandpass filter with $f_r = 10.7$ MHz and requires a BW of 200 kHz. Calculate the required Q for this filter.

3–8–41. The circuit described in Problem 3–8–40 is shown in Fig. 3-22(a). If $C = 0.1$ nF (0.1×10^{-9} F), calculate the required inductor value and the total circuit resistance.

3–9–42. A parallel LC tank circuit has a Q of 60 and coil winding resistance of 5 Ω. Determine the circuit's impedance at resonance.

3–9–43. The circuit described in Problem 3–9–42 has $L = 27$ mH, $C = 0.68$ μF, and a coil winding resistance of 4 Ω. Calculate f_r, Q, Z_{max}, the BW, f_{lc}, and f_{hc}.

3–9–44. Define what is meant by the term "oscillator." Describe the role that an LC circuit might assume in an oscillator.

3–9–45. Describe with words and a sketch what is meant by a damped sine wave. How can it be created, and why does it decay?

```
┌─────────────────────────────┐
│                             │
│            4                │
│                             │
│                             │
│       MAGNETICS             │
│                             │
└─────────────────────────────┘
```

4-1 MAGNETISM

Magnetism is a force of attraction or repulsion that can be introduced most pronouncedly in certain metals by means of an electric current flowing in a nearby wire. Magnets formed in this way are referred to as artificial magnets. Magnets may also be classified as either permanent magnets or electromagnets, depending upon their ability to retain their magnetic properties. *Permanent magnets* are usually made of steel or steel alloys because steel has a tendency to retain its magnetic properties once it has been magnetized. *Electromagnets* are usually made of soft iron or iron alloys because iron tends to lose most of its magnetic properties once the magnetizing influence of electric current flow has been removed. Hence, the effect of magnetism can be controlled by an electric current. Such is the case in many types of electric machines, wherein electromagnets are major components.

Magnetic Effects

(a) *Magnetic Influence*

The influence of magnets in the space surrounding them may be detected in many ways. Experiments have shown that this influence (i.e., the force of attraction and repulsion) varies inversely as the square of the distance

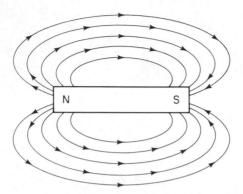

FIGURE 4-1 Magnetic lines of force about a bar magnet.

from the magnet. To account for this influence, a magnet is said to establish a magnetic *field* around itself, which is represented pictorially by directed lines.

Consider the permanent bar magnet of Fig. 4-1. For convenience, the ends of a magnet are arbitrarily referred to as *poles,* the north pole being the end from which the magnetic lines of force leave the magnet and the south pole being the end into which the magnetic lines of force reenter the magnet. The magnetic lines of force pass through the magnet from its south to its north pole. Lines of magnetic force are always closed loops. These lines of force are sometimes referred to as lines of *flux,* or simply magnetic flux.

(b)   *Lines of Force*

If a second bar magnet is placed near the first bar magnet (Fig. 4-2) so that the unlike poles are adjacent, a force of attraction results and some of the magnetic lines of force from the first bar magnet are diverted toward the pole of the second bar magnet. However, if one magnet is reversed so

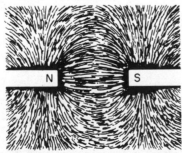

FIGURE 4-2 Magnetic lines of force about two bar magnets of opposite polarity.

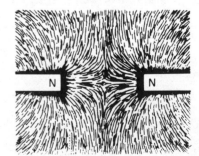

FIGURE 4-3 Magnetic lines of force about two bar magnets of like polarity.

that the like poles are adjacent, a force of mutual repulsion will result, tending to separate the magnets (Fig. 4-3). The magnetic lines of force emanating from each of the adjacent like poles of the two bar magnets will tend to repel each other while simultaneously seeking to reenter the opposite end (opposite pole) of their respective magnets. From these observations it can be seen that unlike poles of magnets attract each other and like poles repel each other. This phenomenon of magnetic attraction and repulsion produces the rotation in many types of electric motors.

(c)   *Current in a Wire*

When an electric current is passed through a wire, a magnetic field is produced around the wire (Fig. 4-4). The direction of the magnetic lines of force, which form concentric circles around the wire, depends upon the

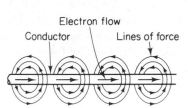

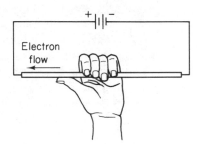

FIGURE 4-4 *Magnetic lines of force about a current-carrying conductor.*

FIGURE 4-5 *Left-hand rule for a current-carrying conductor.*

direction of the electric current through the wire. The direction of the magnetic lines of force about a current-carrying conductor can be determined by using the *left-hand rule* for a current-carrying conductor. This rule states that if a current-carrying conductor is grasped in the left hand with the thumb pointing in the direction of electron flow (negative to positive), the fingers will encircle the wire in the direction of the magnetic lines of force, as illustrated in Fig. 4-5.

Wire Coil Effects

(a)   *Magnetic Field about a Coil*

The magnetic field resulting from the conditions shown in Fig. 4-4, even with high currents, is relatively weak. But if the electrical conductor is formed into a coil, a relatively stronger magnetic field is created with magnetic lines of force running through the center of and perpendicular to each turn of the coil, as shown in Fig. 4-6. The coil current sets up circular

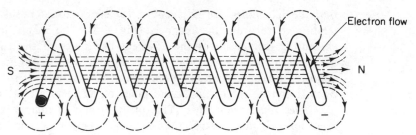

FIGURE 4-6 *Magnetic field about a current-carrying coil.*

magnetic lines of force around each coil turn in accordance with the left-hand rule. Consequently, the magnetic lines of force encircle the upper part of each coil turn in a clockwise direction. In the center of the coil, all the magnetic lines of force run in the same direction, thereby aiding each other to produce a net positive effect. Between adjacent coil turns, the magnetic lines of force cancel each other. Consequently, magnetic lines of force travel the entire length of the coil in order to complete their loops. This makes the coil behave as a magnet with a north and a south pole. By using the left hand, again, the north-pole end of a current-carrying coil can be determined. This time, if the coil is grasped so that the fingers of the left hand encircle the coil in the direction of the electron flow through the coil, the thumb will point to the north pole of the coil.

(b) *Electromagnets*

If a piece of magnetic material, usually soft iron, is placed within a coil through which current is flowing, the magnetic properties of the coil are tremendously increased. The reason for this increase in magnetic strength is due to the greater permeability of the soft iron. *Permeability* is the ease with which magnetic lines of force pass through a substance. A coil wound around a core of magnetic material is called an electromagnet. The coil may be wound with one or more layers of wire from one end to the other and back, provided, of course, that the current flows around the core continuously in the same direction.

(c) *Magnetic Field about a Coil with an Iron Core*

Figure 4-7 illustrates the effect that an iron core has on the magnetic lines of force surrounding a current-carrying coil. In Fig. 4-7(a) notice that the lines of force passing through the coil are confined to the iron core. If the iron core is pulled partially out of the coil as shown in Fig. 4-7(b), the magnetic lines of force will be extended to enter the end of the iron core that is outside the coil. Once the lines have established themselves in the core, they tend to shorten, thereby exerting a force on the core. This force

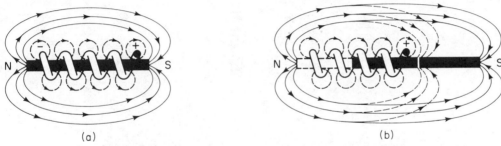

(a)   (b)

FIGURE 4-7   *Effect of an iron core on the magnetic field of a coil.*

tends to pull the core until its center coincides with the center of the coil, as shown by the dashed lines in Fig. 4-7(b). This action has many practical applications in the various types of electrical controlling devices used in industry and in the home. A common example is the solenoid, which is introduced at the end of the chapter.

4-2 INDUCED ELECTROMOTIVE FORCE

Having determined the existence of magnetic lines of force around any wire due to the flow of an electric current, one may conceive of establishing an electric current by means of a magnetic field. Such is the case when an electrical conductor is moved across a magnetic field. As the conductor cuts the lines of magnetic force, an electromotive force (EMF or voltage) is induced in the conductor, which causes an electric current to flow if the conductor is part of a closed loop or circuit. If the conductor is not part of a closed loop, current will not flow through the conductor.

However, there will still be an electromotive force induced in the conductor as long as it cuts across magnetic lines of force. An EMF can still be present in a circuit without the flow of an electric current. An EMF can also be induced in a stationary electrical conductor if the magnetic field is made to move so that its lines of force cut across the conductor. In each of these instances (i.e., where a magnetic field and an electrical conductor are moving relative to each other, causing an induced EMF in the conductor), the magnetic field is assumed to be constant in magnitude. However, a magnetic field that does not move relative to a conductor within its boundaries can also induce an EMF in the conductor. This is done by varying (alternating) the magnitude of the magnetic field with respect to time.

Figure 4-8 shows a loop of wire revolving in a magnetic field. The magnetic field is created by field poles of the kind found in the most elementary type of electric generators or motors. The ends of the loop are connected to slip rings, which revolve with the loop. Stationary brushes are used to collect the current from the rings and deliver it to an external circuit. In Fig. 4-8(a), the white conductor moves to the left while the black conductor moves to the right and the induced EMFs in both conductors may be added together. The total EMF will depend upon the position

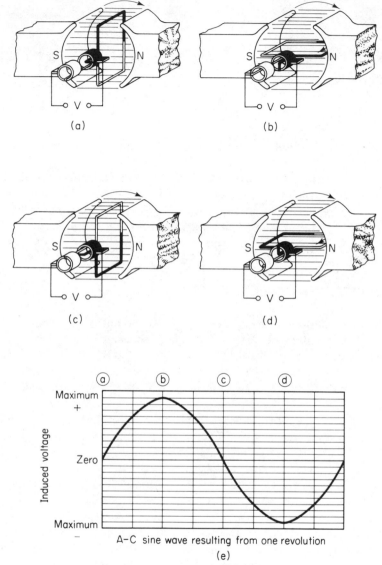

(a)   (b)

(c)   (d)

A–C sine wave resulting from one revolution

(e)

FIGURE 4-8 *Loop of wire rotating in a magnetic field.*

of the loop in the magnetic field. This is true because only that portion of the motion perpendicular to the field is effective in producing EMF. The wave shape that results from one complete revolution of the loop is also shown in Fig. 4-8(e).

(a)   To trace the development of the wave shape, let us start with the loop, as shown in Fig. 4-8(a). In this position each conductor is moving parallel to the magnetic field, the loop is in a neutral position, and the generated EMF is zero. This corresponds to point *a* in the generated voltage in Fig. 4-8(e).

(b) As the loop continues in a clockwise direction, the EMF increases, due to the loop cutting more lines in a given period of time, until it reaches a maximum when the loop is parallel to the field in Fig. 4-8(b). [point *b* in Fig. 4-8(e)].

(c) As the loop continues to rotate to the position shown at Fig. 4-8(c), the EMF decreases until it is again zero [point *c* in Fig. 4-8(e)].

(d) If the loop is turned further through an angle of 90° in the same direction [Fig. 4-8(d)], it will again be cutting lines of force at a maximum rate; however, the EMF and resulting current will be reversed from its previous polarity. The reversal occurs because of the change in the direction in which the conductor passes through the field [point *d* in Fig. 4-8(e)].

(e) As the conductor continues to rotate, the generated EMF decreases until, at the starting point, it is again zero. The wave shape of Fig. 4-8(e) is a sine wave.

4-3   AC GENERATORS

Section 4-2 showed how voltage could be generated by passing a wire through a magnetic field. A device that converts mechanical rotational energy into electrical energy is called a *generator*. In this section we will look at the important features of a generator. The majority of electrical power used in the home or car is created by a generator. The rotational energy for these generators is obtained from falling water, steam (heated by oil or coal) that blows through a turbine causing rotation, or from an automobile engine's rotation.

The generator described in the preceding section (Fig. 4-8) created a sine-wave ac voltage. An ac generator is often termed an *alternator*. In it the coil in which voltage was generated (called the *armature*) was rotated through a magnetic field. A more common configuration is shown in Fig. 4-9. In it the armature coil is wound

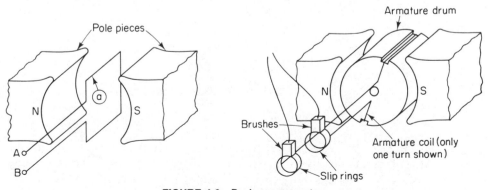

FIGURE 4-9   *Basic ac generator.*

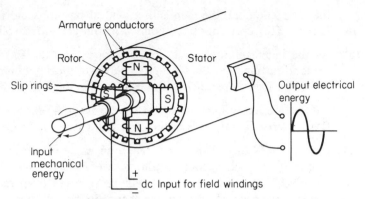

FIGURE 4-10 *Single-phase ac alternator.*

on an armature drum that is rotated through a fixed field. The rotating member of a generator is termed the *rotor*. The stationary portion is termed the *stator,* which in this case includes the magnets used to create the magnetic field. These magnets are generally electromagnets created by coils of wire. These coils are termed the *field* winding.

A more practical alternator is shown in Fig. 4-10. The rotor consists of four field windings that create two north and two south poles. The stator is the armature, which contains a number of conductors, whose induced voltage is brought out as a sine-wave voltage as shown. This alternator requires the input rotational mechanical energy shown and a small amount of dc input for the field windings through the slip rings as shown in Fig. 4-10. The input electrical energy through the slip rings for the field is much less than the output electrical energy. This approach therefore allows for greater slip-ring life than the approach shown in Fig. 4-9 because of the reduced power through the rings. This approach (Fig. 4-10) also allows for lower weight in the rotor, and thus bearing-support problems are minimized.

The frequency of the sinusoidal voltage created by the alternator is provided by the equation

$$f = \frac{P \cdot N}{120} \tag{4-1}$$

where $P =$ number of poles in the field

$N =$ rotor speed, in revolutions per minute (rpm)

$f =$ frequency, in hertz

The output voltage of an alternator typically falls if more current is drawn from it. This occurs because a greater voltage drop within the armature winding resistance occurs as more and more current is drawn through and from it. This is referred to as an *iR* voltage loss *(iR = v).* This voltage loss (drop) within the armature leaves less voltage available at the output terminals and leads to a definition of *voltage regulation:*

$$\text{regulation} = 100\% \times \frac{v_{NL} - v_{FL}}{v_{FL}} \tag{4-2}$$

where v_{NL} = output voltage at no-load conditions (zero current output)

            v_{FL} = output voltage at full-rated output current

The lower the percentage regulation, the better the alternator. Regulation is an indication of how constant the output voltage remains with changing load currents.

EXAMPLE 4-1

The alternator represented in Fig. 4-10 has input rotational mechanical energy at 1800 rpm. Its output voltage at no-load conditions is 130 v, and at rated full load it has a 113-v output. Determine the frequency of its output and its voltage regulation.

Solution:

A reference to Fig. 4-10 shows that the field windings consist of four poles. Thus,

$$f = \frac{P \cdot N}{120} \tag{4-1}$$

$$= \frac{4 \text{ poles} \times 1800 \text{ rpm}}{120}$$

$$= \underline{60 \text{ Hz}}$$

$$\text{regulation} = 100\% \times \frac{v_{NL} - v_{FL}}{v_{FL}} \tag{4-2}$$

$$= 100\% \times \frac{130 \text{ v} - 113 \text{ v}}{113 \text{ v}}$$

$$= \underline{15\%}$$

4-4   DC GENERATORS

It is often desired to provide a dc output from a generator. In those cases it is common to utilize a rotating armature and stationary field winding. Instead of using slip rings to take the output voltage from the armature, a split-ring device as shown

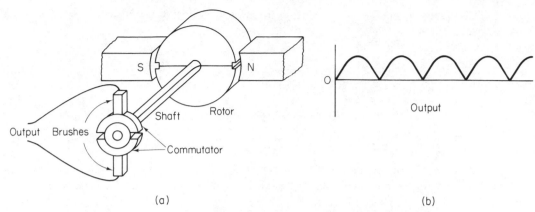

(a)   (b)

FIGURE 4-11   *Dc generator commutator.*

in Fig. 4-11 is used to reverse the normal sine-wave polarity change such that the output shown in Fig. 4-11 is now of a single polarity and therefore is a dc output. The slip-ring device with brushes to pick up the output voltage is called a *commutator*.

To provide a smoother dc output it is common to provide a number of armature windings with separate commutator output sections, as shown in Fig. 4-12. The output is now the solid section shown, which is made up of the peaks of several separate windings. The small output variation still remaining is termed *ripple* and is reduced to almost nothing in commercial units using a large number of windings (and commutator sections).

The principal components of a dc generator are the armature, the commutator,

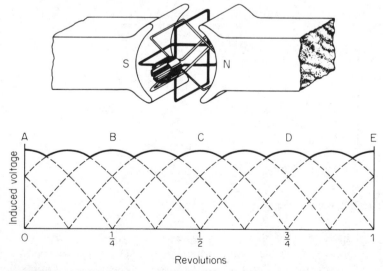

FIGURE 4-12   *Voltage generated by four separate coils connected to an eight-segment commutator.*

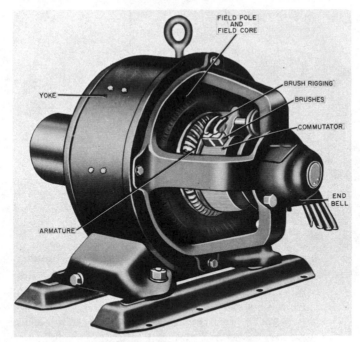

FIGURE 4-13   *Typical dc generator.*

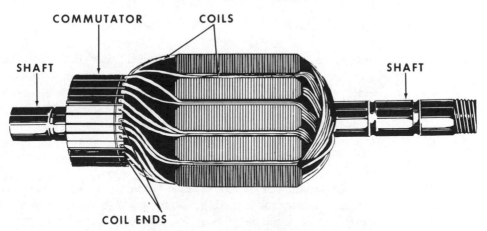

FIGURE 4-14   *Dc generator armature.*

the field poles, the brushes and brush rigging, the yoke or frame, and the end bells or end frames, as shown in Fig. 4-13.

(a) *Armature*

The armature (Fig. 4-14) is the structure upon which are mounted the coils that cut the magnetic lines of force. It is affixed on a shaft which is

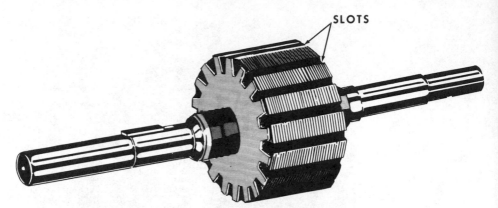

SLOTS

FIGURE 4-15 *Unwound armature core on a shaft.*

suspended at each end of the machine by bearings set in the end bells or end frames. The armature core is circular in cross section and is built up from sheets of soft iron (Fig. 4-15) called a *lamination*. The outer surface edge of the laminated core is slotted to receive the coil windings. The windings are held in place and in their slots by wooden or fiber wedges. Steel bands are sometimes wrapped around the completed armature and, with the wedges, hold the coil windings in place. On small machines the laminations of the armature core are usually pressed onto the armature shaft.

(b) *Commutator*

The *commutator* is that component of the generator that rectifies the generated alternating current to provide direct current output and serves to connect the stationary output terminals to the rotating armature. A typical commutator (Fig. 4-16) consists of commutator bars, which are wedged-shaped segments of hard-drawn copper, insulated from each other by thin strips of mica. These commutator bars are held in place by steel V-rings or clamping flanges, which are bolted to the commutator sleeve by hexagonal cap screws.

The commutator sleeve is keyed to the shaft, which rotates the armature. A mica collar or ring insulates the commutator bars from the commutator sleeve. The commutator bars are usually provided with risers or flanges to which the leads from the associated armature coils are soldered (Fig. 4-16).

The brushes, which make contact with the commutator bars, collect the current generated by the armature coils and, through the brush holders, pass the current to the main terminals. As the commutator bars are insulated from each other, each set of brushes, as it makes contact with the commutator bars, collects current of the same polarity, resulting in a continuous flow

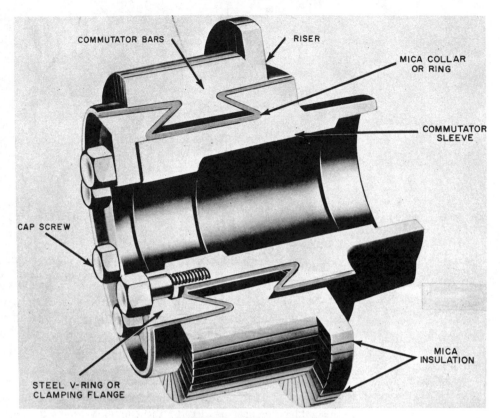

FIGURE 4-16 *Typical commutator.*

of direct current. The finer the division of the commutator bars, the less ripple will be present and the smoother will be the flow of the dc output.

(c) *Field Pole and Frame*

The frame, or yoke, of a generator serves both as a mechanical support for the machine and as a path for the completion of the magnetic circuit. The lines of force which pass from the north to the south pole through the armature are returned to the north pole through the steel frame of the generator.

    1. Field poles (Fig. 4-17) are required to produce the magnetic field, or flux, which passes through the conductors of the armature. The minimum number of field poles required to complete the magnetic circuit is two, a north pole and a south pole. Most commercial generators are made with four or more field poles, depending on the speed of the generator; the slower the speed of the generator,

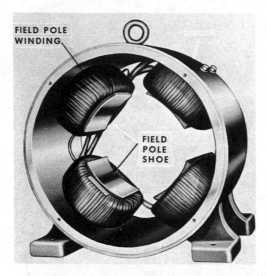

FIELD POLE
WINDING

FIELD
POLE
SHOE

FIGURE 4-17 *Field poles in a dc machine.*

the more poles required to produce the same output. The number of field poles must always be an even number, each set consisting of a north pole and a south pole. Multipole generators are in common use. In dc generators, each set of field poles requires a set of brushes. Thus, a four-pole generator requires four brushes, a six-pole unit requires six brushes, and so on.

2. Field poles are usually fastened to the inside of the frame with screws or bolts, except for some small units which are made with the field poles cast as part of the frame. Most field poles consist of rectangular laminations held together by means of bolts or rivets. That section of the pole away from the frame is generally flared. These flared sections, called shoes, serve to distribute the flux beneath the pole and to hold the coil on the pole.

The preceding detailed description of a dc generator has mechanical similarity to ac generators and to motors as well.

4-5 DC MOTORS

A *motor* is a machine that changes electrical energy to mechanical energy. A motor operates because a force is exerted on a current-carrying conductor placed in a magnetic field. Its construction and operation is similar to the generators described in previous sections except that the desired output is mechanical energy with a motor, whereas the generator output is the electrical energy.

Figure 4-18 shows the force acting on a single-turn coil in a magnetic field. Electrons are flowing out of the left-hand conductor and into the right-hand conductor. This outward electron flow is denoted by a circle with a dot in the center, and

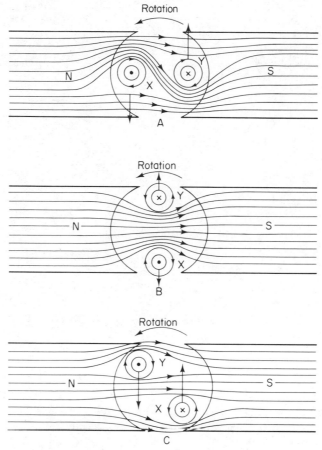

FIGURE 4-18 *dc motor principle.*

inward electron flow is shown by a circle with an X in it. This force tends to move the right-hand conductor up and the left-hand conductor down [Fig. 4-18(a)]. Both forces acting on the coil produce a turning effect. This rotational force is called *torque*. When the coil reaches the position shown in Fig. 4-18(b), the forces tend to spread the coil apart and there is no torque. This position is called *dead center*. If the current in the coil reverses at this point and the coil is also carried beyond the dead center, forces will be developed as in Fig. 4-18(c). The torque tends to continue turning the coil in the same direction as before.

If the direction of current is always reversed at the proper time, a continuous rotating force is developed. A commutator is used to produce the change in current flow, just as in the dc generator.

A single armature coil motor, as shown in Fig. 4-18, is impractical because the torque is pulsating. A large number of coils wound on the armature will produce a much smoother torque. The construction of a motor armature is the same as that

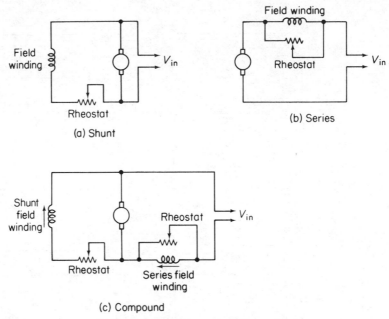

FIGURE 4-19 *Motor connection methods.*

of a generator armature. Additional pairs of poles, as in a generator, are also used.

The rotational force (torque) that turns the motor armature depends on two factors, the armature current and the strength of the magnetic field. The speed of the motor is also dependent on these factors, as well as the amount of load connected to the output shaft.

The circuits and characteristics of generators are also applicable to motors. A generator will run as a motor and convert electrical energy into mechanical energy, provided that the proper voltage is applied to the output terminals. This voltage must be the same as that generated when the machine is rotated at the proper speed. Dc motors, like dc generators, are classified as shunt, series, and compound motors, based on the field windings being in parallel (shunt), series, or both (for compound) with respect to the armature. These conditions are shown schematically in Fig. 4-19. Notice the rheostats in each circuit. They are used to control the motor speed. These same conditions exist in dc generators. In this case, V_{in} (Fig. 4-19) becomes V_{out} and the rheostats control the amount of output electrical energy.

Reversing the polarity of the voltage applied either to the armature or to the field will reverse the direction of any of these motors. Figure 4-20(a) shows the force acting on the armature of a dc motor producing counterclockwise rotation. In Fig. 4-20(b) the polarity of the voltage applied to the field coils has been reversed, reversing the direction of the field. Notice that the rotation is now clockwise.

In Fig. 4-20(c) both the field and the direction of current through the armature have been changed. As a result, the rotation is still counterclockwise, as it was in

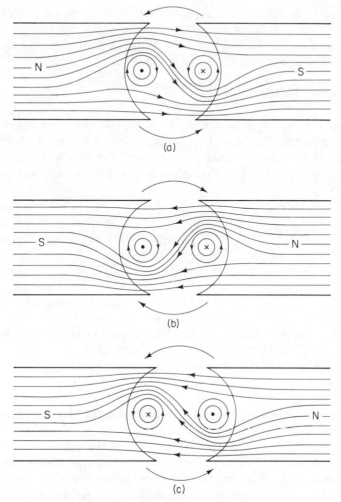

FIGURE 4-20   *Direction of rotation.*

Fig. 4-20(a). Since this is true, it does not matter which side of the dc motor is connected to the positive and which to the negative side of the source voltage regarding rotation direction.

4-6   AC MOTORS

Series Motor

Since changing the polarity of the voltage applied to both the armature and the field does not change the direction of rotation of a series motor, ac may be applied to the terminals rather than dc. Since the field and armature are in series, the field and armature currents are in phase. Therefore, the flux is in phase with the armature current, and the series motor develops about the same torque with ac applied as it does with dc. This is not true with the shunt motor, because the flux there is nearly 90° out of phase with the armature current.

Induction Motor

Because of its ruggedness and simplicity, the induction motor is the most widely used ac motor. Figure 4-21 is an illustration of the basic principle of the induction motor. A horseshoe magnet is placed over a disk of metal so that its magnetic field passes through the disk. The disk is mounted on an axis so that it can rotate freely.

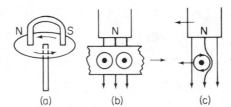

FIGURE 4-21 Basic induction motor principle.

The magnet is rotated in a clockwise direction above the disk (Fig. 4-21). As it moves, the magnetic field cuts through the disk.

This metal disk can be considered to be made up of many conductors side by side. Figure 4-21(b) is a view looking at the edge of the disk below the north pole. Since the results would be the same if the disk were moved in a counterclockwise direction (the relative motion between conductor and field would be the same), the conductors are indicated as moving.

Electron flow is induced in the disk by the cutting of the lines of force. They are shown in Fig. 4-21(b) as coming out of the paper (circles with dots in the center). The lines of force about the conductor are shown in Fig. 4-21(c). Remember that these lines of force are due to the induced current in the conductor. The magnetic field on the right side of the conductor in Fig. 4-21(c) is weakened, while the field to the left is increased. The force acting on the conductor tends to move it to the left. This is the same direction in which we are moving the magnet above the disk.

The disk is rotating in the same direction as the magnet but at a slightly slower rate. It can never rotate as fast as the magnet, because if it were to do so, there would be no relative motion. With no relative motion, there would be no induced voltage and no magnetic field about the disk conductor. There would be, therefore, no reaction between magnetic fields, no torque, and a slowing down of the disk. Once the disk began to slow down, there would be relative motion between the magnet and the disk, resulting in induced currents in the disk, with their accompanying magnetic fields. The disk once more would have a force tending to rotate it. The difference of speed between the magnet and the disk is called the *revolutions slip*.

Instead of a disk, a cylinder may be used. Instead of a horseshoe-shaped magnet, four poles may be rotated within the cylinder. The cylinder is more like the commercial induction motor and works on the same principle as the disk just described.

Rather than rotating a magnet, a rotating field is used in the practical induction motor. The rotating field is due to the currents in multiple windings. There is no mechanical rotation of the fields; they are produced by electrical means. In many cases a capacitor (external to the motor) is connected to maintain a phase difference between two magnetic fields. By switching the capacitor from one winding to the other, the relative phase of their magnetic field may be reversed, thereby reversing the motor's direction of rotation.

An armature placed within the rotating fields created by the multiple field windings will have currents induced in it. The magnetic field due to this current reacts with the rotating field and produces a torque. This torque tends to turn the armature in the direction in which the field is rotating. (This is the same principle that produced the rotation in the simple disk described earlier.)

You will recall that in the simple induction motor we explained that the rotor cannot turn as fast as the rotating field. The speed of the rotating field is called the *synchronous* speed of the motor. The difference in speed between that of the rotating field and the speed of the rotor is called the *revolutions slip*. Usually, the slip is given as a percentage of the synchronous speed. That is, a motor whose synchronous speed is 3600 rpm and whose rotor speed is 3528 rpm has a slip of 2%.

$$[(3600 - 3528) \div 3600] \times 100\% = 2\%$$

This simple induction motor is often called a *squirrel-cage motor* because the armature, or rotor, consists of bars with the ends connected together by end rings. It looks very much like the cage that spins when a squirrel inside it starts running.

As the rotating field moves, the currents induced in the rotor bars create magnetic fields. The interaction of the two fields causes the rotor to turn in the direction of the rotating field. At no load the slip of the squirrel-cage motor is very small, but as the load is increased, the slip increases. To produce the torque required by the load, more current must flow. More lines of flux must be cut per unit time to induce this current. Since the field cannot be changed, the rotor must have a greater relative motion with respect to the rotating field, or a greater slip. The slip increases until the proper torque is produced.

Synchronous Motor

The design of a synchronous motor is almost the same as an alternator. Figure 4-22(a) shows the poles of a synchronous motor field and two of the conductors in the armature. The field is constant, since it is due to direct current flow. The current throughout the armature is alternating. At the instant shown, the torque on the two conductors tends to move them to the left. Since this is alternating current, the next half-cycle will find the current through the armature reversed, and the torque reversed if the wires remain in the same relative position. If the armature moved

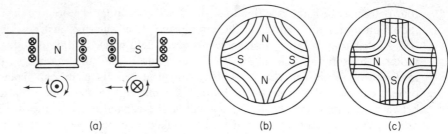

FIGURE 4-22 *Synchronous motor principle.*

far enough so that wire 2 were beneath the north pole and wire 1 below a south pole, the torque would be in the same direction, keeping the armature turning in the same direction as during the first half-cycle.

If each half-cycle sees the armature conductors moved the distance of one pole, the torque will always be in the same direction, resulting in continuous motion. Thus, the synchronous motor must operate at constant speed if the frequency of the applied armature voltage is constant. If the average speed differs from synchronous speed, the average torque becomes zero and the motor stops.

Figure 4-22(b) shows a rotating field for a four-pole synchronous motor. At the instant shown, the two N poles are in a vertical position and the two S poles in a horizontal position. Now let us insert an X-shaped rotor in this field. The flux from the field will move the rotor so that the flux is maximum and there is very little air gap at each pole. The pole pieces of the rotor are locked in with the poles produced by the rotating field in the stator. Now assume that the current flowing in the armature (rotor) produces poles that rotate at the same speed as the poles due to the stator winding. They will lock in, just as though everything were at a standstill [Fig. 4-22(c)]. The rotating poles of the rotating field of the stator pull the rotor around with them at the same speed.

Thus, the speed of a synchronous motor is "locked" to the frequency of the applied voltage. In equation form,

$$N = \frac{120f}{P} \qquad \qquad \textbf{(4-3)}$$

where $N =$ rotor speed, in rpm

　　　　　$f =$ frequency of applied voltage

　　　　　$P =$ number of poles

Synchronous motors find application where an extremely constant speed is a necessity. Examples include clock motors, record turntable drives, and office machine applications.

EXAMPLE 4-2

Determine the speed of a synchronous four-pole motor with standard line voltage (120 V, 60 Hz) applied. Repeat for a 50-Hz applied voltage.

Solution:

$$N = \frac{120f}{P} \qquad\qquad \textbf{(4-3)}$$

$$= \frac{120 \times 60 \text{ Hz}}{4 \text{ poles}}$$

$$= \underline{1800 \text{ rpm}}$$

If the applied voltage were 50 Hz:

$$N = \frac{120 \times 50 \text{ Hz}}{4 \text{ poles}}$$

$$= \underline{1500 \text{ rpm}}$$

4-7 TRANSFORMERS

A *transformer* consists of two wire coils that couple ac electrical energy from one to the other. Depending on the frequency of operation and the particular piece of equipment in which it is installed, a transformer may use an iron, a magnetic alloy, or an air core. The coil that is connected to the ac source is called the *primary winding* and the coil that is connected to the load is called the *secondary winding*. The relative number of turns of the windings depends on whether the voltage induced in the secondary is to be larger or smaller than the voltage applied to the primary; the number of primary turns depends on the magnetic properties of the core and the frequency.

A typical iron-core transformer and its schematic symbol is shown in Fig. 4-23. Notice the vertical lines between the two windings in the schematic symbol.

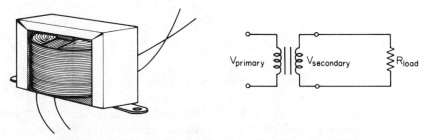

FIGURE 4-23 *Typical transformer and schematic symbol with load.*

They indicate that an iron core transformer—the most common variety—is being used. The most useful property of a transformer is its ability to "transform" ac voltages. Thus, a transformer is used when the 120-v line voltage is actually needed at a 30-v level. The amount of voltage transformation is determined by the *turns ratio*, which is the ratio of primary turns to secondary turns. In equation form,

$$T_r = \frac{v_p}{v_s} = \frac{N_p}{N_s} \tag{4-4}$$

where v_p = primary voltage

v_s = secondary voltage

N_p = turns of wire in primary winding

N_s = turns of wire in secondary winding

T_r = turns ratio

EXAMPLE 4-3

A transformer is to be used so that a 12-v bulb can be operated from a 120-v, 60-Hz voltage source. Determine the required turns ratio, T_r. If the primary contains 360 turns of wire, determine the required number of secondary turns.

Solution:

$$T_r = \frac{v_p}{v_s} \tag{4-4}$$

$$= \frac{120 \text{ v}}{12 \text{ v}}$$

$$= \underline{10}$$

$$\frac{N_p}{N_s} = T_r \tag{4-4}$$

$$N_s = \frac{N_p}{T_r}$$

$$= \frac{360}{10}$$

$$= \underline{36 \text{ turns}}$$

Except for some minor power losses in the transformer due to heat, the power into the primary equals the power output of the secondary. Thus, as an approximation (neglecting the power losses), $P_p = P_s$ and thus

$$v_p i_p = v_s i_s \tag{4-5}$$

It should be noted that the amount of usable output energy has a practical limit based upon the amount of iron core and the size of wire used for the primary and secondary windings. The greater these quantities, the more power output is possible. Thus, a large transformer has a greater power-handling capability than a small one.

EXAMPLE 4-4

The light bulb in Example 4-3 draws 0.5 a at 12 v. Calculate the amount of primary current.

Solution:

$$v_p i_p = v_s i_s \tag{4-5}$$

$$120 \text{ v} \times i_p = 12 \text{ v} \times 0.5 \text{ a}$$

$$i_p = \frac{12 \text{ v} \times 0.5 \text{ a}}{120 \text{ v}}$$

$$= 0.05 \text{ a}$$

$$= \underline{50 \text{ ma}}$$

Electricity Distribution

The electrical utilities make widespread use of transformers in the distribution of electrical energy. The transformers provide an efficient means of changing voltage magnitude to the desired levels. The following example shows why the transformers are so useful in that function. The example shows that it is more economical to transmit electrical energy at high voltage levels and then use a transformer to step it down to the necessary lower levels near the place of usage. Figure 4-24 represents a user 4 miles from a distribution site. Transformer T_1 steps up the line voltage to

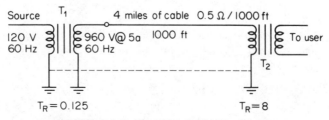

FIGURE 4-24 *Electricity distribution scheme.*

960 v for distribution over 4 miles of cable that has 0.5 Ω resistance per 1000 ft. Notice the grounding technique that allows the earth to serve as the "missing" conductor to complete the circuit. The dashed line represents the earth's connection, and the wire savings that results is a significant savings. Although the "earth" is not a great conductor in small amounts, the large magnitude of cross-sectional area involved between any two points on earth presents approximately zero ohms of resistance. Thus earth, or ground as it is often called, is said to represent a *common* reference point.

EXAMPLE 4-5

Determine the power losses for the electrical distribution system shown in Fig. 4-24 both with and without the transformers.

Solution:

Notice that T_1 is a "step-up" transformer. Its output voltage is eight times its input. The wire connecting T_1 to T_2 has a total resistance of

$$0.5 \ \Omega/1000 \ \text{ft} \times 4 \ \text{mi} \times 5280 \ \text{ft/mi} = 10.56 \ \Omega$$

Thus, the power loss involved is

$$
\begin{aligned}
P &= i^2 R \\
&= (5 \ \text{a})^2 \times 10.56 \ \Omega \\
&= \underline{264 \ \text{w}}
\end{aligned}
\tag{2-2}
$$

The 120-v source is delivering a certain current:

$$v_p i_p = v_s i_s \tag{4-5}$$

$$120 \ \text{v} \times i_p = 960 \ \text{v} \times 5 \ \text{a}$$

$$i_p = \frac{960 \times 5}{120}$$

$$= 40 \ \text{a}$$

If the transformers were not used, the 120 v, 40 a would be transmitted using the wire with 10.56 Ω resistance. The power lost in this transmission cable would now be

$$
\begin{aligned}
P &= i^2 R \\
&= (40 \ \text{a})^2 \times 10.56 \\
&= 16,896 \ \text{W} \qquad \text{or} \qquad \underline{16.896 \ \text{kw}}
\end{aligned}
\tag{2-2}
$$

Example 4-5 clearly shows the desirability of distribution of power at high voltage levels. Large amounts of power are delivered by electrical utilities at several hundred thousands of volts to minimize the size of wire needed to keep losses to an acceptable level.

Automotive Electrical System

At the heart of the automotive electrical system pictorially shown in Fig. 4-25(a), a transformer is utilized. This transformer is the *ignition coil*. Its name is not entirely appropriate in that it is really a transformer and consists of two coils, a primary and a secondary wound on a common core as shown in Fig. 4-25(b). The points

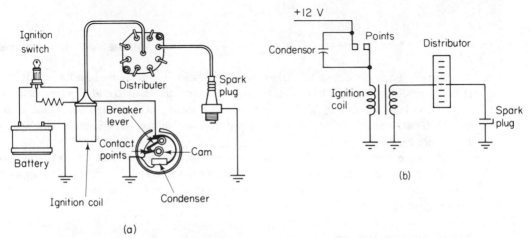

FIGURE 4-25 *Automotive electrical system.*

serve to change the 12 V dc into a pulsating signal by interrupting the dc on a periodic basis. Most of the newer automobiles use electronic means to generate the pulsating signal rather than the points. These are called electronic ignition systems. The ignition coil steps up the primary voltage to the high level necessary to "fire" the spark plug (about 30,000 V). The primary voltage is actually greater than the pulsating (due to the points) 12 V that you might expect because of the so-called *inductive kick*. The primary winding exhibits inductance, and any time the current flow through its inductance is rapidly changed, a voltage now is generated that far exceeds the original level due to inductive kick. In the case here, it is common for the primary to reach a 300-V level instead of the expected 12 V due to this effect.

The capacitor across the points in Fig. 4-25 serves to prolong the life of the points by acting as a short when the current flow is at its highest level. This lowers the current flow through the points, thus minimizing the arcing that occurs as they open. Excessive arcing causes the points to burn out quickly. In automotive circles, a capacitor is usually termed a condensor. The distributor "distributes" the high voltage from the coil's secondary to the various spark plugs on a periodic basis as required for proper engine operation.

Autotransformer

A useful transformer variation is shown in Fig. 4-26 and is called an *autotransformer*. In it a coil of wire is wound on an iron core and a tap is provided to a point along the coil as shown. If an input ac voltage is applied to the input terminals, A and C,

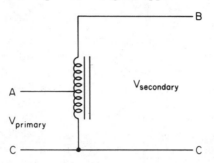

FIGURE 4-26 *Autotransformer.*

the current flow creates a changing magnetic field in the core around which the entire winding is wound. The output voltage between terminals B and C is proportional to the number of turns in BC divided by the number in winding AC. Thus, if BC has 2000 turns and AC has 100, the output voltage in Fig. 4-26 should be 240 v, if 12 v were applied to AC.

If lead B were connected to a sliding contact, the BC output voltage would be variable if the slide were varied. The transformer is then useful in laboratory applications where a variable ac voltage is necessary. One name for commercially available, variable autotransformers is *variac*.

Isolation Transformer

Notice that the C point in Fig. 4-26 is common to both input and output. In regular transformers there is no electrical connection between input and output. That electrical *isolation* is a feature not available with the autotransformer. The isolation available in regular transformers is an electrical safety feature. In some cases a transformer with a 1:1 turns ratio (input equals output voltage) is used to prevent electrical shock. This situation is shown in Fig. 4-27 and the transformer is termed an *isolation transformer*. The person's feet are "grounded" (perhaps by standing on a damp concrete floor) but no electrical shock occurs because of the isolation afforded by the transformer. Shock could occur (perhaps death) if the person touched point A in the primary circuit under these conditions.

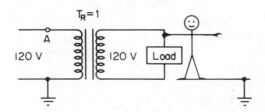

FIGURE 4-27 *Isolation transformer.*

4-8 OTHER MAGNETIC APPLICATIONS

Relays

A *relay* is an electromechanical device where a movable element (the armature) is actuated by magnetic forces from current flowing through a coil. The armature causes *contacts* to open or close when it moves.

   In the most common type of relay, a current change in an electromagnet in one circuit produces an armature movement that opens or closes contacts. These contacts act as switches in another circuit. In Fig. 4-28, when the switch in the low-voltage circuit is closed, current flowing through coil *L* sets up a magnetic field. This magnetic field attracts the soft-iron armature *A,* which moves down and makes contact at *C,* closing the switch in the high-voltage circuit. When the switch in the low-voltage circuit is opened, the field collapses and the tension of spring *S* pulls the armature up, breaking the contact at *C* and opening the high-voltage circuit. This type of relay is called a single-pole, single-throw relay. More complicated relays with multiple sets of contacts are used when a single control over a number of separate circuits is desired.

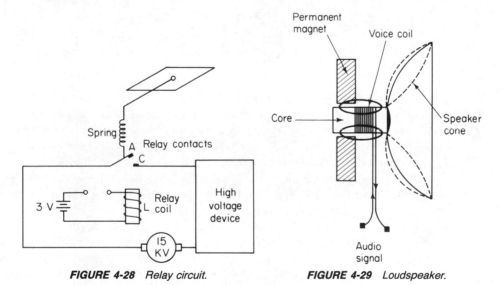

FIGURE 4-28   Relay circuit.   FIGURE 4-29   Loudspeaker.

Loudspeaker

A loudspeaker converts electrical energy into the energy of sound. A device that converts from one form of energy to another is called a *transducer*. The permanent magnet shown in Fig. 4-29 is fixed in position. The electrical sound signal (*audio signal*) is applied to the voice coil. This coil is wound on a core that is attached to the speaker cone. The varying electrical audio signal creates a varying magnetic field that, in conjunction with the permanent magnet's field, causes the core to move in and out in step with the audio input signal. The core movement, in turn, causes the speaker cone to move in and out, causing air to be fluctuated at the audio rate and amplitude. These air variations *(sound waves)* are subsequently picked up by our ears and perceived as sound.

Microphone

The dynamic microphone (also a transducer) shown in Fig. 4-30 does the reverse of the loudspeaker just discussed. The sound waves are impressed upon a flexible cone that is also attached to the core, as shown. The core is moving inside a coil and is surrounded by a permanent magnet. The core movement induces a voltage

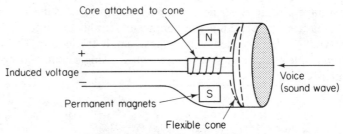

FIGURE 4-30 *Dynamic microphone.*

whose frequency equals the frequency of the input sound waves. The voltage amplitude is proportional to the sound-wave amplitude at any instant of time. The sound "signal" is therefore converted to an electrical signal that otherwise has identical characteristics. Besides this dynamic microphone, other common types include the carbon granular (used in telephones), condensor (capacitor), and crystal microphones.

The dynamic microphone just described is identical in principle to a *magnetic phonograph cartridge,* with one exception. The audio signal is picked up by vibrations of a needle as it traverses the grooves. The grooves in the record have the appropriate physical variations to cause the proper needle vibration. The needle is equivalent to the flexible cone shown in Fig. 4-30; otherwise, the dynamic microphone and magnetic cartridge are identical in function. The audio information contained in the record grooves is converted to an electrical signal by the cartridge. That signal is now *amplified* (made larger and more powerful) by electronic circuitry and then converted to sound energy by a loudspeaker.

Magnetic Tape

The use of magnetic tape is widespread in today's society. It is used to record and play back audio signals, video (i.e., television) signals, and computer information. The basic process is actually quite simple. The single-channel tape head shown in

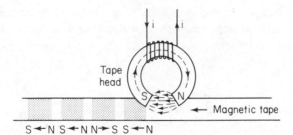

FIGURE 4-31   Magnetic-tape head.

Fig. 4-31 is an electromagnet that has an electrical signal applied to its coil. That signal, as it varies in amplitude and frequency, causes the tape (with its magnetic coating) passing past it to be alternately magnetized with varying north–south polarities. The electrical signal has been converted to a magnetic replica. The strength of the magnets recorded on the tape corresponds to the amplitude of the electrical signal that created them. The rate at which the created magnets change from north to south polarity corresponds to the frequency of the applied electrical signal. The magnetic tape is coated with an iron material that retains its magnetic characteristics.

The "playback" mode involves the reverse process from that just described. The magnetic tape is passed over the tape head at the same speed that the "recording" took place. The magnets on the tape, passing the head, cause a voltage to be induced in the winding on the tape head. That voltage should be a replica of the voltage that created the original magnetic pattern on the tape. It is then suitably processed by electronic circuitry so that it can be used to drive a loudspeaker, TV screen, or other equipment.

Solenoid

Another important electromagnetic device is a *solenoid* as shown in Fig. 4-32. It consists of a movable iron core inside of a coil of wire. When the coil is energized with current, the magnetic field created exerts a force on the movable core so as to "center" its position with respect to the coil. When energizing current is removed, the spring returns the core to its original position. The movable core is then used

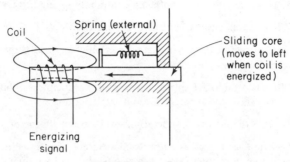

FIGURE 4-32   Solenoid.

to actuate a water valve in a washing machine, a gas valve in a clothes dryer, a bolt in an electric lock, or used any place where a mechanical force is desired as a result of an electrical control signal.

QUESTIONS AND PROBLEMS

4-1-1. Define magnetism. Explain the difference between a permanent magnet and an electromagnet.

4-1-2. What are magnetic lines of force? Explain the lines of force shown in Figs. 4-2 and 4-3.

4-1-3. Explain how electron flow through a coiled conductor creates a much stronger magnetic field than through a straight conductor. Use the left-hand rule to show the direction of magnetic field so created.

4-1-4. What effect does including an iron core within a current-carrying coil have on the lines of force? Explain the effect on such a core not centrally aligned within the current-carrying coil.

4-2-5. Explain how an EMF can be induced in a conductor. How can that EMF also result in a current flow?

4-2-6. Briefly explain the process whereby an EMF is generated in Fig. 4-8. Show why the slip rings are necessary.

4-3-7. Provide another name for an ac generator. Define armature, field, rotor, and stator.

4-3-8. Describe the advantages of a field rotor, armature stator configuration (Fig. 4-10) as opposed to the armature rotor, field stator generator (Fig. 4-9).

4-3-9. Calculate the frequency of generated voltage for an alternator with eight poles and a rotational speed of 900 rpm.

4-3-10. The output of an alternator is 109 V at full-rated load and 124 V at no load. Calculate the regulation.

4-4-11. Explain the necessity for split rings in a dc generator as opposed to slip rings in an ac generator. What is the name commonly applied to the split-ring arrangement?

4-4-12. What techniques are used to minimize the ripple in the output of a dc generator?

4-4-13. Briefly describe the physical construction of a typical dc generator's armature, commutator, and field poles.

4-5-14. What is a motor? Explain the dc motor principle as illustrated in Fig. 4-18. Define torque and "dead center" in your explanation.

4-5-15. Explain the factors that determine the output torque of a motor.

4-5-16. Describe the three basic motor connection schemes. Why do you think the rheostat used to control motor speed is also in the field winding circuit?

4-5-17. Describe how the direction of rotation for a dc motor can be reversed.

4-6-18. Briefly explain the basic induction motor principle. Include the concept of "revolutions slip" in your explanation.

4-6-19. Describe how a capacitor is used to reverse the direction of rotation in an ac induction motor.

4-6-20. Explain the relationship between slip, motor load, and torque for the squirrel-cage (induction) motor.

4-6-21. Explain the basic principle of operation for a synchronous motor. What are some common applications for this motor as a result of its speed/input frequency relationship?

4-6-22. A synchronous motor is operating on 60-Hz line voltage and uses eight poles. Calculate the motor's output speed.

4-7-23. In general terms, describe the function of a transformer. Include the meaning of primary and secondary in your description.

4-7-24. Define the meaning of turns ratio for a transformer. What effect does it have on the output voltage?

4-7-25. A doorbell buzzer requires 18 v, 60 Hz for proper operation. A transformer operating off 120-v, 60-Hz line voltage has 45 turns in its secondary. Calculate the turns required for the primary winding.

4-7-26. The doorbell buzzer from Problem 4-7-25 requires 1.5 W of energy for operation. Calculate the primary and secondary currents.

4-7-27. Explain how *and* why transformers are widely used by electrical utilities for the distribution of electricity.

4-7-28. Describe the operation of the automotive electrical system shown in Fig. 4-25, including an explanation of inductive kick.

4-7-29. Calculate the required turns ratio for an automotive coil that has a 280-v level in the primary due to inductive kick and requires a 25,000-v level to fire the spark plugs.

4-7-30. What is an autotransformer? Explain how a variable output voltage autotransformer is constructed and give another name for such a unit.

4-7-31. What is an isolation transformer? Explain why it offers safety from electrical shock and also why an autotransformer does not afford isolation.

4-8-32. What is the function of a relay? Provide a schematic of a relay that has a 12-v coil that is actuated by a thermostat and is used to turn on a heater motor operating on 120 v.

4-8-33. Describe the process whereby a loudspeaker converts electrical energy into sound energy.

4-8-34. Explain how a dynamic microphone is able to convert from sound energy to electrical energy. How is this process similar and different from the function of a magnetic phonograph cartridge?

4-8-35. Provide a description of how magnetic tape can be used to record and play back electrical signals.

4-8-36. Explain the operation of a solenoid. Provide three applications and include a schematic drawing that illustrates one of those uses.

5

DIODES
AND
POWER SUPPLIES

5-1 *PN* JUNCTION DIODE

A *pn* junction is a combination of two types of semiconductor material. A semiconductor is a material whose electrical properties lie between those of an insulator and a conductor. The most often used semiconductors are germanium and silicon. Crystals of either of these materials are modified to create either a *p*-type or an *n*-type semiconductor. This modification process is referred to as *doping* and is accomplished by adding impurity atoms to pure germanium or silicon to increase the number of free electrons or holes. A *hole* is an electron vacancy in the outer shell of a crystaline structure. The *n*-type semiconductor is doped so that it can donate free electrons to the semiconductor crystal while the *p*-type provides holes to the semiconductor crystal.

The *pn* junction exhibits the interesting and useful characteristic whereby a low resistance to current flow in one direction and a high resistance in the opposite direction is exhibited. A *pn* diode and its symbol are shown in Fig. 5-1. It is shown that conventional current can easily flow from the *p*-type to the *n*-type [Fig. 5-1(b)] material. Thus, the direction of the arrowhead in a diode's schematic symbol functions as a memory aid for the direction of allowable conventional current flow. You should recall that the actual direction of electron flow is in the opposite direction—from the *n*-type to the *p*-type material. This is logical because it is the *n*-type material that has been doped so as to have an excess of free electrons. The *p*-type side of the diode is called the *anode* and the *n*-type side is referred to as the *cathode,* as

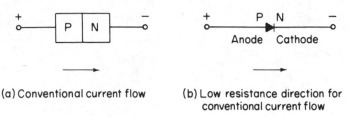

(a) Conventional current flow

(b) Low resistance direction for conventional current flow

FIGURE 5-1   pn junction diode.

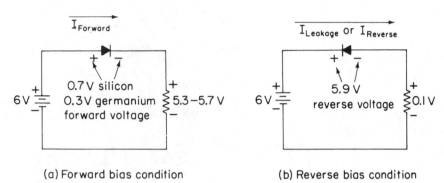

(a) Forward bias condition

(b) Reverse bias condition

FIGURE 5-2   Diode bias conditions.

shown in Fig. 5-1(b). When the diode is polarized for easy conduction (anode positive with respect to cathode), we say that the diode is *forward-biased*.

Figure 5-2(a) illustrates a condition of forward bias. Forward bias results when the diode's *p*-type material is at a more positive level than the *n*-type material, which allows for easy conduction of current through the diode. Ideally, this easy conduction of current would mean zero voltage drop across the diode with the full battery voltage appearing across the resistor. As can be seen from Fig. 5-2(a), however, a small voltage drop of 0.3 V for a germanium diode and 0.7 V for a silicon diode actually occurs. This difference of potential at the junction is called the *barrier potential*. Whenever the battery voltage is large enough to overcome this barrier potential, the diode presents a low resistance to the circuit and significant current flow may result. This is shown graphically by the current–voltage relationships in the upper right-hand section of Fig. 5-3 (positive current and voltage). The forward current is extremely low until the barrier potential is reached (≈ 0.3 V in this case, indicating a germanium diode). As the forward voltage increases above the barrier potential, a rapid increase in current results. The diode is now presenting an extremely low resistance, and the current flow must be limited by external circuit resistance to prevent burnout of the diode. The power dissipated by a diode equals the product of the voltage across it and the current through it. In this case, shown in Fig. 5-3, burnout occurs at 1 A and about 0.5 V. Hence, the power rating of.this diode is 0.5 W *(P = VI)*.

Referring back to Fig. 5-2(a), the 6-V battery forward-biases the diode with all

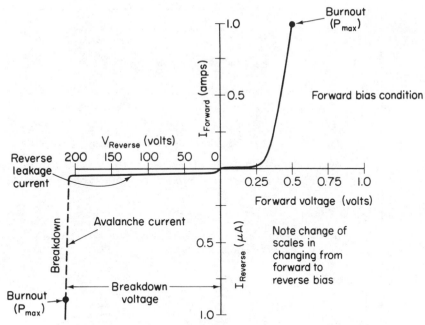

FIGURE 5-3   *Current-voltage relationships for a germanium pn diode.*

the battery's voltage applied across the load, except for the diode's barrier potential of about 0.3 V if germanium, or 0.7 V if silicon. This then leaves either 5.7 or 5.3 V across the resistor when the diode is forward-biased. In Fig. 5-2(b) the diode has been physically reversed, which results in a reverse bias on the diode. A reverse-biased diode has the *n*-type material (the cathode) at a more positive voltage level than the *p*-type material. The battery is now aiding the diode's barrier potential in preventing current flow, and only a very small reverse leakage current flows. This is shown graphically in the lower left-hand section of Fig. 5-3 (negative voltage and current). Notice that the current remains at a low level until the device's *breakdown voltage,* V_B, is exceeded. The breakdown voltage varies, depending on diode type, from a range of several volts up to 1000 V or more. If the breakdown voltage is exceeded, burnout will once again occur, as shown in Fig. 5-3.

Notice that the reverse-biased condition of Fig. 5-2(b) results in almost all the battery's voltage being dropped across the diode, with a very small voltage developed across the load resistor due to the small leakage current. A reverse-biased diode is not a perfect open circuit, as shown in Fig. 5-3, but exhibits some high value of reverse resistance, R_R. The reverse resistance of a diode can be approximated as simply its reverse voltage, V_B, divided by the reverse current, I_R. The manufacturer's data sheet for a specific diode usually supplies the necessary information to enable a calculation of R_R by providing a value of I_R at a specific value of V_B.

EXAMPLE 5-1

The *I–V* curve for a germanium diode is given in Fig. 5-3.

(a) When forward-biased with a 0.25-A current flow, determine its forward voltage drop and the diode power dissipation.

(b) When reverse-biased, it exhibits a leakage current of 0.05 μA. Calculate the power dissipated in the diode when $V_{\text{reverse}} = 100$ V.

(c) If the diode were in a circuit with 100 Ω resistance, calculate the resistor power dissipation for the conditions of parts (a) and (b).

Solution:

(a) From Fig. 5-3, the forward voltage drop is about 0.35 V when $I = 0.25$ A. The power dissipated is

$$P = IV$$
$$= 0.25 \text{ A} \times 0.35 \text{ V}$$
$$= 0.0875 \text{ W} \quad \text{or} \quad 87.5 \text{ mW}$$

(b) The power dissipated is equal to $I \times V$ at any given condition. Thus,

$$P = 0.05 \ \mu\text{A} \times 100 \text{ V}$$
$$= 5 \ \mu\text{W}$$

(c) For the forward-bias condition of part (a),

$$P = I^2 R$$
$$= (0.25 \text{ A})^2 \times 100 \ \Omega$$
$$= 6.25 \text{ W}$$

For the reverse-bias condition of part (b),

$$P = I^2 R$$
$$= (0.05 \ \mu\text{A})^2 \times 100 \ \Omega$$
$$= 0.25 \times 10^{-12} \text{ W} \quad \text{or} \quad 0.25 \text{ pW}$$

5-2 HALF-WAVE RECTIFICATION

When a diode is used to convert ac to dc, it is termed a *rectifier* and the process is called *rectification*. The most elementary method of rectification is the half-wave rectifier circuit shown in Fig. 5-4. When the transformer secondary voltage is on its

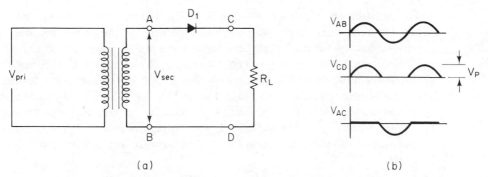

FIGURE 5-4   *(a) Half-wave rectifier circuit; (b) Waveforms.*

positive half-cycle (V_{AB} positive), the diode D_1 is forward-biased and hence presents a very low resistance to the voltage source. This means that almost all the secondary voltage is dropped across the load resistance R_L. The forward-biased diode will typically have a forward voltage drop of 0.5 to 1.0 V if it is silicon or 0.2 to 0.6 V if germanium. To simplify circuit analysis, this voltage drop is often neglected, especially when the supply voltage is high, in which case the diode forward voltage drop will be a small percentage of the resulting output voltage.

Figure 5-4(b) illustrates the performance of a half-wave rectifier. Notice that the output V_{CD} is zero when the transformer voltage V_{AB} is negative. This occurs because the diode has become reverse-biased (anode negative with respect to cathode). Recall that a reverse bias causes the diode to look like a very high resistance—ideally, an open circuit. Hence, by the voltage divider law, the voltage across R_L will approach zero. The average dc voltage (V_{dc}) is equal to 0.318 ($0.318 = 1/\pi$) of the peak value. Many voltmeters are average-reading devices and would thus register 0.318 of the peak voltage for a half-wave rectifier circuit. However, to calculate power, root-mean-square (rms) values must be used. The root-mean-square voltage for the half-wave rectifier circuit is 0.5 of the peak value:

$$V_{dc} = V_{average} = \frac{1}{\pi}\,V_P = 0.318\,V_P \tag{5-1}$$

$$V_{rms} = 0.5\,V_P \tag{5-2}$$

These two different methods of designating voltage can be a source of confusion. Fortunately, this is usually not a concern because the direct current we normally encounter has nearly equivalent root-mean-square and average values. Further detail on this effect is provided in Section 5-4. The current obtained by dividing the load's average voltage by the load resistance is termed the average current, I_0. Hence,

$$I_0 = I_{average} = \frac{V_{average}}{R_L} \tag{5-3}$$

Referring back to Fig. 5-4(b), notice that V_{AC} is the voltage across the diode D_1. When forward-biased, there is a small voltage drop as previously mentioned. When reverse-biased, however, the full peak input voltage is dropped across the diode. This is known as the *peak reverse voltage* (normally abbreviated PRV). Every diode has a maximum allowable PRV rating that should not be exceeded, or device burnout will usually occur. The diode voltage V_{AC} in Fig. 5-4(b) follows V_{AB} while the diode is reverse-biased, since (at this time) the diode has a very high resistance. Notice, also, that while D_1 is forward-biased, the voltage across it (V_{AC}) is not zero but some small positive value. This is the diode's forward voltage drop, which is usually less than 1 V.

EXAMPLE 5-2

Referring to Fig. 5-5 the following data are given:

$$V_{PRI} = 115 \text{ v, } 60 \text{ Hz}$$

$$T_1 \quad \text{has a } 10:1 \text{ turns ratio } (T_R = 10)$$

$$V_{\text{forward}} \text{ of } D_1 = 1 \text{ V (when } D_1 \text{ is forward-biased)}$$

$$R_{\text{reverse}} \text{ of } D_1 = 1 \text{ M}\Omega \text{ (when } D_1 \text{ is reverse-biased)}$$

$$R_L = 100 \text{ }\Omega$$

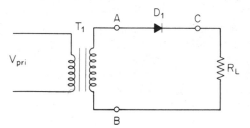

FIGURE 5-5 *Half-wave rectifier circuit for Example 5-2.*

Determine the following:

(a) V_{AB} peak.

(b) V_L (peak, average, and root mean square).

(c) The PRV withstood by D_1.

(d) The power dissipated in R_L.

(e) The peak current through D_1.

(f) The average current (I_0) through D_1.

(g) The peak value of V_L during the negative half-cycle of V_{AB}.

Solution:

(a) Recall from the ac circuit theory in Chapter 2 that the root-mean-square voltage equals the peak voltage divided by $\sqrt{2}$; hence,

$$V_{peak} = V_{rms}\sqrt{2}$$

$$V_{pri} = 115 \text{ v ac}$$

Therefore,

$$V_{pri} \text{ peak} = 115 \text{ V} \times \sqrt{2} = 162 \text{ V}$$

$$V_{sec} = \frac{V_{pri}}{T_R}$$

$$V_{sec} \text{ peak} = \frac{162}{10} = \underline{16.2 \text{ V}} = V_{AB} \text{ peak}$$

(b) D_1 has a forward voltage drop that was given as 1 V.

$$\begin{aligned} V_L \text{ peak} &= V_{sec} \text{ peak} - V_{forward} \text{ of } D_1 \\ &= 16.2 \text{ V} - 1 \text{ V} = 15.2 \text{ V peak} \end{aligned}$$

$$\begin{aligned} V_L \text{ av} &= V_L \text{ peak} \times 0.318 \\ &= 15.2 \text{ V} \times 0.318 \\ &= \underline{4.8 \text{ V}} \end{aligned} \qquad \textbf{(5-1)}$$

$$\begin{aligned} V_L \text{ rms} &= V_L \text{ peak} \times 0.5 \\ &= 15.2 \text{ V} \times 0.5 \\ &= \underline{7.6 \text{ V}} \end{aligned} \qquad \textbf{(5-2)}$$

(c) The peak reverse voltage (PRV) of the diode equals the peak value of the transformer secondary voltage for a half-wave rectifier. Therefore,

$$\begin{aligned} V_{PRV} &= V_{sec} \text{ peak} \\ &= \underline{16.2 \text{ V}} \end{aligned}$$

(d) Recall that the power dissipated in a load equals the product of voltage times current, or I^2R or V^2/R. Remember that root-mean-square values of voltage and/or current must be utilized in making power calculations.

$$\begin{aligned} P &= \frac{V^2 \text{ rms}}{R_L} \\ &= \frac{7.6^2 \text{ V}}{100 \text{ }\Omega} \\ &= \underline{0.58 \text{ W}} \end{aligned}$$

(e)

$$I_{peak} = \frac{V_L\ peak}{R_L}$$

$$= \frac{15.2\ V}{100\ \Omega}$$

$$= \underline{152\ mA}$$

(f)

$$I_{av} = I_0 = \frac{V_L\ av}{R_L} \qquad \qquad \text{(5-3)}$$

$$= \frac{4.8\ V}{100\ \Omega}$$

$$= \underline{48\ mA}$$

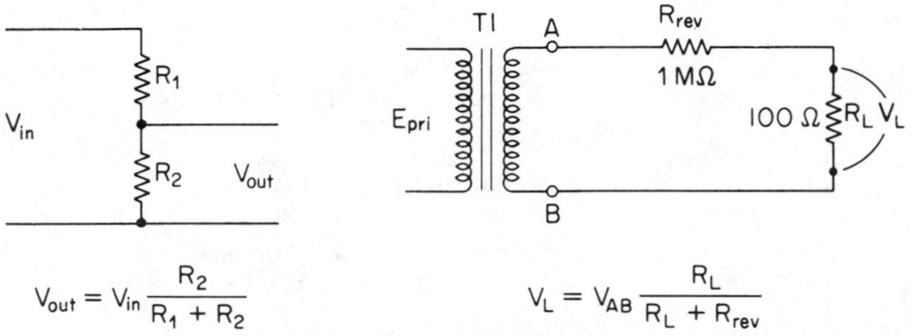

$$V_{out} = V_{in}\frac{R_2}{R_1 + R_2}$$

(a) Voltage divider rule

$$V_L = V_{AB}\frac{R_L}{R_L + R_{rev}}$$

(b) Equivalent circuit of Fig. 2-3 during negative half-cycle

FIGURE 5-6 *Solution to Example 5-2.*

(g) To solve for V_L during the negative half-cycle, the voltage divider rule should be used. Figure 5-6(a) reviews the voltage divider rule in general; part (b) applies this rule to the specific circuit in question:

$$V_L = 16.2\ V \times \frac{100\ \Omega}{100\ \Omega + 1\ M\Omega} \qquad \text{from Fig. 2-4(b)}$$

but since $100\ \Omega + 1\ M\ \Omega \simeq 1\ M\Omega$ ($\simeq$ means approximately equals)

$$V_L = \frac{16.2\ V \times 100\ \Omega}{10^6\ \Omega} = 16.2 \times 10^{-4}\ V$$

$$= \underline{0.00162\ V}$$

5-3   FULL-WAVE RECTIFICATION

A generally more useful and efficient method of converting alternating current to direct current is to recover both the positive and negative portions of the ac input signal. The two circuits used for this purpose are illustrated in Fig. 5-7. This is

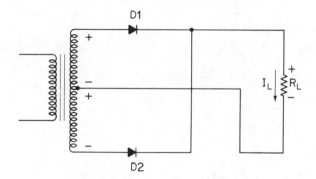

(a) Full-wave center-tap rectifier

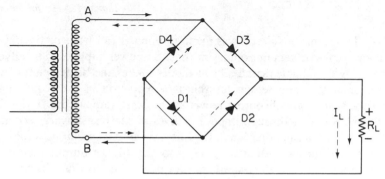

When V_{AB} is positive, solid arrows show current flow
When V_{AB} is negative, dotted arrows show current flow

(b) Full-wave bridge rectifier

FIGURE 5-7 *Full-wave rectifiers.*

known as *full-wave rectification* because the full input wave is utilized in the dc output.

The center-tap version shown in Fig. 5-7(a) uses a secondary winding with a center tap. If, at a given instant, the voltage polarities are as shown, then D_1 is forward-biased and conducting, since its anode is positive with respect to its cathode, while D_2 is reverse-biased and nonconducting. Thus, only D_1 is supplying current to the load.

This situation reverses itself when the polarity of the transformer's secondary voltage reverses on the next ac half-cycle. Then D_1 becomes reverse-biased and D_2 forward-biased, which means that D_2 is then supplying current to the load. Since each diode conducts only half of the time (on alternate half-cycles) a load current double the current rating of the diodes can be provided.

The input and output waveforms are shown in Fig. 5-8. Notice the effective doubling of frequency that takes place between the input and output. This is true because the period T of the output waveform is one-half that of the ac input signal. Recall that frequency is inversely related to the period ($f = 1/T$).

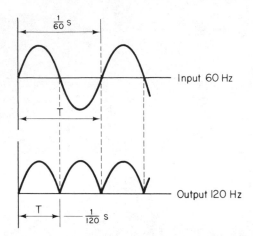

FIGURE 5-8 *Input (60 Hz) and subsequent output (120 Hz) for full-wave rectifiers.*

The center-tap circuit was the most popular full-wave rectifier circuit until silicon diodes replaced vacuum diodes in most dc power supplies. The advent of these low-cost, highly reliable, small silicon diodes makes the bridge circuit the most popular version today. The reason for this is a reduction in the required transformer size for the same available output power as the center-tap circuit. In the center-tap circuit, current is drawn through opposite halves of the transformer secondary on alternate half-cycles. On the other hand, the full-wave bridge circuit uses the entire transformer secondary during both half-cycles, allowing higher output power than the center-tap circuit for equivalent-sized transformers. The obvious disadvantage of the bridge circuit is the need for twice as many rectifiers as the center-tapped version. However, with the availability of low-cost silicon diodes, the full-wave bridge became more economical than the full-wave center-tap circuit, in spite of the need for twice as many diodes.

To analyze the performance of the full-wave bridge circuit, refer to Fig. 5-7(b). The solid arrows (near D_1 and D_3) show the current flow when the voltage V_{AB} is on the positive half-cycle; the dashed arrows denote current flow for the negative half-cycle. Notice that during the positive V_{AB} half-cycle, D_1 and D_3 are conducting, and that during the negative half-cycle, D_2 and D_4 conduct. This results in the full-wave voltage-output waveform shown in Fig. 5-8. Since the diodes conduct only

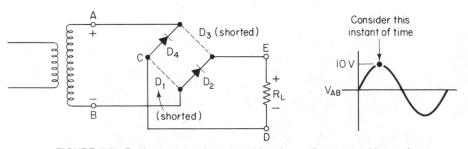

FIGURE 5-9 *Peak reverse voltage considerations—the reverse bias path.*

half the time, the allowable average output current is double the average current ratings of the diodes. Notice that the peak value of the output voltage will be the peak transformer secondary voltage minus the forward drop of two diodes. Notice also that during the time D_1 and D_3 are forward-biased (and hence conducting), D_2 and D_4 are reverse-biased. As shown in Fig. 5-9, when V_{AB} is maximum positive, this voltage appears across the loop $ACDEB$. Writing the loop equation using Kirchhoff's voltage law, $V_{BA} + V_{AC} + V_{DE} + V_{EB} = 0$. Consider the instant that $V_{BA} = -10$ V, as in Fig. 5-9. (If $V_{AB} = +10$ V, then $V_{BA} = -10$ V.) Then $V_{DE} = -10$ V approximately as a result of the conduction of D_1 and D_3. Hence,

$$(-10 \text{ V}) + V_{AC} + (-10 \text{ V}) + V_{EB} = 0$$

and

$$V_{AC} + V_{EB} = 20 \text{ V}$$

Therefore, a total of 20 V is across D_2 and D_4, and it seems logical to assume that they will split this voltage. Therefore, the reverse rating (PRV) of each diode is approximately equal to the peak input voltage of the circuit for a full-wave bridge circuit.

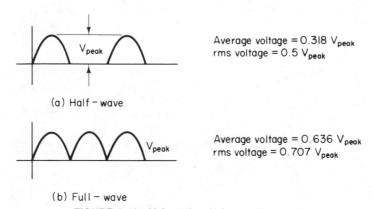

(a) Half – wave

Average voltage $= 0.318\ V_{peak}$
rms voltage $= 0.5\ V_{peak}$

(b) Full – wave

Average voltage $= 0.636\ V_{peak}$
rms voltage $= 0.707\ V_{peak}$

FIGURE 5-10   *Voltage levels for rectifier circuits.*

Full-wave rectification results in a root-mean-square output voltage of $0.707\ V_{peak}$ and an average dc voltage of $0.636\ V_{peak}$. These results are summarized in Fig. 5-10 for full-wave and half-wave rectification. Expressed as equations,

$$V_{dc} = V_{average} = \frac{2}{\pi}\ V_P = 0.636 V_P \qquad \text{(5-4)}$$

$$V_{rms} = 0.707\ V_{peak} \qquad \text{(5-5)}$$

EXAMPLE 5-3

A full-wave rectifier circuit delivers a peak voltage of 24 V to a 48-Ω load. Calculate:

(a) V_{dc} to the load.

(b) V_{rms} to the load.

(c) The load power dissipation.

(d) The average load current (I_0 to load).

(e) The required I_0 rating of the diodes.

(f) The PRV rating of the diodes.

Solution:

(a)
$$V_{dc} = 0.636 V_P \qquad\qquad \textbf{(5-4)}$$
$$= 0.636 \times 24 \text{ V}$$
$$= \underline{15.3 \text{ V}}$$

(b)
$$V_{rms} = 0.707 V_P \qquad\qquad \textbf{(5-5)}$$
$$= 0.707 \times 24 \text{ V}$$
$$= \underline{17 \text{ V}}$$

(c)
$$P = \frac{V^2_{rms}}{R}$$
$$= \frac{17^2}{48 \ \Omega}$$
$$= \underline{6 \text{ W}}$$

(d)
$$I_0 = \frac{V_{dc}}{R}$$
$$= \frac{15.3 \text{ V}}{48 \ \Omega}$$
$$= \underline{0.318 \text{ A}}$$

(e) The required current rating of diodes for both the center-tap and bridge circuits is half the load current or 0.318 A/2 = 0.159 A. This is not a standard rating, but using diodes rated for $I = 0.25$ A (or even 0.5 A or 1 A) would provide a margin of safety and they are standard values. *Incorrect – see supplemental notes*

(f) The required PRV rating for both the center-tap and bridge circuits is approximately equal to the peak load voltage. Thus, a PRV rating greater than 24 V is necessary. The lowest standard PRV rating is 50 V, and this would provide a good margin of safety.

5-4 POWER-SUPPLY FILTERS

The output of the various rectifier configurations we have considered is called pulsating direct current. It is pulsating because it pulsates from a zero level (ground) up to some peak value and back to zero at a steady, repetitious rate. This type of output is sometimes useful to drive a dc motor but not for sophisticated electronic circuits. As a matter of fact, electronic circuits usually require a very steady dc output that approaches the smoothness of a battery's output. A circuit that converts pulsating direct current into a very steady dc level is known as a *filter*, because it "filters" out the pulsations in the output. Refer to Fig. 5-11 for visualization of a filter's function.

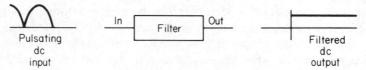

FIGURE 5-11 *Filter function.*

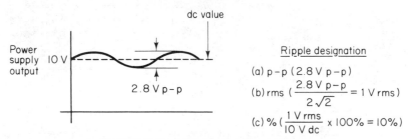

FIGURE 5-12 *Methods of ripple designation.*

Many types of filters have been devised over the years, using capacitors, inductors, resistors, and various combinations of the three. However, the capacitor alone has been most popular for the filters in modern power supplies, because inductors are extremely bulky and expensive compared to the lower-cost capacitors that are available for today's low-voltage circuits. Consequently, we shall concentrate our study of filters in that area.

The amount of pulsation (usually termed the ac component) in a dc power supply output is called *ripple*. Keeping the ripple content low is the function of the filter. In practical applications, ripple is designated by peak-to-peak volts or root-mean-square volts, and sometimes as a percentage of the output. Percentage ripple is defined as follows:

$$\% \text{ ripple} = \frac{\text{rms value of ripple}}{\text{dc output voltage}} \times 100\% \qquad \textbf{(5-6)}$$

Notice that in Fig. 5-12 the power supply has a 10 V dc output with an ac component riding on the dc level. This ac component, the ripple, is designated in three ways—all acceptable:

1. Peak-to-peak value of the ac component.

2. Root-mean-square value of ac component.

3. Percent ripple [Eq. (5-6)].

Recall from your study of capacitance that a capacitor has the ability to store energy. The purpose of the capacitor filter is to smooth the rectifier output by receiving energy during peaks of rectifier output voltage for storage, and then releasing energy to the load during periods of low voltage input to the capacitor. This is illustrated for both the half-wave and full-wave condition in Fig. 5-13. Note that the capacitor's charge curve initially follows the sinusoidal rectifier output, as seen in Fig. 5-13(d) and (f). This occurs while the rectifier output is increasing, but then as it starts decreasing, the capacitor starts releasing the energy it stored initially. Thus, it "holds" the voltage up until the next cycle of rectifier output exceeds the capacitor voltage. At that point the output follows the rectifier voltage and the capacitor is storing energy for release when the rectifier output again starts to rapidly fall. The amount

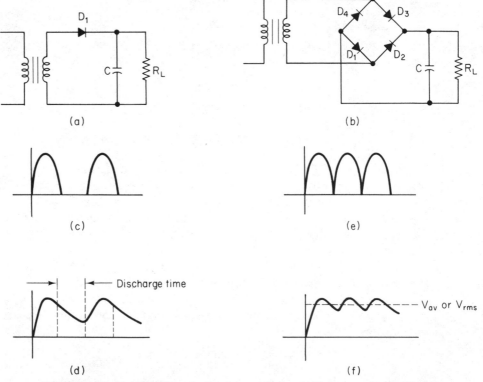

FIGURE 5-13 *Rectifier-filter circuits and waveforms: (a) Half-wave circuit with filter; (b) Full-wave circuit with filter; (c) Half-wave output without filtering; (d) Filtered half-wave output; (e) Full-wave output without filtering; (f) Filtered full-wave output.*

of capacitor and thus load voltage discharge [i.e., see the discharge time in Fig. 5-13(d)] is effectively dependent on the RC time constant between the capacitor and R_L. To minimize discharge voltage, and thus the amount of ripple, the capacitor value should be high. The capacitor is "smoothing" the rectifier output, as pictured in Fig. 5-13(d) for a half-wave circuit and at Fig. 5-13(f) for a full-wave circuit.

It is noteworthy to consider the diode current flow in these circuits. With a capacitor filter, diode forward current flows during short intervals of time. These short conduction intervals correspond to the periods of time when the transformer secondary output voltage is greater than the capacitor voltage, which is also the load or output voltage. Thus, the diode current is of a surging (temporary) nature, giving rise to an important rating of power rectifiers—surge current. This rating should not be exceeded in power-supply design. Given a rectifier's average current rating, the resulting surge current rating is adequate for most designs. The conditions the designer must be wary of are those circuits in which abnormally high values of filter capacitance are being utilized. This situation means that the charge-current duty cycle is extremely short, and hence of high amplitude.

It should be recognized that the magnitude of ripple is less in the full-wave circuit [Fig. 5-13(f)] than in the half-wave circuit [Fig. 5-13(d)] because of the shorter discharge time before the capacitor is reenergized by another pulse of current. For a full-wave circuit, 60-Hz line voltage frequency, and moderate to low ripple factor ($\leq 5\%$), the following approximation can be used to determine % ripple:

$$\% \text{ ripple} \simeq \frac{2.4 \times 10^{-3}}{R_L C} \times 100\% \qquad \textbf{(5-7)}$$

EXAMPLE 5-4

A full-wave bridge power supply is to be designed to operate a 12-V dc tape player. The tape player draws an average current of 0.3 A and can tolerate 0.1 V rms of ripple before audible hum is heard at its output. Determine the required filter capacitor size for this power supply. The ac line voltage is 60 Hz.

Solution:

The effective load resistance of the tape player can be determined by Ohm's Law.

$$R_L = \frac{12 \text{ V}}{0.3 \text{ A}}$$
$$= 40 \ \Omega$$

The % ripple allowable is

$$\% \text{ ripple} = \frac{\text{rms value of ripple}}{\text{dc output voltage}} \times 100\% \qquad \text{(5-6)}$$

$$= \frac{0.1 \text{ V rms}}{12 \text{ V dc}} \times 100\%$$

$$= 0.833\%$$

Now, the value of filter capacitance required can be calculated using Eq. (5-7).

$$\% \text{ ripple} = \frac{2.4 \times 10^{-3}}{R_L C} \times 100\% \qquad \text{(5-7)}$$

$$0.833\% = \frac{2.4 \times 10^{-3}}{40 \text{ }\Omega \times C} \times 100\%$$

The % on each side of the equation cancel, and rearranging,

$$C = \frac{2.4 \times 10^{-3} \times 100}{40 \times 0.833}$$

$$= \underline{7200 \text{ }\mu\text{F}}$$

You will recall from Section 5-3 that the unfiltered dc voltage had different average and rms values [see, for instance, Eqs. (5-1) and (5-2)]. The addition of a filter makes the difference between the rms and average value almost negligible. This condition is shown in Fig. 5-13(f). Of course, if the ripple were zero (pure dc), the rms and average voltages are exactly the same.

5-5 REGULATION

The power supplies that we have so far discussed are of the unregulated variety. A regulated power supply has more than just the basic transformer, rectifiers, and filter discussed in the previous section. A *regulated* power supply has a means of keeping the dc output voltage relatively constant when:

1. The ac input voltage to the transformer varies (variations in the 115-v 60-Hz line from 100 to 130 v are common).

2. The load is varied.

The term that gives an indication of the amount of deviation for load changes is *percentage load regulation,* sometimes referred to simply as the power supply's *load regulation:*

$$\% \text{ load regulation} = \frac{(V_{\text{NL}} - V_{\text{FL}}) \times 100\%}{V_{\text{FL}}} \qquad \textbf{(5-8)}$$

where V_{FL} and V_{NL} refer to the power-supply output under rated full-load and no-load conditions, with constant input voltage.

The term that gives an indication of the power-supply output deviation for input line changes is the *percentage line regulation,* or simply *line regulation:*

$$\% \text{ line regulation} = \frac{(V_{\text{high}} - V_{\text{low}}) \times 100\%}{V_{\text{low}}} \qquad \textbf{(5-9)}$$

where V_{high} refers to the dc output at the highest specified input voltage and V_{low} refers to the dc output at the lowest input voltage at full-rated load current.

The term that predicts a supply's output variations with respect to both line and load variations is the *percentage regulation,* or simply *regulation:*

$$\% \text{ regulation} = \frac{(V_{\text{max}} - V_{\text{min}}) \times 100\%}{V_{\text{min}}} \qquad \textbf{(5-10)}$$

where V_{max} is the maximum dc output voltage for variations in both line and load and normally occurs when the ac input is at its maximum value and when the load is lightest (R_L maximum). V_{min} is the minimum dc output voltage and normally occurs under minimum input voltage ($V_{\text{IN}_{\text{min}}}$) and heaviest load conditions.

For highly regulated power supplies (i.e., very small dc output changes), another designator commonly replaces percentage regulation. Instead of an unwieldy regulation specification, such as 0.005%, it is more convenient to simply say $\pm$0.25 mV. Thus, a 10 V dc power supply with 0.005% regulation could also be designated by the $\pm$0.25-mV regulation expression. Notice that 0.005% of 10 V is 0.5 mV:

$$0.005\% \times \frac{1}{100\%} \times 10 \text{ V} = 0.5 \text{ mV}$$

The term 0.005% in this case indicates a 0.5-mV envelope within which this power supply will remain for all line and load variations, as does the term $\pm$0.25 mV.

An unregulated power supply's output will vary directly with the input line voltage variation. If the line voltage changes from 115 V by 10% to 126.5 V, the dc output will also change by 10%. For instance, a 10-V dc power supply would go up to 11 V dc. This 10% load voltage change is fairly substantial and should give the reader an indication of the need for some form of voltage regulator for most electronic systems.

EXAMPLE 5-5

A regulated power supply (rated at 105 to 125 v rms input and 1 A load current) is checked out in the laboratory and the following data are obtained;

$v_{in} = 105$ v rms	$v_{in} = 125$ v rms
$I_L = 1$ A	$I_L = 0$
$V_{out} = 9.75$ V dc	$V_{out} = 10.05$ V dc
$v_{in} = 125$ v rms	$v_{in} = 105$ v rms
$I_L = 1$ A	$I_L = 0$
$V_{out} = 9.93$ V dc	$V_{out} = 10.01$ V dc

Determine line, load, and overall regulation for this power supply.

Solution:

$$\text{\% line regulation} \atop \text{(at 1 A load)} = \frac{(V_{high} - V_{low}) \times 100\%}{V_{low}} \qquad (5\text{-}9)$$

$$= \frac{9.93 \text{ V} - 9.75 \text{ V}}{9.75 \text{ V}} \times 100\%$$

$$= \underline{1.85\%}$$

$$\text{\% load regulation} = \frac{V_{NL} - V_{FL}}{V_{FL}} \times 100\% \qquad (5\text{-}8)$$

At 125 V ac input,

$$= \frac{10.05 \text{ V} - 9.93 \text{ V}}{9.93 \text{ V}} \times 100\%$$

$$= \underline{1.21\%}$$

$$\text{\% regulation} = \frac{V_{max} - V_{min}}{V_{min}} \times 100\% \qquad (5\text{-}10)$$

$$= \frac{10.05 \text{ V} - 9.75 \text{ V}}{9.75 \text{ V}} \times 100\%$$

$$= \underline{3.08\%}$$

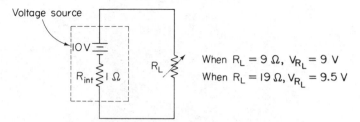

FIGURE 5-14 *Nonideal voltage source and load.*

An ideal power supply would exhibit no output voltage change when the load changes. The characteristic of a power supply that causes an output change is its *internal resistance, R_{int}.* The ideal supply has an internal resistance of zero ohms and would deliver a constant voltage regardless of load.

Unfortunately, all practical sources of voltage always have some internal resistance (often called output impedance) associated with them. Figure 5-14 shows such a source, a 10-V battery with a R_{int} of 1 Ω. If R_L were 9 Ω, the voltage across it by the voltage divider rule would be

$$V_{R_L} = 10 \text{ V} \left(\frac{9 \text{ } \Omega}{1 \text{ } \Omega + 9 \text{ } \Omega} \right) = 9 \text{ V}$$

If the load were changed to 19 Ω, the load voltage would increase to

$$V_{R_L} = 10 \text{ V} \left(\frac{19 \text{ } \Omega}{1 \text{ } \Omega + 19 \text{ } \Omega} \right) = 9.5 \text{ V}$$

Thus, it is seen that the voltage delivered to a load from a nonideal voltage source is dependent on the size of the load. The more current drawn from a practical voltage source, the lower will be the voltage at the load, owing to the increased voltage drop across the source's internal resistance.

EXAMPLE 5-6

Determine the output voltage of a power supply when $I = \frac{1}{2}$ A, which delivers 20 V at no load ($I = 0$) and has an internal resistance of 0.25 Ω.

Solution:

The equivalent circuit for this power supply is shown within dashed lines in Fig. 5-15. The load current of 0.5 A also flows through the internal resistance, R_{int}. Thus, the voltage drop $V_{R_{int}}$ is

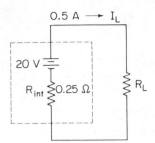

FIGURE 5-15 *Circuit for Example 5-6.*

$$V_{R_{int}} = I_L \times R_{int}$$
$$= 0.5\ \text{A} \times 0.25\ \Omega$$
$$= 0.125\ \text{V}$$

By Kirchhoff's Voltage Law (KVL), the output voltage, V_{R_L}, is

$$20\ \text{V} = V_{R_{int}} + V_{R_L}$$

$$V_{R_L} = 20\ \text{V} - V_{R_{int}}$$
$$= 20\ \text{V} - 0.125\ \text{V}$$
$$= \underline{19.875\ \text{V}}$$

To calculate R_L,

$$R_L = \frac{V_{R_L}}{I_L}$$

$$= \frac{19.875\ \text{V}}{0.5\ \text{A}}$$

$$= \underline{39.75\ \Omega}$$

5-6   ZENER DIODE REGULATOR

Now that we have briefly talked about the characteristics of regulation in general, we shall study the circuitry of some regulators. A zener diode is similar to a normal diode except that its reverse breakdown voltage is put to practical use. In addition, the point of reverse breakdown is carefully controlled in the manufacturing process, and this zener voltage is available anywhere from 2 or 3 V up to 200 V. Figure 5-16 illustrates the zener diode characteristic curve with the pertinent currents and voltages labeled. I_{ZK} is the minimum reverse diode current to put the diode into its regulating (breakdown) region, and I_{ZM} is the maximum reverse current the diode can withstand without exceeding its power rating. Notice that between I_{ZK} and I_{ZM} the zener voltage changes only slightly. The slope of the line between I_{ZK} and I_{ZM} is known as Z_Z, the *zener impedance.* The *slope* of a curve is given by the change along the horizontal axis divided by the change along the vertical axis. A change

$$\text{Slope} = \frac{\Delta y}{\Delta x}, \text{ not what he writes}$$

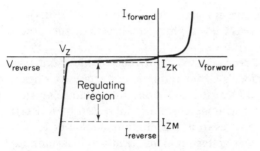

FIGURE 5-16 *Zener diode characteristic curve.*

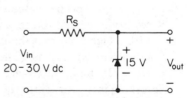

FIGURE 5-17 *Simple shunt regulator.*

in voltage, current, or other variables is represented by preceding it with the Greek capital letter delta, Δ. Thus, the slope for the curve represented in Fig. 5-16 is given by $\Delta V/\Delta I$.)

By operating a zener diode between I_{ZK} and I_{ZM}, the voltage across the diode will remain relatively constant. This principle is put to practical use in the circuit shown in Fig. 5-17. An unregulated source of dc voltage from 20 to 30 V is applied to the circuit. The polarity is such as to reverse bias the zener diode and, it is hoped, to supply enough reverse current so that I_{ZK} will be surpassed and hence will put the diode into the *regulating region* shown in Fig. 5-16. The output voltage will then remain relatively constant, even as the input varies from 20 to 30 V. The zener diode is thus regulating the output voltage and is in shunt (parallel) with the output and the circuit; it is therefore termed a *shunt regulator*. Notice that $V_{out} = V_Z$.

The value of R_S in Fig. 5-17 must be properly selected so as to always provide enough current to keep the zener regulating and to supply the necessary load current. The necessary reverse zener current is I_{ZK}, and the load is usually variable between certain specific limits. If the load were to vary from 0 to 100 mA and I_{ZK} for the zener diode were 10 mA, it would be possible to calculate a suitable value for R_S. The current through R_S will be a minimum when the voltage across R_S is minimum. Therefore, R_S must be small enough to supply the maximum load current of 100 mA and the zener diode's I_{ZK} in order to allow proper circuit operation under all conditions. Thus,

$$R_S = \frac{V_{R_S\text{min}}}{I_{R_L\text{max}} + I_{ZK}}$$

$$= \frac{V_{IN\text{min}} - V_Z}{I_{R_L\text{max}} + I_{ZK}} = \frac{20\,\text{V} - 15\,\text{V}}{100\,\text{mA} + 10\,\text{mA}} \tag{5-11}$$

$$= \frac{5\,\text{V}}{110\,\text{mA}} = 45.5\,\Omega$$

If an R_S greater than 45.5 Ω were used, the zener would not be provided with enough reverse current (I_{ZK}) when the unregulated input is at its minimum value (20 V) and the load is drawing its maximum amount of current (100 mA). As a safety factor, it is therefore wise to select R_S slightly smaller than the exact calculated value.

Once the proper value has been selected for a zener diode shunt regulator, it is necessary to ensure the zener diode's power rating is not exceeded when R_S supplies it with a higher current than it does at the minimum input voltage and/or maximum load current conditions. (The zener diode current is increased by a decrease in load current.) If the input voltage increases from the minimum value of 20 V, the voltage across R_S increases, since the zener voltage remains relatively constant. (Apply KVL as an aid in understanding that $V_{in} = V_{R_S} + V_{out}$). Similarly, if the load changed so as to draw less than 100 mA, that drop in load current would be forced through the zener diode. Thus, we see that the maximum power dissipation in the zener diode of a shunt regulator occurs when the input voltage goes to its highest possible value and when the load current is at its minimum value. For the circuit of Fig. 5-17, then, with an R_S of 45.5 Ω, the maximum zener current can be calculated as

$$I_{Z_{max}} = \frac{V_{RS_{max}}}{R_S} - I_{L_{min}} \qquad (5\text{-}12)$$

Since $V_{RS_{max}}$ is 30 to 15 V and $I_{L_{min}}$ was specified as zero, we have

$$I_{Z_{max}} = \frac{30\text{ V} - 15\text{ V}}{45.5\ \Omega} - 0\text{ mA}$$
$$= 330\text{ mA}$$

The power dissipated by a zener diode is equal to the product of its voltage and current:

$$P_Z = V_Z I_Z$$

The maximum power dissipated by the zener diode occurs when I_Z is maximum. Therefore $P_Z = 15\text{ V} \times 330\text{ mA} = 4.95\text{ W}$.

EXAMPLE 5-7

Calculate an appropriate value for R_S and the required power rating of D_1 for the circuit shown in Fig. 5-18. The input voltage varies from 15 to 20 V and $I_{ZK} = 10$ mA.

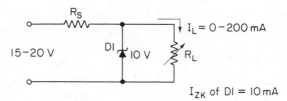

FIGURE 5-18   *Circuit for Example 5-6.*

Solution:

Noting from Fig. 5-18 that the load current can vary from 0 to 200 mA, the maximum allowable value of R_S can be calculated from Eq. (5-11):

$$R_S = \frac{V_{R_{S_{min}}}}{I_{R_{L_{max}}} + I_{ZK}} \qquad \text{(5-11)}$$

$$= \frac{15\,\text{V} - 10\,\text{V}}{200\,\text{mA} + 10\,\text{mA}} = \frac{5\,\text{V}}{210\,\text{mA}}$$

$$= 23.8\,\Omega$$

As a safety factor, select $R_S = 22\ \Omega$, a standard value of resistance slightly lower than the calculated value of R_S. The maximum current through the zener will occur when the load current is minimum (0 mA) and the input voltage is maximum (20 V). From Eq. (5-12),

$$I_{Z_{max}} = \frac{V_{R_{S_{max}}}}{R_S} - I_{L_{min}} \qquad \text{(5-12)}$$

$$= \frac{20\,\text{V} - 10\,\text{V}}{22\,\Omega} - 0\,\text{mA}$$

$$= 455\,\text{mA}$$

Thus, the required zener power rating can be calculated.

$$P = V \times I$$
$$= 10\,\text{V} \times 455\,\text{mA}$$
$$= 4.55\,\text{W}$$

EXAMPLE 5-8

It is desired to operate a 9-V portable cassette tape player ($I_{L_{max}} = 150$ mA) from an automobile electrical system. The available zener diode is rated at 9 V with an I_{ZK} of 1 mA. Calculate the appropriate value for R_S and the required power ratings for the zener diode and R_S. The voltage from the automobile's electrical system varies from 11 V up to 15 V.

Solution:

A minimum load current was not given, but in this case the circuit should be designed for no-load conditions for when the tape player is turned off or disconnected with the shunt regulator still energized. Thus,

$$R_S = \frac{V_{R_{S_{min}}}}{I_{R_{L_{max}}} + I_{ZK}} \tag{5-11}$$

$$= \frac{11\ V - 9\ V}{150\ mA + 1\ mA}$$

$$= 13.25\ \Omega$$

As a safety factor we will use a 12-Ω resistor. Now

$$I_{Z_{max}} = \frac{V_{R_{S_{max}}}}{R_S} - I_{L_{min}}$$

$$= \frac{15\ V - 9\ V}{12\ \Omega} - 0\ mA$$

$$= 250\ mA \quad = 500\ mA$$

Therefore,

$$P_Z = 9\ V \times 250\ mA \quad 500\ mA$$

$$= 2.25\ W \quad = 4.5\ w$$

To determine the maximum resistor power, we use

$$P = I^2 R$$

$$= (I_{R_{S_{max}}})^2 \times R$$

The maximum I_{R_S} is (15 V − 9 V)/12 Ω or 250 mA (500 mA). Thus, $P_{R_{S_{max}}} = (250\ mA)^2 \times 12\ \Omega, = 0.75\ W$ (3 w).

Although shunt regulators are effective in maintaining a constant load voltage as the source voltage changes and/or the load changes, they are relatively inefficient. Large amounts of power are wasted in the zener and R_S in comparison with the amount of power actually delivered to the load. Example 5-8 illustrates this situation, because the zener diode and resistor dissipation can be as high as 2.25 and 0.75 W, respectively. The load power is only a maximum of 9 V × 150 mA = 1.35 W by comparison. The series regulator covered in the next section is much more efficient in this respect.

5-7 SERIES REGULATORS

In contrast to the shunt regulator, the control element of the series regulator is in series with the load. The block diagram shown in Fig. 5-19 is useful in analyzing the performance of such a circuit. The first two blocks are the familiar elements of a basic unregulated power supply. The series pass element is the control element—the element that turns the "electron spigot" up and down to maintain a constant

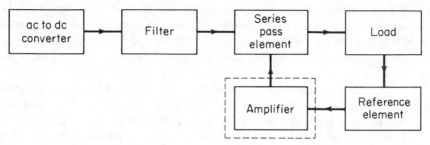

FIGURE 5-19 *Block diagram for a series regulator power supply.*

load voltage while the load current may be changing, or to compensate for changes in the input voltage. The reference element is a constant-voltage device (usually a zener diode) that compares the output voltage with its zener voltage and produces an "error" signal if the load voltage is not correct. This signal is made larger by the amplifier shown in Fig. 5-19 and then the amplifier's output is used to control the series pass element. The series pass element is usually a transistor that controls the current flow to the load so as to keep a constant load voltage.

The series regulator just described is a *feedback control system*. It senses its own output and, if incorrect, sends an error signal to an amplifier, which subsequently is used to correct this condition. Since you have not yet been introduced to basic transistor amplifiers, we shall not try to provide details on the inner workings of a series regulator. However, all of the circuitry required for the reference element,

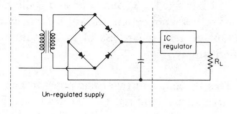

Un-regulated supply

FIGURE 5-20 *IC regulated power supply.*

amplifier, and series pass element in Fig. 5-19 is available in a single three-leaded device, and we are now prepared to study its performance.

The technique of integrating a great deal of electronic circuitry within a small package allows complex functions to be performed at minimal cost and in a very reliable fashion. These circuits are called *integrated circuits* (ICs) and in slang terms are often referred to as "chips," because they are fabricated out of tiny chips (or wafers) of silicon. A large number of regulator ICs are available to easily convert an unregulated supply into a regulated unit. This situation is illustrated in Fig. 5-20. The IC regulator chip might typically cost $1 and provide extremely good regulation.

The manufacturer's specifications for a popular IC regulator, the LM309, are provided in Fig. 5-21. They are included with the permission of the manufacturer, National Semiconductor. At first glance you may be somewhat overwhelmed by the many strange terms and symbols. Do not be dismayed; we will not discuss all of them now, but let us consider some of the more important parts of this regulator's specifications. The following descriptions refer to the various headings of Fig. 5-21.

General description: This section provides the general characteristics of the device and explains the major applications. In this case we find out that it is intended mainly as a fixed 5-V dc output device for digital logic circuits (Chapter 10 will provide an introduction to these circuits). It is also noted that this regulator IC is essentially blowout-proof. This means that if too much current is drawn through the chip, causing overheating, it will limit the current draw and/or shut itself down until it cools.

Schematic diagram: As can be seen, this IC contains a large amount of complex circuitry. The many three-terminal devices shown in the schematic with Q designators are transistors, but also shown are resistors and one capacitor. The use of capacitors in ICs is minimized because they take up a lot of space in comparison to the other circuit elements. Keep in mind that this circuit is contained on a silicon wafer less than ⅛ inch square! If the circuit were duplicated using individual (discrete) components, the cost might be $40, as opposed to about $1 in integrated form.

Typical applications: The first application shown is a high-stability regulator that can provide better than 0.01% regulation. It uses the LM309 in conjunction with another IC, the LM308A, and some other discrete components. Additional applications shown include the basic fixed 5-V regulator and a method to obtain an adjustable output regulator. The last application shown is for a current regulator. This circuit provides a constant current regardless of load or input changes. It is often termed a *constant current* supply.

Connection diagrams: This IC comes in two basic physical packages, as shown in Fig. 5-21. If low output currents are required (< 200 mA), the TO-5 package is suitable while output currents up to 1 A are possible with the TO-3 unit. These

Voltage Regulators

LM309 five-volt regulator
general description

The LM309 is a complete 5V regulator fabricated on a single silicon chip. It is designed for local regulation on digital logic cards, eliminating the distribution problems associated with single-point regulation. The device is available in two common transistor packages. In the solid-kovar TO-5 header, it can deliver output currents in excess of 200 mA, if adequate heat sinking is provided. With the TO-3 power package, the available output current is greater than 1A.

The regulator is essentially blow-out proof. Current limiting is included to limit the peak output current to a safe value. In addition, thermal shutdown is provided to keep the IC from overheating. If internal dissipation becomes too great, the regulator will shut down to prevent excessive heating.

Considerable effort was expended to make the LM309 easy to use and minimize the number of external components. It is not necessary to bypass the output, although this does improve transient response somewhat. Input bypassing is needed, however, if the regulator is located very far from the filter capacitor of the power supply. Stability is also achieved by methods that provide very good rejection of load or line transients as are usually seen with TTL logic.

Although designed primarily as a fixed-voltage regulator, the output of the LM309 can be set to voltages above 5V, as shown below. It is also possible to use the circuit as the control element in precision regulators, taking advantage of the good current-handling capability and the thermal over-load protection.

To summarize, outstanding features of the regulator are:

- Specified to be compatible, worst case, with TTL and DTL
- Output current in excess of 1A
- Internal thermal overload protection
- No external components required

schematic diagram

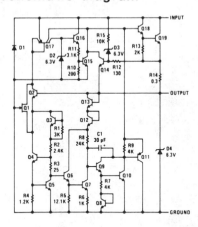

typical application

High Stability Regulator*

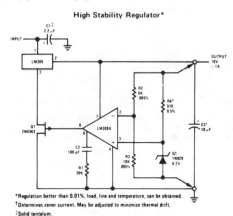

*Regulation better than 0.01%, load, line and temperature, can be obtained.
†Determines zener current. May be adjusted to minimize thermal drift.
‡Solid tantalum.

connection diagrams

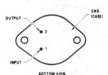

TO-5 (H)

BOTTOM VIEW

Order Number LM309H
See Package 9

TO-3 (K)

BOTTOM VIEW

Order Number LM309K
See Package 18

FIGURE 5-21 *Courtesy of* National Semiconductor.

absolute maximum ratings

Input Voltage	35V
Power Dissipation	Internally Limited
Operating Junction Temperature Range	$0°C$ to $125°C$
Storage Temperature Range	$-65°C$ to $150°C$
Lead Temperature (Soldering, 10 sec)	$300°C$

electrical characteristics (Note 1)

PARAMETER	CONDITIONS	MIN	TYP	MAX	UNITS
Output Voltage	$T_j = 25°C$	4.8	5.05	5.2	V
Line Regulation	$T_j = 25°C$ $7V \leq V_{IN} \leq 25V$		4.0	50	mV
Load Regulation LM309H LM309K	$T_j = 25°C$ $5\,mA \leq I_{OUT} \leq 0.5A$ $5\,mA \leq I_{OUT} \leq 1.5A$		20 50	50 100	mV mV
Output Voltage	$7V \leq V_{IN} \leq 25V$ $5\,mA \leq I_{OUT} \leq I_{max}$ $P < P_{max}$	4.75		5.25	V
Quiescent Current	$7V \leq V_{IN} \leq 25V$		5.2	10	mA
Quiescent Current Change	$7V \leq V_{IN} \leq 25V$ $5\,mA \leq I_{OUT} \leq I_{max}$			0.5 0.8	mA mA
Output Noise Voltage	$T_A = 25°C$ $10\,Hz \leq f \leq 100\,kHz$		40		μV
Long Term Stability				20	mV
Thermal Resistance Junction to Case (Note 2) LM309H LM309K			15 3.0		$°C/W$ $°C/W$

Note 1: Unless otherwise specified, these specifications apply for $0°C \leq T_j \leq 125°C$, $V_{IN} = 10V$ and $I_{OUT} = 0.1A$ for the LM309H or $I_{OUT} = 0.5A$ for the LM309K. For the LM309H, $I_{max} = 0.2A$ and $P_{max} = 2.0W$. For the LM309K, $I_{max} = 1.0A$ and $P_{max} = 20W$.

Note 2: Without a heat sink, the thermal resistance of the TO-5 package is about $150°C/W$, while that of the TO-3 package is approximately $35°C/W$. With a heat sink, the effective thermal resistance can only approach the values specified, depending on the efficiency of the sink.

typical applications(con't)

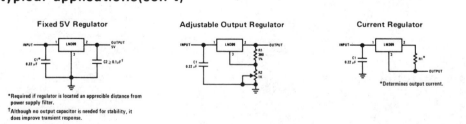

Fixed 5V Regulator

*Required if regulator is located an appreciable distance from power supply filter.
†Although no output capacitor is needed for stability, it does improve transient response.

Adjustable Output Regulator

Current Regulator

*Determines output current.

FIGURE 5-21 *Continued.*

146

typical performance characteristics

FIGURE 5-21   Continued.

current levels require the use of a *heat sink*. This means that the IC is to be physically attached to a heat dissipating unit such as a metal chassis or other surface that can minimize the heat buildup within the chip. The heat buildup is minimized since the heat sink "draws" it away from the IC.

Absolute maximum ratings: These specifications are the absolute limits for the IC. They should not be exceeded under any circumstance.

Electrical characteristics: These characteristics are provided as an aid for the IC user. Notice that for some parameters a minimum, typical, and maximum specification is provided. For example, when used at 25°C, the IC manufacturer claims a minimum output voltage of 4.8 V and a maximum of 5.2 V with a typical value of 5.05 V. In terms of line regulation, the maximum output variation is 50 mV when the input varies from 7 to 25 V. Expressed as a percentage,

$$\% \text{ line regulation} = \frac{50 \text{ mV}}{5 \text{ V}} \times 100\%$$
$$= 1\%$$

Typically, however, the line regulation would be only 4 mV. In terms of load regulation, we find that the LM309H version offers better performance than the LM309K version (typically 20 mV versus 50 mV).

There are still some parts of these specifications that you will not understand—that will come as you progress through this book. Hopefully, you will now have a basic knowledge of this IC regulator.

One final point to consider with respect to regulators is their effect on ripple. Ripple input to a regulator is greatly minimized in the output. The regulator is maintaining a relatively constant output for input voltage variations. Ripple input to the regulator is simply a repetitive input voltage variation. Thus, the regulator serves to greatly reduce ripple and acts as a filter. The IC regulator just described typically reduces ripple by a factor of 10,000 to 1. Thus, if the unregulated input has 3 V p-p ripple, the regulated output ripple would be 3 V p-p/10,000 = 0.3 mV p-p!

We have discussed shunt and series regulators. In recent years a third major type of regulator has emerged. They are known as *switching regulators*. The series regulator is more efficient than the shunt regulator, and the switching variety is even better. Being more efficient means that less power is dissipated within the power supply and generally allows for a smaller and lighter package.

5-8 LIGHT-EMITTING DIODES

In a regular diode, the process of forward conduction results in a certain amount of heat being given off. This is true for both silicon and germanium diodes. It has been found that certain other semiconductor materials emit light when the *pn* junction is forward-biased. This is true for both gallium arsenide and gallium phosphide diodes. By combining various materials and by varying the doping levels, these diodes can be made to emit different wavelengths (colors) of light. These diodes are appropriately termed *light-emitting diodes* (LEDs). The most common color is red, but yellow and green are also widely used.

The brightness of light given off by the LED is dependant on the number of photons released by the recombination of charge carriers inside the LED. The higher the forward bias voltage, the greater will be the brightness of the light emitted. Typically, LEDs have a maximum forward voltage rating from 1 to 2 V and a 50 to 100 mA current maximum. The light output is nearly linear (proportional) with the amount of forward current.

The circuit shown in Fig. 5-22 shows the schematic symbol for an LED and a

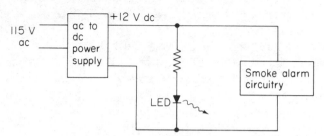

FIGURE 5-22 *LED schematic symbol and application.*

possible application. The LED is serving as a pilot light for the smoke alarm. It will be "on" as long as the smoke alarm is receiving line voltage power and its ac to dc power supply is delivering 12 V dc. Notice the resistor in series with the LED. Its purpose is to limit the current and voltage to allowable levels for the LED.

Another widespread application for LEDs is to combine a whole array of them to provide visual alphanumeric (letter and number) readout displays in calculators, clocks, and instruments (see Fig. 2-16). In these applications, digital circuitry (see Chapters 11 and 12) is used to drive various segments of the display to give the proper readout. In applications where battery life is critical, a different type of unit, the liquid crystal display (LCD), is often used. The LCD requires very low power compared to the LED display and hence LCD popularity in digital wristwatches and other applications having limited battery power. The LCD does *not* use diodes but rather depends upon electrical excitation of certain crystalline structures. The LCD consumes μW of power compared to mW for LEDs.

Since the start of widespread use of LEDs in the early 1970s, they have replaced standard light sources (incandescent bulbs, fluorescent displays, neon bulbs) for a number of reasons:

1. They have a much longer expected life than other light sources.

2. They operate from very low voltages, making them very compatible with most integrated circuit applications.

3.   They possess extremely fast response times (typically several nanoseconds).

4.   Low cost.

The fast response time listed in number 3 makes them useful in high-frequency pulsed applications. Thus, an electrical communication signal that varies at a 100-MHz rate can be impressed upon an LED. The LED light output can then be applied to a fiber of glass (*fiber optics* or *light pipe*) and transmitted from one end of an airplane to another (for example) and then detected and converted back to an electrical signed by a photodiode or phototransistor (see Chapter 13). The fiber optics communication system just described is becoming increasingly popular. The transmission of information over glass fibers rather than copper wire is proving less costly and offers significant weight reduction (a factor in airplanes). Additionally, a single glass fiber can contain much more information than a single copper wire (a factor in many new metropolitan phone systems).

A final application for LEDs is under development. It is a solid-state color TV picture display made up of arrays of LEDs emitting green, red, and blue light in the proper combination to depict full-color scenes. Once perfected, a flat TV set will result that is hung on the wall like a picture or photograph.

5-9   SILICON-CONTROLLED RECTIFIER

A *silicon-controlled rectifier* (SCR) can be thought of as a rectifier with a third lead used to control its conduction. The SCR is a four-layer semiconductor device as shown in Fig. 5-23(a). As can be seen, it contains three *pn* junctions. Conduction from anode to cathode is controlled by the signal applied to the gate. The schematic symbol for the SCR is shown at Fig. 5-23(b).

The major attribute of the SCR is its ability to control a large amount of power to a load with a very low power gate signal. Gate-to-cathode voltages of 1 or 2 V at 10 or 20 mA will typically be able to control loads at hundreds of volts and

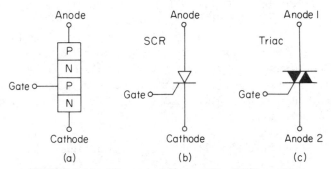

FIGURE 5-23   *Silicon-controlled rectifier (SCR) and triac.*

amperes. The ratings of SCRs range from loads of several hundred mA up to about 1000 A.

The circuit shown in Fig. 5-24(a) is used to control the power delivered to the load, R_L. The regular diode, D_1, rectifies the ac input such that the current to the gate, i_G, is pulsating dc, as shown at Fig. 5-24(c). The level of gate current is determined by the resistance of the R_1 potentiometer and R_2 combination. The load voltage in Fig. 5-24(d) is shown to turn-on at time a and off at time b. An SCR turns off whenever the input anode–cathode voltage and current approach zero [Fig. 5-24(b)]

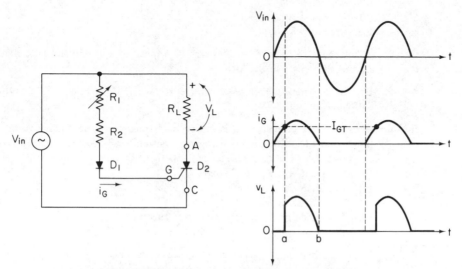

FIGURE 5-24   *SCR lamp dimmer circuit and waveforms.*

or goes negative. The amount of gate current required to trigger or turn the SCR on [I_{GT} in Fig. 5-24(c)] is typically 10 mA and is controlled by R_1's setting in Fig. 5-24(a). The load-voltage waveform in Fig. 5-24(d) can thus be controlled by R_1's value. If R_1 is increased, the gate current will reach the SCR's trigger value I_{GT} at a higher value of v_{in}, so that the SCR will trigger at a later point in the v_{in} half-cycle. The power to the load is thereby decreased.

This type of circuit is used in light dimmer circuits to provide a continuously variable light output. It can also be used to control the speed of dc motors and the output of heater elements. In some cases, SCRs are used to simply turn a device on or off, as opposed to the continuous control shown here.

When full ac is required for a load, a device known as a *triac* is used. It is effectively two SCRs connected in inverse parallel so that the full ac voltage can be controlled to a load. The schematic symbol for the triac is shown at Fig. 5-23(c). They are widely used in computer control systems when a small computer output signal is used to control ac loads (see Chapters 12 and 13).

QUESTIONS AND PROBLEMS

5–1–1. Explain the basics of a *pn* junction in terms of construction and characteristics.

5–1–2. A *pn* junctions for electrons can be compared to a one-way water valve. Explain the analogy between the two.

5–1–3. Provide brief definitions for the following terms:
(a)   Forward bias
(b)   Barrier potential
(c)   Reverse bias
(d)   Breakdown voltage

5–2–4. A half-wave rectifier circuit delivers 17 V peak to a 40-Ω load. Calculate V_{dc}, V_{rms}, I_o, and the power in the load.

5–2–5. Explain the meaning of peak reverse voltage (PRV). How is it determined for a half-wave circuit?

5–2–6. A half-wave rectifier circuit is driven by a transformer secondary voltage of 17.7 V peak. The silicon diode has a reverse resistance of 700 kΩ and a forward voltage drop of 0.7 V. The load resistance is 60 Ω. Calculate the following:
(a) V_L (peak, average, and rms).
(b)   The PRV for the diode.
(c)   The power in R_L.
(d)   The diode peak, average, and rms current.
(e)   The load voltage when the diode is reversed-biased.

5–3–7. Draw a schematic diagram of a full-wave center-tap circuit and explain the basic operation.

5–3–8. Draw a schematic diagram of a full-wave bridge circuit and explain the basic operation.

5–3–9. With respect to the center-tap and bridge rectifier circuits:
(a)   Why are they both "frequency doublers"?
(b)   Why is the bridge version now more popular even though it requires four diodes versus two for the center-tap circuit?
(c)   What is the PRV for each circuit with respect to the peak input voltage?

5–3–10. A full-wave rectifier circuit delivers a peak voltage of 30 V to a 3.3-kΩ load. Calculate:
(a)   The load voltage (dc and rms).
(b)   The average load current.
(c)   The load power.
(d)   The required diode current and PRV ratings.

5–3–11. It is desired to provide 15 V dc to a motor that uses an average current of 0.5 A. Using a full-wave bridge setup, determine the required transformer secondary voltage and its turns ratio if the primary voltage is 115 v ac.

5–4–12. Explain what pulsating dc is. What is the function of a filter in a power supply?

5–4–13. Define ripple and list three ways in which it is commonly designated.

5–4–14. Explain how a capacitor has the ability to "filter" a pulsating dc output.

5–4–15. When a capacitor is used as a power-supply filter, explain why the diode current flow is of short duration with high-valued peaks.

5–4–16. A full-wave rectifier scheme using 60-Hz line voltage is used to drive a 200-Ω load. The % ripple is to be 1%. Calculate the required filter capacitor value. If the load resistance went up, what happens to the ripple?

5–4–17. Explain why, in a well-filtered power supply, the difference between the rms and average voltage is negligible.

5–5–18. Discuss the difference between a regulated and unregulated power supply.

5–5–19. Explain the meaning of *and* how to calculate the following:
(a)   % load regulation
(b)   % line regulation
(c)   % regulation

5–5–20. A regulated power supply has an output of from 4.97 to 5.03 V for input line voltage changes and varies from 4.94 to 5.04 V for rated load changes. Calculate its % line, % load, and % regulation.

5–5–21. Define internal resistance for a power supply and explain its significance for load regulation.

5–5–22. A power supply with internal resistance of 0.2 Ω has an output of 12.0 V at no load current. Determine its output at 1 A load current. Calculate the % load regulation for a rated load range of 0 to 1 A.

5–6–23. Explain the special characteristics of a zener diode. Include the terms I_{ZK}, I_{ZM}, breakdown voltage, and regulating region in your discussion.

5–6–24. Draw a schematic of a zener diode shunt regulator. Explain how it is able to provide regulation.

5–6–25. Design a zener diode shunt regulator that is able to provide 15 V dc output at loads from zero to 400 mA. The unregulated input varies from 19 to 26 V dc. Calculate a value of R_S and the required zener diode voltage rating. Calculate the necessary power ratings for R_S and the zener diode.

5–7–26. Draw a block diagram of a series regulator and briefly explain the function of each block. Include the concept of a feedback control system in your discussion.

5–7–27. Describe several advantages of the series regulator as compared to the shunt regulator.

5–7–28. What is an integrated circuit? Explain several advantages of an IC as compared to standard "discrete" circuitry.

5–7–29. With respect to the LM309 regulator IC, what is the
(a)   Typical output voltage?
(b)   Absolute maximum input voltage?
(c)   Typical line regulation?
(d)   Maximum output current?
(e)   Maximum line regulation?

5–7–30. Draw the schematic of a fixed 5-V regulator, variable output regulator, and current regulator using the LM309 regulator IC. Explain the function of a current regulator.

5–7–31. Describe the effect whereby a regulator is able to greatly reduce the ripple of its unregulated input. If 1.8-V p-p ripple were the input to a LM309 regulator IC, what is the expected level of ripple in the output?

5–7–32. Describe the major advantage of a swiching regulator as compared to other types.

5–8–33. What is an LED? Explain how it works and list some applications for this device.

5–8–34. Show a circuit diagram for a zener diode shunt regulator power supply that uses an LED to indicate the presence of output voltage.

5–9–35. What is an SCR? Provide a schematic symbol and briefly explain how it functions and some applications of this device.

5–9–36. Explain the operation of the circuit shown in Fig. 5-24(a). Include the term I_{GT} in your explanation.

5–9–37. What is a triac? Describe its relationship to an SCR and list some applications.

6

TRANSISTORS

6-1 TRANSISTOR BASICS

The *bipolar junction transistor* (BJT) was developed in 1948. It triggered an electronics revolution that continues to this day. The first transistors did not perform as well as the vacuum tubes they started replacing, but the transistor's small size and low power dissipation were keys to their early acceptance. The rapid improvement in transistor performance and cost reductions that followed led to the elimination of most tube applications by 1960. The diodes introduced in Chapter 5 and the BJT transistor led the way in the electronics revolution. These new devices were the first in a long list of *solid-state* devices that have changed the course of mankind. These new devices are called solid-state because the predecessor devices (vacuum tubes) involved electron travel through a vacuum as compared to solid semiconductor materials such as germanium or silicon.

Transistors are made by forming two semiconductor junctions in close proximity to one another. A *pnp* transistor is formed with *n*-type material between two pieces of *p*-type material, as illustrated in Fig. 6-1(a). The one region of *p*-type material is called the *collector* and the other *p*-type region the *emitter*. The *n*-type region in the center is termed the *base*. Figure 6-1(b) shows the *npn* counterpart, and beneath both structures is shown their corresponding schematic symbol. Notice that the arrowhead on the emitter lead changes direction depending on the transistor polarity— *npn* or *pnp*. As with the diode, the arrowhead points in the direction of easy conventional current flow between the emitter and base.

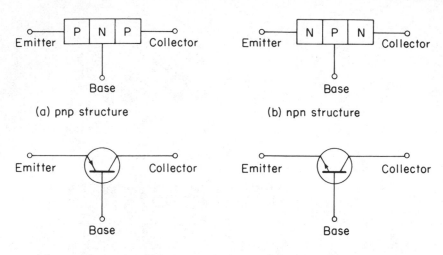

(a) pnp structure   (b) npn structure

(c) pnp schematic symbol   (d) npn schematic symbol

FIGURE 6-1 *Transistor structure and schematic symbols.*

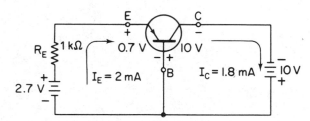

FIGURE 6-2 *Common-base transistor connection.*

We are now prepared to get to the very heart of transistor action. If the base–emitter junction is *forward*-biased and the base–collector junction *reverse*-biased, the transistor can be used to perform a very useful function. Figure 6-2 shows these bias conditions for a *pnp* transistor. A resistor has been included to limit current flow in the base circuit. The dc emitter current I_E can be calculated as 2.7 V minus the V_{EB} of 0.7 V (for silicon) divided by R_E or 2 V/1 kΩ = 2 mA. The base–collector bias is a 10-V reverse bias. A transistor connected as in Fig. 6-2 is known as the common–base (CB) configuration, since the base is common to both sides of the circuit. A set of characteristic curves for the CB circuit is provided in Fig. 6-3. Notice that the curves relate information concerning the collector current as a function of applied V_{CB} for various values of input emitter current. From these curves the collector current of 1.8 mA can be determined from the 2-mA emitter current and 10 V reverse bias on the base–collector junction.

An important relationship exists between the collector and emitter currents in a CB circuit. Since the collector current is the output current and the emitter current is the input current, the ratio I_C/I_E is termed the dc CB forward current gain. It is

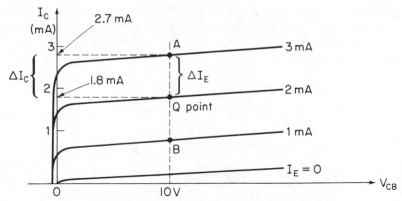

FIGURE 6-3 *Common-base characteristic curves.*

given the symbol alpha (α) or h_{FB}. We shall use h_{FB} throughout the remainder of this book to designate forward current gain of the CB configuration:

$$h_{FB} = \alpha = \frac{I_C}{I_E} \qquad \textbf{(6-1)}$$

The h_{FB} for the circuit of Fig. 6-2 is 1.8 mA/2.0 mA, or 0.9. Typical values of h_{FB} range from 0.9 to about 0.99, but they are always less than 1.

    If an *ac* signal were applied to the input side (emitter) of the circuit in Fig. 6-2, the ac value of CB forward current gain, h_{fb}, is of interest (note the lowercase subscripts for the ac h_{fb}). It is defined as

$$h_{fb} = \frac{\Delta I_C}{\Delta I_E} \Bigg|_{V_{CB} = \text{constant}} \qquad \textbf{(6-2)}$$

Figure 6-4 shows the CB circuit of Fig. 6-2 with an ac input current of 2 ma p-p. The 2-ma p-p current would cause the emitter current to vary around a specific dc bias level. That bias level (2 ma in this case) is called the *quiescent current* and is

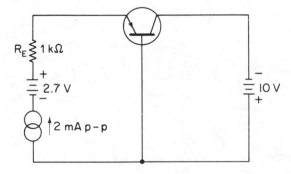

FIGURE 6-4 *Common-base circuit with ac input signal.*

usually abbreviated as the *Q-point*. This is shown in Fig. 6-3, where the emitter current varies between 3 and 1 ma around a 2-ma quiescent level (points *A* and *B* on Fig. 6-3). The ac current gain *(h_fb)* is calculated from the curves of Fig. 6-3 as follows:

$$h_{fb} = \frac{\Delta I_C}{\Delta I_E}\bigg|_{V_{CB}=\text{constant}} \tag{6-2}$$

$$\simeq \frac{2.7\ \text{mA} - 1.8\ \text{mA}}{3\ \text{mA} - 2\ \text{mA}}\bigg|_{V_{CB}=10\,\text{V}}$$

$$= 0.9$$

The *h_fb* was calculated between point *A* of Fig. 6-3 and the *Q*-point in this case, but could just as well have been calculated between point *B* and the *Q*-point. Note that in this instance the dc and ac current gains are equal. In actual practice, the two are not usually exactly equal. A CB circuit provides a current gain of slightly less than 1 but is capable of substantial voltage gain.

Another important transistor characteristic results when the emitter is made the common lead between input and output. As might be expected, this is termed the *common-emitter* (CE) configuration. The base is the input lead, whereas the output is at the collector. The CE forward current gain is the single most important transistor characteristic and is designated by the Greek letter beta (β) or h_{FE}. The latter term will be used throughout the remainder of this text. By definition, the dc CE forward current gain is

$$h_{FE} = \beta = \frac{I_C}{I_B} \tag{6-3}$$

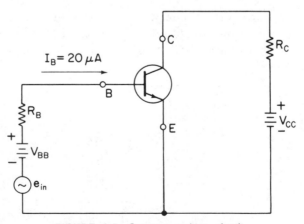

FIGURE 6-5 *Common-emitter circuit.*

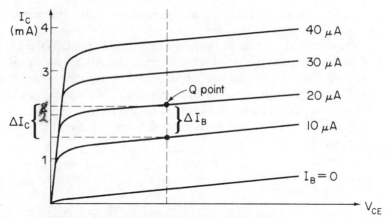

FIGURE 6-6 *Common-emitter characteristic curves.*

Figure 6-5 shows a CE circuit with proper biasing, that is, with the base–emitter junction forward-biased and the base–collector junction reverse-biased. Figure 6-6 shows a set of typical characteristic curves for the CE circuit of Fig. 6-5.

The Q-point conditions of $I_B = 20$ μA and $I_C = 2.2$ mA allow a calculation of h_{FE} as

$$h_{FE} = \frac{I_C}{I_B} \tag{6-3}$$

$$= \frac{2.2 \text{ mA}}{20 \text{ μA}} = 110$$

As with h_{FB}, an ac current gain for the CE circuit is often used. It is defined similarly as

$$h_{fe} = \frac{\Delta I_C}{\Delta I_B} \bigg|_{V_{CE} = \text{constant}} \tag{6-4}$$

For the transistor represented in Fig. 6-6, the ac current gain between the Q-point and point A is calculated as

$$h_{fe} = \frac{\Delta I_C}{\Delta I_B} \bigg|_{V_{CE} = \text{constant}} \tag{6-4}$$

$$= \frac{2.2 \text{ mA} - 1.5 \text{ mA}}{20 \text{ μA} - 10 \text{ μA}}$$

$$= \frac{0.7 \text{ mA}}{10 \text{ μA}} = 70$$

As can be seen, the ac current gain, h_{fe}, is significantly different from the dc current gain, h_{FE}. This is often the case.

The reader should see at this point that (unlike the CB) the CE circuit is capable of providing a greater than 1 current gain, h_{fe}, to an ac signal. It also provides significant voltage gain, as will be shown. The ability to take a small signal and convert it into a large signal of the same shape is the most useful characteristic of the transistor. This process is known as *amplification*.

It can be easily shown by Kirchhoff's current law that a transistor's currents are related by

$$I_B = I_E - I_C \qquad\qquad (6\text{-}5)$$

The base current is quite small with respect to the emitter and collector currents and is equal to their difference. For most calculations it is safe to assume that the emitter and collector currents are equal to one another. Equations (6-2), (6-4), and (6-5) may be combined to eliminate the transistor currents and show the relationship between h_{fb} and h_{fe}. This is left as an exercise for the student at the end of the chapter. The result is

$$h_{fe} = \frac{h_{fb}}{1 - h_{fb}} \qquad\qquad (6\text{-}6)$$

As we shall see, the term $(h_{fe} + 1)$ is often more useful than h_{fe}. As such, it is given the special designation, h'_{fe}. Thus,

$$h'_{fe} = h_{fe} + 1 \qquad\qquad (6\text{-}7)$$

Equations (6-6) and (6-7) may be rearranged to show that

$$h'_{fe} = \frac{h_{fe}}{h_{fb}} = \frac{1}{1 - h_{fb}} \qquad\qquad (6\text{-}8)$$

The ac forms of h_{fe} and h_{fb} were used in Eqs. (6-6) through (6-8), but the equations are also valid for the dc forms (h_{FE} and h_{FB}).

EXAMPLE 6-1

A transistor has a dc base current of 1 mA and emitter current of 50 mA. Calculate I_C, h_{FB}, h_{FE}, and h'_{FE}.

Solution:

$$I_B = I_E - I_C \tag{6-5}$$
$$\therefore \quad I_C = I_E - I_B$$
$$= 50 \text{ mA} - 1 \text{ mA}$$
$$= \underline{49 \text{ mA}}$$

$$h_{FB} = \frac{I_C}{I_E} \tag{6-1}$$
$$= \frac{49 \text{ mA}}{50 \text{ mA}}$$
$$= \underline{0.98}$$

$$h_{FE} = \frac{h_{FB}}{1 - h_{FB}} \qquad \frac{0.98}{1 - 0.98} \tag{6-6}$$
$$= \underline{49}$$

or

$$h_{FE} = \frac{I_C}{I_B} \tag{6-3}$$
$$= \underline{49}$$

$$h'_{FE} = h_{FE} + 1 \tag{6-7}$$
$$= 49 + 1$$
$$= \underline{50}$$

or

$$h'_{FE} = \frac{1}{1 - h_{FB}} \tag{6-8}$$
$$= \frac{1}{1 - 0.98} = \frac{1}{0.02}$$
$$= \underline{50}$$

EXAMPLE 6-2

The circuit from Example 6-1 has an ac input (base) current of 1 ma p-p, which results in an ac collector current of 40 ma p-p. Calculate h_{fe}, h'_{fe}, and h_{fb}.

Solution:

$$h_{fe} = \frac{\Delta I_C}{\Delta I_B} \bigg|_{V_{CE} = \text{constant}}$$

$$= \frac{40 \text{ mA}}{1 \text{ mA}} \qquad\qquad\qquad \textbf{(6-4)}$$

$$= \underline{40}$$

$$h'_{fe} = h_{fe} + 1$$

$$= 40 + 1 \qquad\qquad\qquad \textbf{(6-7)}$$

$$= \underline{41}$$

$$= \frac{h_{fe}}{h_{fb}}$$

$$\therefore \quad h_{fb} = \frac{h_{fe}}{h'_{fe}} = \frac{40}{41} \qquad\qquad \textbf{(6-8)}$$

$$\approx \underline{0.976}$$

Notice that the ac gains are somewhat lower than the dc gains calculated in Example 6-1.

6-2 AMPLIFICATION

Amplification is defined as an enlargement of an input signal by an amplifier. If during this enlargement process the output signal does not maintain the exact same shape as the input, we say that the output has been *distorted*. Distortion is illustrated in Fig. 6-7(b). Note the distinction between an amplifier that just reproduces a signal with no amplification [Fig. 6-7(a)] and one that not only reproduces but amplifies as well [Fig. 6-7(c)]. Amplifiers may provide voltage, current, and power gains in varying combinations.

Transistors can be used to provide amplification because the amount of collector current flowing can be controlled by the much smaller base current. Figure 6-3 shows a curve of collector current versus base current for a typical transistor; Fig. 6-8(a) shows the schematic of a transistor amplifier. Notice in Fig. 6-8(b) that there are two areas of nonlinearity, one for very low currents and one for high currents. In between is an area where $I_C = KI_B$, and this constant of proportionality K is equal to h_{FE}. Remember that $I_C = h_{FE}I_B$ but only where the I_C versus I_B curve is a straight line. If the transistor were used as an amplifier in this linear region, reproduction (no distortion) should occur. Operating at either (nonlinear) extreme will cause distortion in the output. For this particular transistor, h_{FE} is 20 in the linear region, since the collector current is always 20 times the base current.

If a pure sinusoidal current were made to cause the base current to vary between 1 and 3 mA (i.e., a sinusoidal current of 2 ma p-p superimpressed on a dc level of

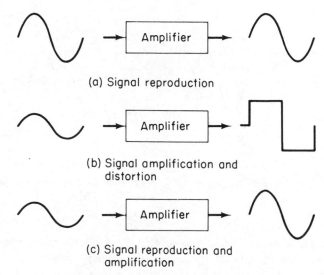

(a) Signal reproduction

(b) Signal amplification and distortion

(c) Signal reproduction and amplification

FIGURE 6-7   Amplification and distortion.

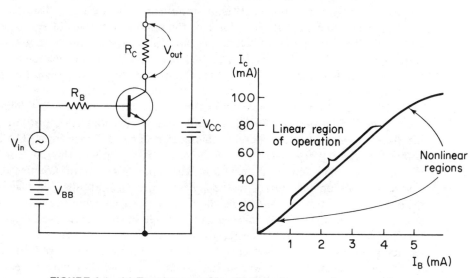

FIGURE 6-8   (a) Transistor amplifier; (b) Collector versus base current curve.

2 mA), we would find a collector current variation of 20 to 60 mA. the collector current variation would also be a pure sinusoid, and hence is an exact replica of the input. This current amplification is illustrated in Fig. 6-9. Notice that both ac currents are riding on a dc level. These dc levels are referred to as the *bias levels* and are necessary to set up the proper dc operating point. For instance, without the 2-mA dc level in the base circuit provided by the battery V_{BB}, the negative excursion of the ac input signal would reverse bias the base–emitter junction, and transistor

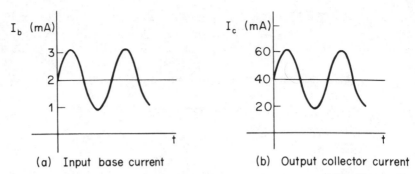

(a) Input base current   (b) Output collector current

FIGURE 6-9 *Input and output current versus time for the amplifier illustrated in Figure 6-8.*

action would cease. These bias levels, as previously mentioned, are also termed the *quiescent* operating point (*Q*-point). The student should now realize that a *Q*-point (i.e., dc bias levels) must be initially set up to allow proper amplification of an ac signal. Without proper biasing, the transistor would be unable to amplify both the positive and negative extremes of the ac input signal, resulting in distortion.

It should be noted that operation into the nonlinear regions of Fig. 6-8(b) would cause distortion of an output signal. In those areas the output is *not* equal to h_{FE} times the input. Since the graph is curved (nonlinear), the proportionality between input and output is variable. This causes the output to not be a replica of the input.

One more important concept can be visualized with the help of Figs. 6-8 and 6-9. The input current *does not* supply current for the output current! That is, only 2 mA p-p need be supplied by the input source to provide an output current of 40 mA p-p. One may wonder where all this output current comes from. The answer is simply that the battery in the collector circuit *(V$_{CC}$)* supplies this current. In other words, the small ac source impressed on the input of the transistor has resulted in a large ac draw on the battery in the collector (output) circuit. The transistor here has acted as a "valve" in controlling the flow of electrons from the collector battery. The transistor has effectively taken a source of dc current *(V$_{CC}$)* and drawn from it a sinusoidal current! For the device to do this, however, it was driven by a much smaller sinusoidal current source into its base (input) lead. Amplification has occurred—an increase in ac power at the expense of some dc power. Energy has not been created, only transformed from dc to ac.

Amplifiers require four terminals for operation, because the input signal requires two leads and the output also requires two leads. Since transistors are only three-terminal devices, it is necessary to use one of the three leads as a common or shared lead between input and output. This presents no special problem, and often the common lead is grounded such that the input signal at point *A* in Fig. 6-10 is measured with respect to ground (terminal *C*) as is the output signal at *B*.

Transistor amplifiers exist in three possible configurations, depending on which one of the transistor's three leads is used as the common terminal between input

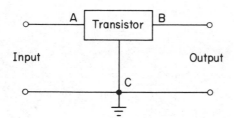

FIGURE 6-10 *Three-terminal amplifier.*

and output. The amplifier's designation is determined by the common terminal, as illustrated in Fig. 6-11. The performance of the amplifier is also determined by which lead is common. The CB and CE configurations have previously been demonstrated. Figure 6-11 shows the third possibility—the common collector (CC).

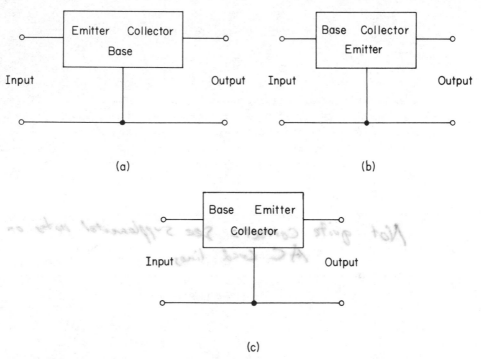

(a)   (b)

(c)

FIGURE 6-11 *Transistor amplifier classifications: (a) Common base (CB); (b) Common emitter (CE); (c) Common collector (CC).*

6-3 COMMON-EMITTER AMPLIFIERS

The most commonly encountered bipolar junction transistor (BJT) amplifier type is the *common-emitter* (CE) configuration. Figure 6-12 shows the most often used circuit layout for the CE amplifier. The ac input signal is coupled into the amplifier via

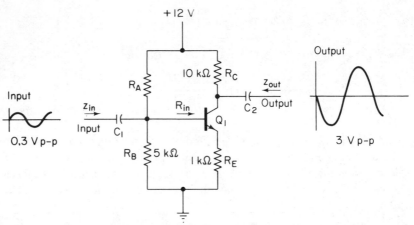

FIGURE 6-12   Circuit for Example 6-3.

coupling capacitor, C_1. The purpose of C_1 is to block (isolate) dc levels between the amplifier and the preceding circuitry or signal driver. Recall that a capacitor looks like an open circuit to dc. The capacitor value is chosen such that it will virtually look like a short circuit to the lowest ac frequency.

The resistors in Fig. 6-12 are used to set up the dc levels in the circuit. In general, it is desirable to have the transistor's dc output (collector for a CE stage) voltage at about one-half the dc supply level. This enables the ac output voltage, which rides on the dc level, to be as large as possible before reaching the limits of either 0 V or the supply voltage (12 V in this case).

Not quite correct = see supplemental notes on AC Load lines

EXAMPLE 6-3

The dc level at the transistor's collector in Fig. 6-12 is to be 6 V. Determine the proper value for R_A if Q_1 is silicon and has h_{FE} equal to 100.

Solution:

If the collector is at 6 V (with respect to ground), the voltage across R_C must be 12 V − 6 V, or 6 V. Therefore,

$$I_C = \frac{V_{RC}}{R_C} = \frac{6\text{ V}}{10\text{ k}\Omega} = 0.6\text{ mA}$$

Since $h_{fe} = 100$,

h_{FE}

$$I_B = \frac{I_C}{h_{FE}}$$ (6-3)

$$= \frac{0.6 \text{ mA}}{100} = 6 \text{ } \mu\text{A}$$

$$\therefore \quad I_E = I_C + I_B$$ (6-5)
$$= 0.606 \text{ mA}$$

Now the voltage across R_E can be calculated as

$$V_{RE} = I_E \times R_E = 0.606 \text{ mA} \times 1 \text{ k}\Omega$$
$$= 0.606 \text{ V}$$

Since Q_1 is silicon, the forward-biased base–emitter junction voltage should be about 0.7 V. Thus, the voltage at the base of Q_1 is the sum of V_{R_E} and V_{BE}, or

$$V_B = V_{R_E} + V_{BE} = 0.606 \text{ V} + 0.7 \text{ V}$$
$$= 1.306 \text{ V}$$

Now the voltage at the base of Q_1 (with respect to ground) is also the voltage across R_B and, therefore,

$$I_{R_B} = \frac{V_{R_B}}{R_B} = \frac{1.306 \text{ V}}{5 \text{ k}\Omega} = 0.26 \text{ mA}$$

Applying Kirchhoff's Current Law (KCL) at the junction between the base, R_A and R_B, we have

$$I_{R_A} = I_B + I_{R_B}$$
$$= 6 \text{ } \mu\text{A} + 0.26 \text{ mA}$$
$$= 0.266 \text{ mA}$$

The voltage across R_A is $12 \text{ V} - V_{R_B}$, or $12 \text{ V} - 1.306 \text{ V} = 10.7 \text{ V}$. Therefore,

$$R_A = \frac{V_{R_A}}{I_{R_A}} = \frac{10.7 \text{ V}}{0.266 \text{ mA}}$$
$$= 40.2 \text{ k}\Omega$$

Example 6-3 shows that the dc analysis of an amplifier stage can be rather tedious. Making two approximations greatly reduces the analysis and generally yields acceptable results. The two approximations are to assume that I_E equals I_C and that I_B is negligible with respect to I_{R_A} and I_{R_B}.

EXAMPLE 6-4

Repeat Example 6-3 assuming that $I_E = I_C$ and that I_B is $<<$ (much less than) I_{R_A} or I_{R_B}.

Solution:

As before,

$$I_C = \frac{6\text{ V}}{10\text{ k}\Omega} = 0.6\text{ mA}$$

Then

$$I_E \simeq I_C = 0.6\text{ mA}$$
$$\therefore \quad V_E = I_E \times R_E = 0.6\text{ mA} \times 1\text{ k}\Omega$$
$$= 0.6\text{ V}$$

$$\therefore \quad V_B = V_E + V_{BE}$$
$$= 0.6\text{ V} + 0.7\text{ V}$$
$$= 1.3\text{ V}$$

Now if $I_B << I_{R_A}$ or I_{R_B}, by the voltage divider law,

$$V_B = 12\text{ V} \times \frac{R_B}{R_A + R_B}$$

$$1.3\text{ V} = 12\text{ V} \times \frac{5\text{ k}\Omega}{R_A + 5\text{ k}\Omega}$$

$$\therefore \quad R_A = \underline{41.2\text{ k}\Omega}$$

The result of Example 6-4 is very close to the more exact analysis (41.2 kΩ vs. 40.2 kΩ) and is thereby justified.

Having taken a close look at the dc analysis of the CE amplifier, we are now prepared to consider the ac characteristics. That is the main goal of the amplifier, ac amplification. As an approximation, the ac voltage gain of the CE amplifier shown in Fig. 6-12 is R_C/R_E. The ac voltage gain is represented by G_V, and therefore

$$G_V \simeq \frac{R_C}{R_E} \qquad \text{(6-9)}$$

The ac input resistance (R_{in}) looking into the base of a transistor is of interest to ac analysis and is approximately equal to

$$R_{\text{in}} \simeq h'_{fe} R_E \tag{6-10}$$

The overall input impedance (Z_{in}) for the CE amplifier of Fig. 6-12 is

$$Z_{\text{in}} = R_A \| R_B \| R_{\text{in}} \tag{6-11}$$

The output impedance (Z_{out}) is simply

$$Z_{\text{out}} = R_c \tag{6-12}$$

The circuit's overall current gain, G_i, can be calculated as

$$G_i = G_v \times \frac{Z_{\text{in}}}{Z_{\text{out}}} \tag{6-13}$$

EXAMPLE 6-5

For the CE amplifier of Fig. 6-12, determine the ac output voltage, the voltage gain (G_v), the impedance looking into the base of Q_1 (R_{in}), the overall input impedance (Z_{in}), the output impedance (Z_{out}) and the overall current gain (G_i). The ac input voltage is 0.3 v p-p and $h'_{fe} = 100$.

Solution:

The ac voltage gain is

$$G_v \simeq \frac{R_c}{R_E} \tag{6-9}$$

$$= \frac{10 \text{ k}\Omega}{1 \text{ k}\Omega} = \underline{10}$$

The ac output voltage is simply the 0.3 v p-p input times the gain or 0.3 v p-p $\times$ 10 = $\underline{3 \text{ v p-p}}$. The impedance looking into the base, R_{in}, is

$$R_{\text{in}} \simeq h'_{fe} R_E \tag{6-10}$$
$$= 100 \times 1 \text{ k}\Omega$$
$$= \underline{100 \text{ k}\Omega}$$

The overall input impedance, Z_{in}, is

$$Z_{\text{in}} = R_A \| R_B \| R_{\text{in}} \tag{6-11}$$
$$= 40.2 \text{ k}\Omega \| 5 \text{ k}\Omega \| 100 \text{ k}\Omega$$
$$= \underline{4.26 \text{ k}\Omega}$$

The output impedance, Z_{out}, is equal to R_C:

$$Z_{\text{out}} = R_C$$
$$= \underline{10 \text{ k}\Omega} \hspace{3cm} \textbf{(6-12)}$$

The overall current gain, G_i, is

$$G_i = G_v \times \frac{Z_{\text{in}}}{Z_{\text{out}}}$$

$$= 10 \times \frac{4.26 \text{ k}\Omega}{10 \text{ k}\Omega} \hspace{2cm} \textbf{(6-13)}$$

$$= \underline{4.26}$$

Notice that the 3-v p-p output signal in Fig. 6-12 goes negative when the input goes positive, and vice versa. The output is, therefore, said to be *out of phase* with the input—180° out-of-phase (or phase inversion) to be more precise. The CE amplifier is said to be *phase-inverting* because of this characteristic. To understand why phase inversion occurs, consider the transistor currents. When the input voltage and thus base current increase positively, the output collector current increases also. This increased collector current causes an increased voltage across R_C. If V_{R_C} goes up, the voltage from collector to ground (V_C) must go down. Thus, the ac output voltage (V_C) is decreasing as the input is increasing. The input/output characteristics for this amplifier are summarized in Fig. 6-13. Notice that the collector voltage includes the 3 v p-p ac component riding on the 6 V dc Q-point voltage. The output coupling

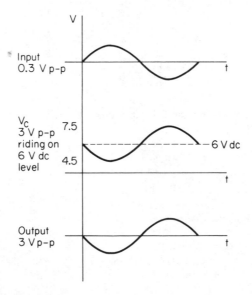

FIGURE 6-13 *Voltage waveforms for amplifier in Figure 6-12.*

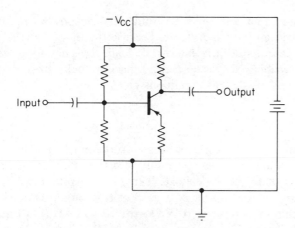

FIGURE 6-14 *CE amplifier using a pnp transistor.*

capacitor in Fig. 6-12, C_2, blocks the dc level so that the output voltage is just the desired 3 v p-p ac signal.

The preceding CE amplifier analysis was made using an *npn* transistor. The analysis using a PNP transistor is essentially identical. Figure 6-14 shows the same circuit as Fig. 6-12 except that a PNP transistor is used. The only difference is that the power-supply polarity must be reversed for the *pnp* transistor. Notice that in Fig. 6-12 the "shorthand" method for showing a dc supply is used, whereas in Fig. 6-14 the full connection scheme is provided.

6-4 COMMON-COLLECTOR AMPLIFIER

So far, you have been introduced to the CE and CB amplifiers. The CB variety is seldom used and so has not been emphasized. The basic common-collector (CC) amplifier is shown in Fig. 6-15. The input is at the base and the output is at the emitter. The common lead is the collector, but that may not be readily apparent to

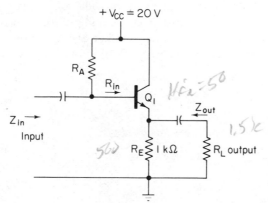

FIGURE 6-15 *Basic CC amplifier.*

you. It is common to both input and output through the dc power supply. This configuration is often termed an *emitter follower,* because the output at the emitter "follows" the input at the base. Notice the external ac load, R_L. The output coupling capacitor keeps dc from reaching R_L because the capacitor looks like an open to dc. Thus, R_L has no effect on the dc bias conditions of Q_1.

DC Analysis

The resistors R_A and R_E in Fig. 6-15 are used to set up the amplifier's required dc conditions. It is normally desirable to have the dc voltage at the output lead, the emitter, at about one-half the dc supply voltage. Thus, if the emitter voltage were $\frac{1}{2} \times V_{CC}$, 10 V should exist at the emitter with respect to ground. Thus, V_E (V_{R_E} also) is 10 V and the dc emitter current is 10 V/R_E, or 10 V/1 kΩ = 10 mA. The value of R_A in Fig. 6-15 should be selected to make this V_E = 10 V condition happen.

To determine the proper value for R_A, assume that Q_1 is silicon (as almost all BJTs are) and that it has h'_{FE} = 100. Since the voltage V_E is 10 V, the voltage at the base, with respect to ground, V_B, should be 10.7 V. This is true since the base–emitter junction must be forward-biased and thus V_{BE} is approximately equal to 0.7 V for a silicon device. With V_B of 10.7 V and V_{CC} of 20 V, the voltage across R_A must be the difference, 20 V − 10.7 V, or 9.3 V. Now since I_E must be 10 mA, so that V_E is 10 V, the base current is I_E/h'_{FE}, or 10 mA/100 = 100 μa. The base current must flow through R_A, and thus I_{R_A} = 100 μa. Now the current through (100 μa) and voltage across (9.3 V) R_A are known, so that its required value can be calculated as 9.3 V/100 μa = 93 kΩ.

AC Analysis

The CC amplifier does *not* provide any voltage gain. We shall assume a voltage "gain" of 1, but in fact the actual "gain" is typically 0.9 to 0.99. A fair question might be why bother with an amplifier configuration that does not amplify? Well, even though an emitter follower stage does not provide voltage gain, it does provide significant current gain and is used to prevent loading of an ac signal source. The ease with which a CC stage can be biased is another consideration in its popularity.

For the emitter-follower circuit (Fig. 6-15), the following ac relationships can be assumed.

$$G_v \simeq 1 \tag{6-14}$$

$$R_{in} \simeq h'_{fe}(R_E \| R_L) \tag{6-15}$$

$$Z_{in} \simeq R_A \| R_{in} \tag{6-16}$$

$$G_i \simeq \frac{Z_{in}}{R_L} \tag{6-17}$$

The expression for Z_{out} is rather complicated and is not presented here. It is, however, considerably lower than the value of R_E.

EXAMPLE 6-6

A signal source of 5 v p-p with an internal resistance of 600 Ω is to drive a 500-Ω load. Determine the load voltage with a direct connection and using the emitter-follower circuit of Fig. 6-15 (assume that $h'_{fe} = 200$ for Q_1).

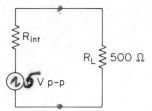

FIGURE 6-16   Direct-connection for Example 6-6.

Solution:

The direct-connection method is shown in Fig. 6-16. The signal source will be heavily loaded down, as can be shown by applying the voltage divider to this circuit.

$$V_{R_L} = 5 \text{ v p-p} \times \frac{500 \ \Omega}{500 \ \Omega + 600 \ \Omega}$$

$$= \underline{2.27 \text{ v p-p}}$$

A much better solution is to use the emitter-follower circuit of Fig. 6-15 between the signal source and R_L as a "buffer." In that case, with $R_L = 500 \ \Omega$,

$$R_{in} \approx h'_{fe} \, (R_E \| R_L) \qquad \qquad \textbf{(6-15)}$$
$$= 200 \, (1 \text{ k}\Omega \| 500 \ \Omega)$$
$$= 66.7 \text{ k}\Omega$$
$$Z_{in} \approx R_A \| R_{in} \qquad \qquad \textbf{(6-16)}$$

From the previous analysis, $R_A = 93 \text{ k}\Omega$, so

$$Z_{in} = 93 \text{ k}\Omega \| 66.7 \text{ k}\Omega$$
$$= 38.8 \text{ k}\Omega$$

Now the signal source "sees" an impedance of 38.8 kΩ, so that its loading can be determined using the voltage divider law as

$$\text{voltage input to amp} = 5 \text{ v p-p} \times \frac{Z_{in}}{Z_{in} + R_{int}}$$

$$= 5 \text{ v p-p} \times \frac{38.8 \text{ k}\Omega}{38.8 \text{ k}\Omega + 600 \text{ }\Omega}$$

$$= 4.92 \text{ v p-p}$$

Since

$$G_v \simeq 1 \qquad\qquad (6\text{-}14)$$

the load voltage is equal to the input, so that $V_{R_L} = \underline{4.92 \text{ v p-p}}$.
The current gain can be determined using

$$G_i \simeq \frac{Z_{in}}{R_L} \qquad\qquad (6\text{-}17)$$

$$= \frac{38.8 \text{ k}\Omega}{500 \text{ k}\Omega}$$

$$= \underline{77.6}$$

Example 6-6 shows the extreme usefulness of the emitter-follower stage. It was able to make a relative low resistance load look like a high resistance (38.8 kΩ versus an actual 500 Ω) and thus prevent excessive loading of the signal source.

6-5 JUNCTION FIELD-EFFECT TRANSISTORS

The *field-effect transistor* (FET) has come into its own in today's electronic circuits. Although basic research on this device was taking place simultaneously with or before that of the bipolar transistor, it did not come into general usage until the 1960s. Field-effect transistors come in two general categories, the *junction field-effect* (JFET) and the *metal oxide semiconductor field-effect transistor* (MOSFET). The meaning of all this will become apparent as our discussion progresses. Field-effect transistors are also sometimes referred to as *unipolar* transistors because conduction is by majority charge carriers only. This is in contrast to *bipolar* junction transistors, in which both majority and minority charge carriers are involved in the conduction process.

A major advantage of FETs over bipolar transistors is their extremely high input impedance. Input impedances of 10^8 to 10^{14} Ω are available as compared to 10^2 to 10^6 Ω for bipolar transistors—a considerable advantage for many applications. The FET also offers advantages in electrical noise characteristics (see Chapter 9) when working from high impedances, with regard to temperature effects and simpler biasing techniques. The biasing of MOSFETs operating in the enhancement mode is uniquely simple. One final advantage of extreme importance to today's circuits is the fabrication simplicity and lower cost of integrated circuits using MOSFETs. At this point, the student is entitled to ask why FETs have not completely supplanted bipolar transistors. The bipolar transistor is capable of high voltage gains, better high-frequency response, and greater power dissipations. The selection between unipolar (FETs) or bipolar transistors is often a complex technical–economic decision depending on many factors, with many of today's circuits being a combination of both varieties.

JFET Theory of Operation

The JFET is formed as an *n*-type or *p*-type *channel* and with an opposite polarity semiconductor material known as the gate. The physical construction is shown in Fig. 6-17(a). Figure 6-17(b) shows the appropriate symbols and required voltage polarities for *n*-channel and *p*-channel operation. The names of the three leads of a FET

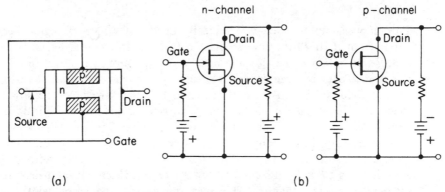

FIGURE 6-17 *(a) JFET n-channel structure; (b) JFET symbols and necessary voltage polarities.*

are also noted. The drain and source leads are often interchangeable and are determined by the polarity of the supply voltage to which they are connected. The drain is the lead connected to the positive side of the battery for the *n*-channel device and to the negative side for the *p*-channel device.

A rough analogy exists between the JFET and bipolar transistor with respect to gate and base, drain and collector, and source and emitter. However, the JFET's conduction is controlled by the reverse bias voltage on the gate–source junction and the electric field it creates. The bipolar transistor's conduction control is by the base current in the forward-biased base–emitter junction. This accounts for the JFET's high input impedance, since its input (the gate) is a reverse-biased junction.

Figure 6-18 provides a curve of drain current, I_D, versus the drain-to-source voltage, V_{DS}. The bias voltage or voltage from gate to source, V_{GS}, is equal to zero in this instance. In the first area, from 0 to *A*, the *ohmic region,* the JFET acts as an ordinary resistor. The current increases linearly as the applied voltage is increased.

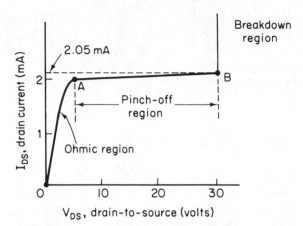

FIGURE 6-18 *Curve of I_D versus V_{DS} with $V_{GS} = 0$.*

This happens up to 1 to 3 V typically until the pinch-off region begins. At this point (from *A* to *B*) the reverse bias on the gate-to-channel *pn* junction has caused a *depletion region* in the channel next to the junction, as shown in Fig. 6-19. Until the reverse bias is large enough to reach the pinch-off region, the depletion layers are not wide enough to meet. When enough bias is provided to allow the layers to touch, the channel is depleted of charge carriers and is *pinched off.* Be careful here—"pinch off" *does not* mean "current off." Pinch off is an area of almost *constant* current flow for wide changes in V_{DS}. Refer to Fig. 6-18 and note that the curve is almost horizontal in this region from *A* to *B*. As the channel width is decreasing to zero, it tends to make the current flow decrease, which causes a drop in the voltage gradient along the channel. This then reduces the depletion width, which causes the current to increase. The final result is a stabilized value of current that is in a state of equilibrium.

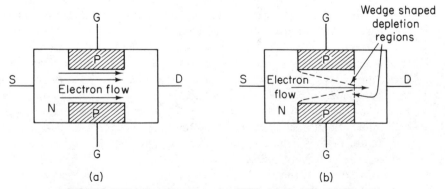

FIGURE 6-19 *Cross section of a JFET: (a) No bias; (b) Biased.*

The *dynamic drain resistance, r_{ds},* is defined as the slope of the curve in the pinch-off region:

$$r_{ds} = \frac{\Delta V_{DS}}{\Delta I_D} \bigg| V_{GS} = \text{constant} \qquad \text{(6-18)}$$

It is a very high value, and for Fig. 6-18 between points *A* and *B* is

$$r_{ds} \approx \frac{30\,\text{V} - 5\,\text{V}}{2.05\,\text{mA} - 2\,\text{mA}} = \frac{25\,\text{V}}{0.05\,\text{mA}} = 500\,\text{k}\Omega$$

This is the ac resistance of the channel, since it is the resistance to changes in V_{DS}. On the other hand, the static, or dc, resistance of the channel (R_{DS}) at a given V_{DS} is a low resistance and is given simply by the ratio of V_{DS} to I_{DS}. Hence, for Fig. 6-18, at $V_{DS} = 20$ V,

$$R_{DS} = \frac{V_{DS}}{I_D} \approx \frac{20\,\text{V}}{2\,\text{mA}} = 10\,\text{k}\Omega$$

The breakdown region for voltages beyond point *B* is also shown in Fig. 6-18. Breakdown results when the reverse-biased gate-channel *pn* junction undergoes avalanche breakdown. It results in a rapid increase in current flow for a small change in V_{DS}. It is similar to the constant voltage region of a zener diode and is nondestructive if the FET's power rating is not exceeded. It is interesting to note that increasing values of V_{DS} cause the JFET to appear first as a resistor (ohmic region), then as a constant current source (pinch-off region), and finally as a constant voltage source (breakdown region).

In the preceding discussion, the gate and source were at the same potential $(V_{GS} = 0)$. If a plot of I_D versus V_{DS} were made at different values of V_{GS}, the *drain characteristic curves,* as shown in Fig. 6-20, would result. It is customary to not show the breakdown (zener) region on these curves, since JFETs are seldom operated in this region. Notice that various (increasingly negative) bias potentials (V_{GS}) result in decreasing values of drain current in the pinch-off region. This is the basis for an FET's amplifying ability, and its *transconductance (g_{fs})* is a measure of the change in drain current for a change in V_{GS}:

$$g_{fs} = \frac{\Delta I_D}{\Delta V_{GS}} \bigg| V_{DS} = \text{constant} \qquad \text{(6-19)}$$

The transconductance is variable to a certain extent, as shown by the unequal spacing of I_D for I-V changes in bias voltage in Fig. 6-20. The variation in g_{fs} from device to device for a specific type of FET is very high also, with max–min ratios

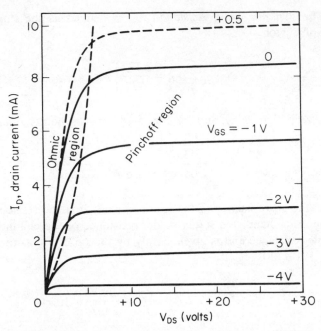

FIGURE 6-20 *Drain characteristic curves of an n-channel JFET.*

of 3:1 typical. Since voltage gain is proportional to g_{fs}, this variability can be trouble-some.

The transconductance of the device represented by the characteristics of Fig. 6-20 for a change of V_{GS} from -1 to -2 V is

$$g_{fs} = \frac{\Delta I_D}{\Delta V_{GS}} \approx \frac{5.5 \text{ mA} - 3 \text{ mA}}{1 \text{ V}} = 2500 \ \mu\text{℧}$$

Note also from Fig. 6-20 that operation with a small forward gate–source bias is permissible as long as the gate–channel junction is not allowed to conduct (become forward-biased). Since FETs are normally made of silicon material, this would imply a positive voltage no greater than about 0.5 V. Conduction would, of course, mean a significant current flow and probably cause destruction of the device due to heating effects in the *pn* junction.

The JFET can either be *n*-channel or *p*-channel, but the *n*-channel is normally preferred. It offers conduction via electrons, whereas the *p*-channel device operates via hole conduction. Since electrons are more mobile than holes, the *n*-channel device allows a higher frequency of operation, all other factors being equal. It also turns out that *n*-channel devices introduce less *noise* (unwanted electrical disturbances) into a circuit than do their *p*-channel counterparts. The *n*-channel characteristic curves of Fig. 6-20 are equally applicable to a *p*-channel device. However, all voltage polarities would have to be reversed.

6-6 MOSFETS

The gate of a MOSFET is a very small, high-quality capacitor and conduction through the channel is controlled by the voltage applied between the gate and the source. The MOSFET input current is thus the leakage current of the capacitor, as opposed to an input current for the JFET, which is the leakage current of a reverse-biased *pn* junction. Thus, the MOSFET input impedance is usually orders of magnitude greater than the JFET's (often 10^6 times as large).

A thin layer of insulating material, silicon dioxide, is deposited over the channel. Over this silicon dioxide layer is a metallic material—often aluminum. Thus, the input capacitor has one metallic plate, a silicon dioxide dielectric, and a semiconductor plate: hence the abbreviation MOS. Figure 6-21 shows the construction features of a typical MOSFET.

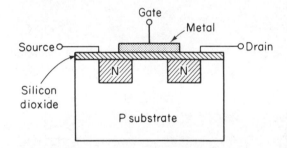

FIGURE 6-21 *Typical MOSFET structure.*

MOSFETs are available as either depletion-mode or enhancement-mode devices, depending on their construction details. Figure 6-22(a) illustrates the characteristics of an *n*-channel MOSFET. In the depletion mode the channel is normally a conductor, whereby current flow may be reduced or even cut off by applying sufficient gate voltage, just as with the JFET. In the enhancement mode the channel is normally cut off, and may be turned on or enhanced (hence, enhancement) and controlled by application of a gate voltage. As it turns out, the depletion-mode MOSFET can also be operated in the enhancement mode, as illustrated in Fig. 6-22(d).

The capacitance of the gate is very low, and therefore the input resistance very high; hence, the gate is very easily charged to a voltage that might break down the capacitor's narrow silicon dioxide dielectric. Typically, a 50-V level can cause this breakdown, and this could easily be induced by the static charges resulting from normal handling. For this reason these devices are often packaged with the leads shorted together. Care should be exercised in handling them by using a grounded tip soldering iron and guarding against static charges.

The frequency response of the MOSFET is generally better than the JFET because

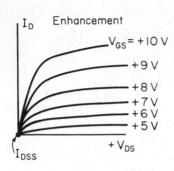

(a) Enhancement mode n-channel
characteristics

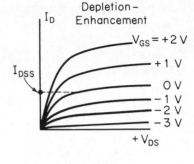

(d) Depletion mode n-channel
characteristics

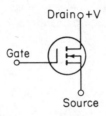

(b) n-channel enhancement
mode symbol

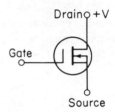

(e) n-channel depletion
mode symbol

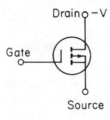

(c) p-channel enhancement
mode symbol

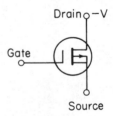

(f) p-channel depletion
mode symbol

FIGURE 6-22 *MOSFET characteristics and symbols.*

the inherent capacitances between gate and drain and between gate and source (that limit high-frequency response) are generally lower for the MOSFET. Temperature effects are minimal with MOSFETs, whereas the JFET has an input leakage current that rises exponentially with temperature, as does the leakage current of any reverse-biased *pn* junction. MOSFETs are the least temperature-sensitive semiconductor devices generally available today.

6-7 FET CIRCUIT OPERATION

There are many methods to set up an FET to prepare it for amplification of an ac signal. In this section we explore the most important of these methods with an eye toward the most practical circuits. For our discussion we shall consider the 2N5458 JFET. It is a general-purpose low-cost *n*-channel device, which can be purchased for less than $1.00 in small quantities and less than $0.50 in lots of 100 or more. It has the following ratings:

1. Transconductance: g_{fs} = 1500 to 5500 $\mu\mho$.

2. Zero gate–source voltage drain current: I_{DSS} = 2 to 9 mA.

3. Gate leakage current: I_{GSS} = 1 nA max.

The setup shown in Fig. 6-23 is the simplest common-source biasing scheme. If we let R_G = 100 MΩ, then even at the maximum I_{GSS} (leakage current) of 1 nA,

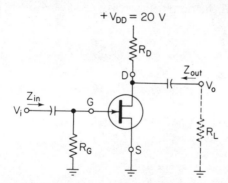

FIGURE 6-23 *Elementary JFET biasing scheme.*

we have just a 0.1-V drop across R_G. This can be reasoned by applying KVL to the loop that includes R_G and the gate-source junction. By KVL, the voltage V_{R_G} must equal V_{GS}. It is therefore safe to assume that $V_{GS} \simeq 0$ and the drain current will, therefore, equal I_{DSS}. If R_D were made equal to 2 kΩ, the operating point voltage V_D would range from

$$V_D = V_{DD} - V_{R_D}$$
$$= 20\text{ V} - (2\text{ mA} \times 2\text{ k}\Omega) = 16\text{ V}$$

at the minimum value of I_{DSS} of 2 mA up to

$$V_D = 20\text{ V} - (9\text{ mA} \times 2\text{ k}\Omega) = 2\text{ V}$$

at I_{DSS} = 9 mA. Obviously, this method of biasing has caused a very wide Q-point range, but this may be quite adequate for low-input ac signal levels. However, the

181

input signal must be limited to not exceed about $+0.5$ V on the gate-channel junction to prevent forward bias and the resulting input current flow and probable device destruction. This then implies a maximum ac input voltage of about 1 v p-p. Despite these limitations, the circuit of Fig. 6-23 could practically be put to use in the lab as a simple one-of-a-kind, low-signal-level, high-input-impedance amplifier. In addition, most FET amplifiers are used as an input stage to an amplifier system to offer a high input impedance from a high impedance signal source. This classification of signals is generally very low level—millivolts and less. Since the voltage gain of FET amplifiers is usually 10 or less, the output signal is still small enough to allow operation with the 2- to 16-V bias level range this design results in. Thus, the circuit is applicable to many production designs.

The voltage gain for the common-source FET amplifier shown in Fig. 6-23 is

$$G_v = g'_{fs} R_D \tag{6-20}$$

where g'_{fs} is the transconductance at zero bias ($V_{gs} = 0$) conditions. If an external load resistor is connected at the output (shown with dashed lines in Fig. 6-23), the voltage gain is

$$G_v = g'_{fs}(R_D \| R_L) \tag{6-21}$$

Thus, for the 2N5458 JFET using the circuit in Fig. 6-23, the gain varies from

$$G_V = g'_{fs} R_D = 2 \text{ k}\Omega \times 1500 \ \mu\mho$$
$$= 3$$

up to

$$G_V = 2 \text{ k}\Omega \times 5500 \ \mu\mho$$
$$= 11$$

through its full possible transconductance range. The input impedance for this circuit is equal to R_G in parallel with the impedance seen looking into the gate. The gate impedance is usually much higher than R_G, so that as a good approximation,

$$Z_{\text{in}} \simeq R_G \tag{6-22}$$

The purpose of R_G is to provide a path for gate leakage current to flow. Both the reverse-biased gate-channel junction for the JFET and the capacitance input of the MOSFET result in a small leakage current. A glance at Fig. 6-23 shows that R_G provides a path to ground for this current. Without that path, the leakage current could build up a static charge (voltage) at the gate that can change the bias or even destroy the FET. Typical values for R_G vary from 500 kΩ up to 100 MΩ.

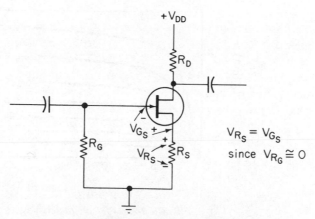

FIGURE 6-24 *Self-bias scheme.*

This resistance value essentially becomes the input impedance of the amplifier shown in Fig. 6-23.

The output impedance is equal to the value of R_D. The FET amplifier works from a very high input impedance (R_G) into a relatively low output impedance (R_D). It is, therefore, very effective in providing large-scale impedance transformations.

The circuit shown in Fig. 6-24 is also a common-source amplifier but has a resistor, R_S, from source to ground. This addition reduces the voltage gain but makes it more constant for transconductance variations. For this circuit,

$$G_v = \frac{g_{fs} R_D}{1 + g_{fs} R_s} \tag{6-23}$$

The input and output impedance relationships are the same as before—R_G and R_D, respectively.

The dc biasing conditions are affected by the inclusion of R_S. Since the drain and source current are essentially equal, whatever value of I_D that results from the biasing will also flow through R_S. This will cause a dc voltage drop of the polarity shown in Fig. 6-24 across R_S. Using Kirchhoff's voltage law around the loop that includes R_S, R_G, and the gate–source junction, the gate–source voltage is seen to approximately equal V_{RS} and is of the polarity shown. This is true since the gate leakage current is very low and causes only a minimal voltage drop across R_G. Hence, V_{RG} is essentially zero and $V_{GS} = V_{RS}$.

EXAMPLE 6-7

The characteristic curves for a particular JFET are provided in Fig. 6-25. With a V_{DD} of 20 V and $R_D = 3$ kΩ, determine the necessary value of R_S to provide a 10-V Q-point at the drain.

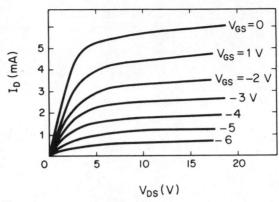

FIGURE 6-25 *FET characteristics.*

Solution:

I_D can be calculated as $3\frac{1}{3}$ mA (10 V/3 kΩ). This corresponds to V_{GS} = −2 V from the characteristic curves. Therefore, V_{R_S} must equal 2 V (V_{R_S} = V_{GS}) and will have $3\frac{1}{3}$ mA flowing through it. Hence,

$$R_S = \frac{2 \text{ V}}{3\frac{1}{3} \text{ mA}} = \underline{600 \ \Omega}$$

EXAMPLE 6-8

Determine the voltage gain of the circuit in Example 6-7. If a new FET were used with double the transconductance, calculate the change in voltage gain.

Solution:

To calculate voltage gain, the transconductance must be known. That can be determined graphically from Fig. 6-25. At V_{GS} = −2 V, $I_D \simeq 3$ mA; at V_{GS} = −3 V, $I_D \simeq 2\frac{1}{3}$ mA.

$$g_{fs} = \frac{\Delta I_D}{\Delta V_{GS}} \qquad \text{(6-19)}$$

$$= \frac{3 \text{ mA} - 2\frac{1}{3} \text{ mA}}{-3 \text{ V} - (-2 \text{ V})}$$

$$= \frac{0.67 \text{ mA}}{1 \text{ V}}$$

$$= 670 \ \mu\mho$$

Thus,

$$G_v = \frac{g_{fs}R_D}{1 + g_{fs}R_S} \qquad (6\text{-}23)$$

$$= \frac{670\ \mu\mho \times 3\ k\Omega}{1 + 670\ \mu\mho \times 600\ \Omega}$$

$$= \frac{2}{1 + 0.4}$$

$$= \underline{1.43}$$

If g_{fs} were doubled,

$$G_v = \frac{2 \times 670\ \mu\mho \times 3\ k\Omega}{1 + (2)670\ \mu\mho \times 600\ \Omega}$$

$$= \frac{4}{1.8}$$

$$= \underline{2.22}$$

Thus, doubling the transconductance changed the voltage gain from 1.42 up to 2.22.

The preceding examples show the method whereby operation at nonzero bias ($V_{GS} \neq 0$) can be attained. It simply requires the use of a resistor in the source. Less voltage gain is possible this way, but the gain is less variable from changes in g_{fs} that can occur from one FET to the next. Additionally, larger ac input voltages are possible, since positive inputs of greater than about 0.5 V would forward-bias the gate-source junction under zero-bias conditions but are permissible when a source resistor is used.

The circuit operation details of this section have used JFETs as examples. The MOSFET can be substituted in all these cases, however. The enhancement-mode MOSFET offers a special case of operation not possible with the JFET.

The bias simplicity for the enhancement-mode MOSFET provides some interesting applications. Figure 6-26 shows some typical characteristic curves and a likely biasing scheme. Notice that the bias voltage (V_{GS}) is of the same magnitude as the Q-point voltage V_D. This is because the gate of a MOSFET draws almost no current (10^{-14} A is typical) through R_G, resulting in zero potential difference between gate and drain, since $V_{RG} = 0$.

The minimal bias circuitry makes MOSFETs ideal for integrated circuits since MOSFETs turn out to be cheaper than resistors in ICs because of their much smaller size. In fact, MOSFETs are often used as resistors in integrated circuits. We discuss these aspects in greater detail in Chapter 8.

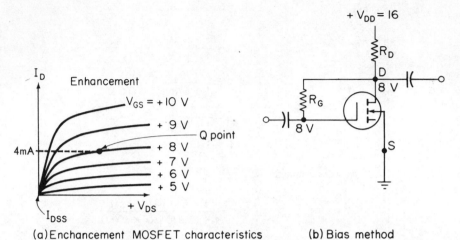

(a) Enchancement MOSFET characteristics   (b) Bias method

FIGURE 6-26 *Enhancement mode bias of MOSFET.*

EXAMPLE 6-9

Calculate the proper value for R_D in Fig. 6-26(b). If $g_{fs} = 1000\ \mu\mho$, determine the circuit's voltage gain.

Solution:

The voltage at the drain, V_D, is 8 V with respect to ground. Thus, one side of R_D is at 8 V and the other is at 16 V. The voltage across it is, therefore, the difference, 16 V − 8 V, or 8 V. The curves in Fig. 6-26(a) show that I_D is 4 mA when $V_{GS} = 8$ V. Thus,

$$R_D = \frac{V_{R_D}}{I_D} = \frac{8\text{ V}}{4\text{ mA}}$$

$$= \underline{2\text{ k}\Omega}$$

The voltage gain is

$$G_v = g_{fs}R_D$$
$$= 1000\ \mu\mho \times 2\text{ k}\Omega$$
$$= \underline{2}$$

QUESTIONS AND PROBLEMS

6–1–1. What is meant by the expression "solid-state" device? How do they differ from vacuum-tube devices?

6–1–2. Explain the basic construction of a transistor. Proper transistor operation requires that the two *pn* junctions be properly biased. Explain those conditions.

6–1–3. Provide definitions for the terms h_{FB} and h_{fb}. Repeat for h_{FE}, h_{fe}, and h'_{fe}.

6–1–4. A transistor has been properly biased (base–emitter junction forward-biased and base–collector junction reverse-biased). It has a dc emitter current of 10 mA and dc base current of 0.2 mA. Calculate I_C, h_{FB}, h_{FE}, and h'_{FE}.

6–2–5. Define amplification and distortion.

6–2–6. A certain transistor has an I_C versus I_B curve as shown in Fig. 6-27.
    (a) Determine the range of base and collector currents through which this transistor will exhibit a linear relationship.
    (b) What is the name of the constant of proportionality between I_C and I_B?
    (c) Calculate the value of that constant for this transistor.

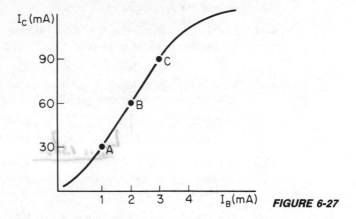

FIGURE 6-27

6–2–7. (a) An ac sinusoidal current of 1 ma p-p is applied to the base of the transistor utilized in Problem 6–2–6. The dc base current (*Q*-point) is at point *B* of Fig. 6-27. Sketch the resulting collector current.
    (b) The same current is now applied around a *Q*-point at point *C* of Fig. 6-27. Sketch the resulting collector current. (*Note:* The collector current is no longer a pure sinusoid, and hence the amplifier has introduced distortion into the signal.)

6–2–8. (a) State which of the following characteristics were exhibited by the amplifier in Problem 6–2–7(a): (1) reproduction; (2) amplification; (3) reproduction and amplification.
    (b) Repeat for Problem 6–2–7(b).

6–2–9. In an amplifier, the ac input base current of 1 ma p-p is amplified to 50 ma p-p in the collector output. Calculate h_{fe} and explain where that new current comes from.

6–2–10. What are the three possible amplifier configurations for a bipolar junction transistor (BJT)? Explain why it is necessary to have a common lead in these amplifiers.

6–3–11. For the CE amplifier shown in Fig. 6-12, explain the function of C_1, C_2, R_A, and R_B.

6–3–12. In general, why is it desirable to have the output lead (collector) of the CE amplifier at about one-half of the dc supply voltage?

6–3–13. The CE amplifier shown in Fig. 6-12 has its collector resistor changed from 10 kΩ to 5 kΩ. If $h_{FE} = 100$, calculate the proper value for R_A.

6–3–14. Calculate G_v, R_{in}, Z_{in}, Z_{out}, and G_i for the circuit analyzed in Problem 6–3–13.

6–3–15. Explain why the CE amplifier is termed "phase-inverting." Provide sketches of ac input and output voltage to help explain this condition.

6–3–16. Describe the required circuitry changes when changing from an NPN to a PNP BJT.

6–4–17. Provide another common name for the common-collector (CC) configuration and explain the significance of this name.

6–4–18. For the circuit shown in Fig. 6-15, calculate the proper value of R_A to make the voltage from emitter to ground equal 10 V if R_E were changed to 500 Ω and $h'_{FE} = 50$.

6–4–19. What is the typical voltage gain for an emitter-follower circuit? What characteristics make it such a popular configuration?

6–4–20. The circuit from Problem 6–4–18 is driven by signal source of 4 v p-p with internal resistance of 1.5 kΩ. If R_L is 1 kΩ, calculate the ac load voltage, R_{in}, Z_{in}, and G_i.
R_e is 1.5kɅ

6–4–21. If the signal source of Problem 6–4–20 directly drove the 1.5-kΩ load, determine the load voltage and compare it with the result of Problem 6–4–20. Explain the usefulness of the emitter follower on the basis of these results.

6–5–22. Describe the basic construction of a JFET. Identify and discuss its three areas of operation.

6–5–23. Define the transconductance of a FET. Calculate the transconductance (g_{fs}) for the FET represented by the characteristic curves in Fig. 6-20 between $V_{GS} = 0$ and -1 V.

6–5–24. Explain the preference for *n*-channel versus *p*-channel JFETs.

6–6–25. Describe the basic construction of a MOSFET. Explain the significance of the terms "enhancement mode" and "depletion mode."

6–6–26. Compare some of the characteristics of JFETs and MOSFETs.

6–7–27. For the circuit shown in Fig. 6-23, what is the dc voltage across R_G? If $R_D = 3$ kΩ, $I_{DSS} = 3$ mA, and $g'_{fs} = 4000$ μʊ, determine the drain to ground voltage (V_D) and the voltage gain.

6–7–28. If an external ac load, R_L, of 2 kΩ were added to the circuit in Problem 6–7–27, calculate the new value of voltage gain.

6–7–29. The FET amplifier shown in Fig. 6-24 has $V_{DD} = 15$ V and $R_D = 2$ kΩ. Determine the value of R_S necessary to provide a 7-V Q-point at the drain. The characteristic curves for the FET are provided in Fig. 6-25.

6–7–30. Using the results of Problem 6–7–29, determine the voltage gain of the circuit. What is the circuit's input impedance?

6–7–31. Calculate the required value of R_D in Fig. 6-26 if V_{DD} is changed to 12 V and a 6-V Q-point is desired. The MOSFET transconductance is 1500 $\mu \mho$ and R_D is 3 kΩ. Calculate the voltage gain of the circuit.

7

AMPLIFIERS

7-1 REGULATED POWER SUPPLY

In Chapter 6 you were introduced to transistors and basic amplifier considerations. This chapter provides a look at some additional amplifier topics.

You will recall the basics of regulated power supplies from Chapter 5. At that time, the zener diode shunt regulator and a linear integrated circuit (LIC) series regulator were presented. Now that you have transistor basics behind you, a series regulator using a transistor (as an amplifier) can be presented.

The block diagram in Fig. 7-1 shows the form of a series regulated power supply. The first two blocks are the familiar elements of a basic unregulated power supply. The series pass element is the control element—the element that turns the "electron

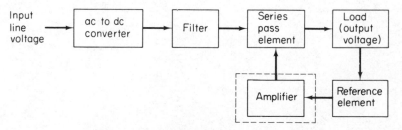

FIGURE 7-1 *Block diagram for a series regulator.*

spigot" up and down to maintain a constant voltage at the load, a load that may be changing, or it compensates for changes in the input line voltage. The series pass element is normally a power transistor performing this control function. The reference element is a constant voltage device that compares the output voltage with itself and sends out an "error" signal to the control element when necessary. This error signal is passed on to an amplifier, which then relays it to the series pass element. In the simpler circuits, a separate amplifier is not used—the reference element feeds the "error" signal directly to the series pass element, which also functions as the amplifier.

What has just been described is a *feedback control system*. This is a system that senses its own output and, if the output is not correct, sends an error signal back to the amplifier to be amplified and then to the control element to correct this condition. In our particular system, the output voltage can instantaneously change from the desired level, which, when compared with the reference element, results in an error signal. Figure 7-2 illustrates a simple but effective series regulator circuit. The zener diode, D_1, is the reference element and Q_1 is the amplifier and series pass element. The voltages V_{BE}, V_Z, and V_{OUT} form a closed loop. Kirchhoff's voltage law states that the algebraic sum of the voltages in a closed loop is zero. Therefore, it stands to reason that if V_{BE} and V_Z are relatively constant, as indeed they should be in a properly working circuit, V_{OUT} should also be a constant voltage. This is obviously the desired result for a voltage regulator, and that is exactly how this circuit works, at least in very simple terms.

Let us now take a look at the circuit operation in a different way and in greater detail. R_S supplies current for the base of Q_1 and for D_1 to keep it operating on the zener breakdown area (regulating region). Care must be taken that R_S has a low enough value to be able to always supply the needs of these two devices. The worst-case condition in this respect occurs at the minimum input voltage and maximum output current. This is because the voltage across R_S is then minimum, making I_{R_S} minimum, and I_E and I_C of Q_1 are maximum, which means that R_S must then supply the highest value of base current. The zener diode "absorbs" any of the excess current

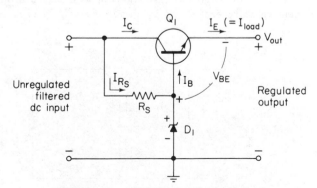

FIGURE 7-2 *Simple series regulator circuit.*

from R_S not needed for the transistor's base, but requires some minimum value to keep it operating in its regulating region (I_{ZK}—refer to Fig. 5-16).

If, for instance, the load on this power supply were suddenly doubled (i.e., I_L doubled), the output voltage would instantaneously tend to drop, because the unregulated input would drop as more current is drawn from it. However, the constant voltage effect of the zener diode causes a larger forward bias (V_{BE}) on Q_1; hence, I_B of Q_1 will increase and cause a greater I_C flow and thus a greater I_E flow. This increase in bias occurs because the load voltage has instantaneously dropped while the zener voltage will, it is hoped, remain constant. Since V_{BE} is the only other voltage drop in the loop with V_Z and V_L, it stands to reason that it must increase instantaneously. This results in a larger forward bias and, hence, greater load (emitter) current. This greater load current then tends to bring the load voltage back up to its original value. Thus, Q_1 is acting as a "valve" in controlling current flow in order to obtain a constant voltage output. An increase in output voltage caused by a change to a lighter load is compensated for in a similar but naturally reverse fashion. A circuit of this type can typically offer a 5% regulation figure for line and load changes.

The speed with which this regulator "reacts" to input line changes or output load changes is determined by the small internal junction capacitances within the transistor. It is normally safe to assume transistor reaction times of microseconds or quicker. Now consider the speed with which the unregulated input voltage fluctuates because of ripple—typically 120 Hz or a complete cycle every $\frac{1}{120}$ s (8.3 ms). Thus, the regulator input ripple fluctuations are slow in comparison to its reaction time. It should now be clear that the regulator serves to reduce ripple and hence functions as a sophisticated "filter." A good regulator serves to reduce ripple by a factor of 100 or more.

EXAMPLE 7-1

You are to design a voltage regulator as indicated in Fig. 7-3. The load current varies between 0 and 1 A, and the unregulated dc input varies from

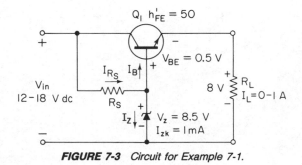

FIGURE 7-3 *Circuit for Example 7-1.*

12 to 18 V for all line and load changes. The 8.5-V zener diode requires at least 1 mA of current to stay in its regulating region ($I_{ZK} = 1$ mA).

(a) Determine the value of R_S to ensure proper circuit operation.

(b) Determine the worst-case power dissipation of all circuit elements.

Solution:

(a) As previously stated, the worst-case conditions dictate that the value of R_S must be low enough to supply current to both the base of Q_1 and the zener diode. This means that under worst-case conditions, R_S must supply at least the $I_{ZK} = 1$ mA rating of D_1 plus the maximum base current:

$$I_{B_{max}} = \frac{I_{E_{max}}}{h'_{FE}} = \frac{I_{L_{max}}}{h'_{FE}} = \frac{1\ \text{A}}{50} = 20\ \text{mA}$$

$$I_{R_S} = I_{ZK} + I_{B_{max}} = 1\ \text{mA} + 20\ \text{mA} = 21\ \text{mA}$$

Now this 21 mA must be supplied by R_S under all conditions of input voltage variation—even when the input falls to 12 V, which causes the minimum voltage across R_S and hence the lowest value of current it will be able to supply. Therefore,

$$R_S = \frac{12\ \text{V} - 8.5\ \text{V}}{21\ \text{mA}} = \frac{3.5\ \text{V}}{21\ \text{mA}} = 166\ \Omega$$

Should the value of R_S become any larger, the power supply would go out of regulation at $V_{IN} = 12$ V and $I_L = 1$ A, since the zener diode would not be supplied with the 1 mA it requires to regulate at 8.5 V. Therefore, R_S should be 166 Ω or slightly less to provide a margin of safety.

(b) The maximum power dissipation of R_S occurs when the voltage across it is maximum. Hence,

$$P_{R_{S_{max}}} = \frac{(V_{R_{S_{max}}})^2}{R_S} = \frac{(V_{IN_{max}} - V_Z)^2}{R_S}$$

$$= \frac{(18\ \text{V} - 8.5\ \text{V})^2}{166\ \Omega} = 0.544\ \text{W}$$

Next, consider the power dissipated in the zener diode. The voltage across it should remain at close to 8.5 V. The current varies from a low of 1 mA to a value not yet determined. The maximum value will occur when V_{IN} is maximum and the load current is minimum—in this example I_L goes down to zero, which means that I_E ($I_E = I_L$) and, hence, I_B will be zero. If I_B is zero, all the current through R_S will pass through the diode:

$$I_{Z_{max}} = I_{R_{S_{max}}} = \frac{V_{IN_{max}} - V_Z}{R_S}$$

$$= \frac{18\text{ V} - 8.5\text{ V}}{166\ \Omega} = 57.2\text{ mA}$$

$$P_{Z_{max}} = V_Z I_{Z_{max}} = 8.5\text{ V} \times 57.2\text{ mA} = \underline{0.486\text{ W}}$$

The maximum power dissipation in Q_1 will occur when its collector-emitter voltage (V_{CE}) is maximum and when its emitter current is maximum. Its V_{CE} is equal to the input voltage minus V_L (Kirchhoff's Voltage Law):

$$V_{CE_{max}} = V_{IN_{max}} - V_L$$
$$= 18\text{ V} - 8\text{ V} = 10\text{ V}$$
$$P_{Q1_{max}} = V_{CE_{max}} \times I_{E_{max}} = 10\text{ V} \times 1\text{ A} = \underline{10\text{ W}}$$

The simple series regulator described in this section is able to give good performance at minimal cost. If better regulation than about 5% is desired, additional amplification is necessary. That means more transistors between the reference element and the series pass element. In these cases, the use of a LIC voltage regulator is probably the best choice (see Chapter 5).

7-2 POWER AMPLIFIERS

The amplifiers we have thus far considered were generally of the small-signal variety. A broad definition of a *small-signal amplifier* is that the ac input and output voltages and currents are small with respect to the quiescent dc voltage and current levels. In a *large-signal* amplifier, either the ac voltage or current (or both) levels are approaching their maximum possible value. That maximum level is determined either by the power supply employed or the transistor voltage and/or current ratings. In cases where appreciable amounts of power are involved (for instance, 1 W or more), the amplifier is termed a *power amplifier*. Thus, not all large-signal amplifiers are necessarily power amplifiers, but it is safe to say that most are.

Power amplifiers (PAs) are similar to those amplifiers previously studied in many ways. For instance, a schematic diagram of a single-stage small-signal amplifier (SSA) and a power amplifier might appear identical. Despite the similarities, many new aspects must be considered when dealing with PAs. It is these aspects that form the basis for this section.

It is the function of an amplifier to drive a load. The type of load being driven determines the required level of voltage and current. If the load is a high-fidelity speaker with an 8 Ω impedance and with 50 W of drive necessary for full volume certain requirements of the amplifier may be determined. The rms output voltage is calculated from the given values of P and R as

$$P = \frac{v^2}{R}$$

$$\therefore \quad v^2 = P \times R$$
$$= 50 \text{ W} \times 8 \ \Omega = 400$$

$$v = 20 \text{ v rms}$$
$$= 20 \times 2 \times \sqrt{2} \simeq 56 \text{ v p-p}$$

The ac output current is calculated as

$$i = \frac{v}{R} = \frac{56 \text{ v p-p}}{8 \ \Omega}$$
$$= 7 \text{ a p-p}$$

Obviously, this is a power amplifier, and the load being driven effectively dictates the power output requirement.

If the load is a cathode-ray tube (CRT) that requires a 200-v p-p signal for horizontal deflection, one might be tempted to say that it requires a PA for drive. However, since very little current is drawn by the CRT, the power level is very low, requiring a large-signal amplifier but not a PA.

Since the cost of transistors is roughly proportional to their power rating, it is customary to use power transistors at close to their rated power level. For instance, it would be uneconomical to use the transistor with the characteristics shown in Fig. 7-4 in a lower-power application because a lower-cost smaller device could be

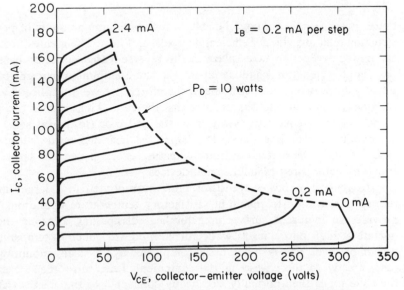

FIGURE 7-4 *High-voltage power transistor characteristics.*

found to do the job. Note the dashed line indicating the maximum power dissipation level of 10 W in Fig. 7-4. Any area to the right or above that line $(V_{CE} \times I_C)$ exceeds the device's power rating. Power transistors have a maximum permissible junction temperature (T_j), as do all other semiconductors. The power the transistor dissipates $(\approx V_{CE}I_C)$ determines the amount of temperature rise above the temperature of the transistor's environment *(ambient temperature)*. Operation at junction temperatures above about 110°C for germanium or 175°C for silicon devices results in device destruction. The amount of device dissipation is determined by the product of I_C and V_{CE} at all times. With a 10-W limit, this describes the hyperbolic dashed line shown in Fig. 7-4.

Of special interest in Fig. 7-4 is the high-voltage rating of the particular device shown. Note the 300-V maximum rating for V_{CE}. The present power transistor state-of-the-art value for maximum voltage levels is about 1000 V. Power transistors with allowable collector currents of about 250 A are at the upper limit with respect to current. Thus, there are three basic limitations on power transistors: voltage, current, and power. Exceeding any one of these ratings normally results in device destruction.

Since it is normal and economical to operate a power transistor close to its absolute limits, distortion of the output signal becomes an important consideration. At both very high and very low values of collector current, the transistor's h_{fe} is reduced. Operation in these nonlinear areas results in distortion, which is either tolerated or compensated for by reducing the gain (which limits distortion). Figure 7-4 also illustrates this effect by showing a smaller spacing between the $I_B = 0.2$ mA per step characteristics at both high and low collector currents.

Heat Dissipation

Since power transistors must be able to dissipate large amounts of power, their construction must provide for efficient dissipation of heat. The greatest fraction of transistor power dissipation takes place in the reverse-biased collector–base junction (the base–emitter junction conducts about the same amount of current but has a much lower voltage drop). It is customary, therefore, to have the collector electrically and physically connected to the metal transistor case. This enables the generated heat to be efficiently conducted away from the collector (to a heat sink) and then into surrounding air or free space. The larger the heat-dissipating surface area of both case and sink, the better the heat-dissipating effect; hence, the larger size of power transistors compared to small-signal devices.

In general, a power transistor must be mounted to a larger metallic surface termed a *heat sink* to maintain satisfactory temperature operation. The heat sink provides an increased surface area for heat dissipation, allowing the transistor to operate at high power levels while still keeping the junction temperature (T_j) at an acceptable level. This is sometimes accomplished by physically mounting the transistor case directly to the chassis of the equipment. Care must then be exercised, since the collector is then normally electrically connected to the chassis (which is usually at ground potential) unless insulated in some fashion. Mica washers between the

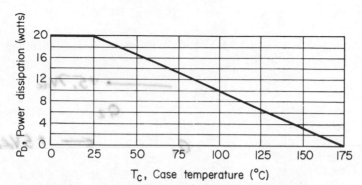

FIGURE 7-5   *Power derating curve for a 2N3766 transistor.*

transistor and metal chassis are often used for such collector isolation at low collector voltages, since they offer good thermal conductivity but very high electrical resistance. In some cases, special heat sinks are used instead of just attaching the transistor to the chassis. These are usually made of aluminum, which offers excellent thermal conductivity with low weight. The aluminum is often extruded or corrugated to offer the maximum surface area possible and may also be painted black to maximize its heat-dissipating ability.

The power rating of a power transistor is only valid when the transistor-case temperature is held to 25°C ($\simeq$ room temperature). This temperature is difficult to maintain even with a heat sink, and it is therefore necessary to derate the transistor according to the expected maximum case temperature, and *derating* information is supplied by the transistor manufacturer. Figure 7-5 shows the power-versus-tempera-ture *derating curve* for a medium-power 2N3766 transistor. It has a 20-W rating, but if used above 25°C case temperature, it must be derated 0.133 W/°C. Notice that the power rating falls to 0 W at 175°C, since this is the maximum allowable junction temperature. Any power dissipation at 175°C would raise the junction tem-perature above this maximum of 175°C. It is presumed that a junction temperature above 175°C causes device failure.

Class A Power Amplifier

Consider the circuit shown in Fig. 7-6. It is a two-stage amplifier with Q_1 a CE configuration and Q_2 a CC (emitter follower). Both of these stages are operating in what is called a *class A* mode. Class A may be defined as a method of bias that allows for a complete replica of the input signal of an amplifier at the output. Thus, the transistor neither saturates or cuts off during the complete range of the input signal. All the amplifiers thus far discussed have been of this category, and therefore class A operation is nothing new to you—other than its name.

The class A amplifier in Fig. 7-6 can deliver a theoretical maximum ac load voltage of 10 v p-p since the dc supply is 10 V. For that to be possible, the dc voltage at the output (emitter of Q_2) must be 5 V. The dc current flow in the output

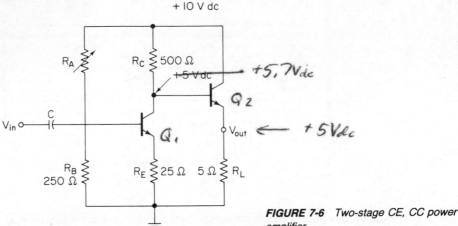

FIGURE 7-6   Two-stage CE, CC power amplifier.

stage (Q_2) is therefore 1 A, since 5 V dc exists across the load R_L of 5 Ω. We are now in a position to determine the efficiency of the amplifier. The *efficiency* (η) of an amplifier stage is defined as the ratio of ac power out to dc power in:

$$\text{efficiency} = \eta = \frac{P_{\text{out}}\,\text{ac}}{P_{\text{in}}\,\text{dc}} \qquad (7\text{-}1)$$

As a percentage,

$$\eta = \frac{P_{\text{out}}\,\text{ac}}{P_{\text{in}}\,\text{dc}} \times 100\% \qquad (7\text{-}2)$$

We can now calculate η for the ~~amplifier~~ Second Stage in Fig. 7-6. This will be the maximum theoretical efficiency, since an output of 10 v p-p is a maximum theoretical value.

$$\eta = \frac{P_{\text{out}}\,\text{ac}}{P_{\text{in}}\,\text{dc}} \times 100\%$$

$$= \frac{(10\text{ v p-p}/2\sqrt{2})^2 \div R_L}{10\text{ V} \times 1\text{ A}} \times 100\% \qquad (7\text{-}2)$$

$$= \frac{(100/8) \div 5\ \Omega}{10\text{ W}} \times 100\% = \frac{2.5\text{ W}}{10\text{ W}} \times 100\%$$

$$= 25\%$$

Notice that the 10-v p-p ac output voltage was divided by $2\sqrt{2}$ to convert into the necessary rms voltage for power calculation. The dc power delivered to the output stage is simply the dc supply voltage (10 V) times the dc current supplied (5 V/R_L = 1 A).

This 25% efficiency is the maximum obtainable in a class A amplifier. In reality this figure will be less, since the peak-to-peak output voltage cannot fully equal the dc supply voltage without introducing clipping. This is due to the saturation voltage of a transistor, which might typically be $V_{CE\ sat} = 1$ V. Thus, even if perfect biasing existed for the remaining 9 V that are usable, a maximum 9 v p-p signal would be the absolute limit. To ensure that clipping of the output signal did not occur, it would be wise to not let the signal exceed 8 v p-p, which would mean an efficiency of

$$\eta \simeq \frac{(8\ \text{v p-p}/2\sqrt{2})^2/5\ \Omega}{10\ \text{W}} \times 100\% = 16\%$$

This is a typical class A circuit efficiency. At the power levels being dealt with (several watts), this may be acceptable, but when dealing with higher powers (say, 10 W or more), the cost of the wasted power becomes too great to tolerate, and class B circuitry (see the next section) is then used.

The power rating required of Q_2 in Fig. 7-6, at no ac signal conditions, is approximately equal to $V_{CE}I_C$, which in this case averages out to 5 V $\times$ 1 A = 5 W. At full ac signal conditions, the average power dissipated is unchanged; thus, a 5-W transistor is required when a maximum 2.5 W ac output power is obtained. This ideal case for class A amplifiers requires a transistor power rating double the ac load power. In a practical situation a good rule of thumb is to maintain a 2.5:1 ratio or higher. Thus, if 1 W ac output power is required, about 2.5 W is dissipated by the transistor.

The power amplifier thus far considered delivers both dc and ac power to the load. In fact, the ideal case delivered 2.5 W of ac power to the load at full signal and dissipated 5 W in the transistor. Since the battery was supplying 10 W of power, this leaves 2.5 W of dissipation unaccounted for. That 2.5 W is dissipated as dc power in the 5-Ω load—a situation that is often unacceptable. A good example is a loudspeaker in which the dc current may destroy the speaker due to excessive heat or, at best, will cause poor speaker performance. A solution to this problem is to capacitively couple the load, as shown in Fig. 7-7. In this circuit, Q_2's emitter resistor allows a dc bias voltage to be set up, and the coupling capacitor keeps dc from entering the load. An additional feature of this circuit is an increase in efficiency, since there is no longer any dc power dissipation in the load. The maximum theoretically attainable efficiency η is now 50% if $R_E = \infty$, but in practice 30 to 40 percent is typical, since a finite (low) value of R_E is required.

Another possible class A power amplifier circuit is shown in Fig. 7-8. The major difference here is that the power stage, Q_2, is now a CE circuit instead of an emitter follower. This allows for voltage and current gain in both stages, which in some cases might be an advantage, but several drawbacks over the previous circuits are introduced. First, notice that direct coupling between stages is no longer possible. If the base of Q_2 were directly connected to the collector of Q_1, the forward bias on Q_2 would cause it to saturate. It is convenient to direct couple a CC stage to

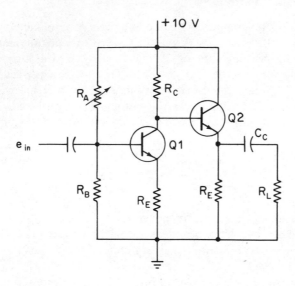

FIGURE 7-7 Power amplifier with ac load coupling.

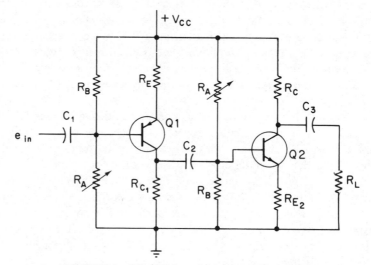

FIGURE 7-8 CE, CE class A power amplifier circuit.

the output of a CE amplifier, but it is not normally possible in a PA to connect a CE stage to another CE stage directly. The inability to directly couple means that Q_2 must have its own bias resistors, and a coupling capacitor is necessary to isolate the dc levels between the two stages. Thus, three additional passive components are necessary in Fig. 7-8 (two resistors and one capacitor), and the coupling capacitor C_2 must be a high-valued electrolytic unit if good low-frequency response is necessary. An emitter resistor R_{E_2} for stability in the power stage is also required. This results in added power dissipation and thus lower overall efficiency. In general, the use of an emitter-follower circuit for the output of a class A amplifier is the logical choice over a CE stage.

　　One final note on the amplifier of Fig. 7-8. Since a *pnp* transistor was used for stage 1, it was necessary to "flip" it over, schematically speaking, to allow for proper voltage polarities. Q_1 is still a CE stage, since the output is taken at the collector and the input is at the base. Various combinations of *pnp* and *npn* transistors are often utilized in multistage amplifiers, requiring careful schematic analysis to allow comprehension of the circuit.

7-3　PUSH-PULL OPERATION

Push-pull amplifiers operating class B make up the largest majority of linear (output is replica of input) power amplifiers. The transistors in *class B push-pull* amplifiers operate 50% of the time. Figure 7-9 shows the result of applying a sine wave to a single transistor amplifier operating class B. Amplification of the positive half of the input signal has occurred, and rectification has also taken place. A lack of any bias on a transistor will yield this result, as shown in Fig. 7-10. During the positive half-cycle of the input signal, the transistor base–emitter junction becomes forward-biased, and the pictured out-of-phase output signal is obtained. The negative half-cycle of the input signal reverse biases the base–emitter junction, and hence cuts off the transistor and no ac output signal is obtained. This amplifier by itself then has little value because of this extreme distortion that it introduces into the output signal. However, if two transistors are properly connected, their combined outputs

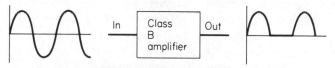

FIGURE 7-9　*Class B operation.*

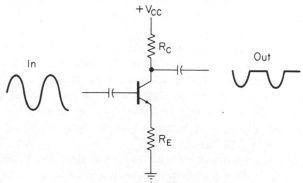

FIGURE 7-10　*Class B transistor amplifier.*

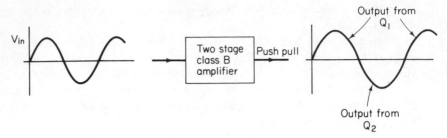

FIGURE 7-11   *Two-stage push-pull class B amplifier representation.*

will result in a good replica of the input, as shown in Fig. 7-11. Operation of two class B transistors in this fashion is known as *push-pull operation.*

The major advantage of push-pull operation is simply a matter of more power to the load with less power dissipated in the active devices and hence greater efficiency. With no ac input signal applied to the amplifier of Fig. 7-10, the only current being drawn from the dc supply is a very small leakage current. This contrasts with a class A stage, which is always biased at some significant level of quiescent (dc) current. When an input signal is applied to the class B amplifier, the current flow through the transistor increases as the transistor voltage drop (V_{CE}) is decreasing. Thus, class B transistor power dissipation is minimized, since when its V_{CE} is highest, the current flow is lowest (a leakage current), and when the current flow reaches a maximum, V_{CE} falls to some minimum value since power equals $V_{CE} \times I_C$, the dissipated transistor power in either case is minimized. The maximum theoretical efficiency of a class B push-pull amplifier is 78%, at which point the power rating of each transistor must be only one-tenth of the ac output power. Thus, for the theoretical case, a 100-W output power requirement would dictate two 10-W transistors for class B operation or one 200-W transistor for class A operation! Not only would a massive heat sink be required for the class A design, but a more expensive, higher-current dc power source would be necessary. Another important consideration is that virtually zero power is dissipated under no-signal conditions for the class B design, but the full 200 W is being dissipated by the class A transistor under no-signal conditions. For these reasons, virtually all PAs with over several watts output operate class B. Some special applications allow and warrant the use of *class C* or *class D* designs. These designs are not covered here but do offer even better efficiencies than class B operation.

Recall that a practical class A design requires a transistor power rating of about 250% of the output power to allow for a safety factor and the somewhat less than ideal operating conditions that all amplifiers operate in. In a similar fashion, the class B push-pull amplifier transistor ratings require padding. Instead of using transistors with 10% of the required ac output power as their rating, a good rule of thumb is to use a 15 to 20% factor.

Figure 7-12 shows two of the many possible class B push-pull circuit configurations. In Fig. 7-12 a transformer coupling is utilized at the input and output. The effect of the center-tap input is to provide phase reversal of the signals applied to

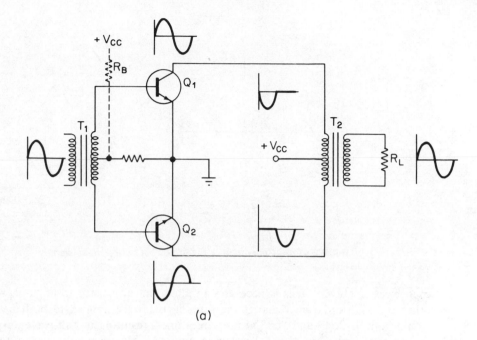

(a)

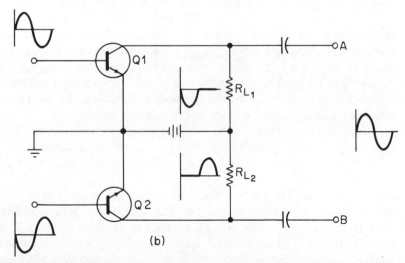

(b)

FIGURE 7-12 *Possible class B configurations: (a) Transformer coupling; (b) Transformerless coupling.*

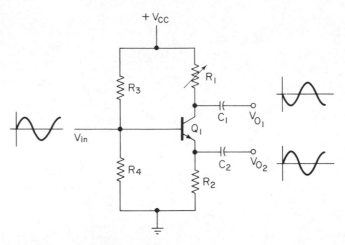

FIGURE 7-13 *Generation of two equal but out-of-phase signals.*

each base, as shown. This is necessary to allow one transistor to be "on" while the other is "off," and vice versa. If the same signal were applied to both bases, they would both be "on" and "off" at the same time—resulting in half-wave output. The output transformer in Fig. 7-12(a) then combines the two half-wave outputs into the complete signal. This type of design will be found in many older circuits but is now seldom used due to the high cost and weight of transformers.

The same circuit [Fig. 7-12(a)] could easily be converted to a class A push-pull design by the resistor addition shown with dashed lines. Thus, R_B would provide forward bias to each transistor through the secondary winding of T_1. This technique is occasionally used in designs requiring extremely low distortion levels. The class A push-pull design has the inherent advantage of canceling distortion components, which does not occur for the class B-design. However, the low efficiency of class A operation is still in effect and thereby minimizes its use.

The circuit of Fig. 7-12(b) shows a transformerless design, if some means of generating two equal but out-of-phase input signals can be accomplished. The resulting output from A to B must be left floating with respect to ground, which is a disadvantage in some situations.

The most widely used method of obtaining the out-of-phase inputs required for the circuit of Fig. 7-12(b) is shown in Fig. 7-13. Circuits that perform this function are termed *paraphase amplifiers, phase splitters,* or *phase inverters.* The circuit shown provides the 180° phase reversal, since the collector output inverts its input signal while the emitter output does not. If R_1 is made approximately equal to R_2, the two outputs will be equal in magnitude, since the ac current through each of about the same $(i_e \simeq i_c)$. The voltage gain will then be about 1, since $G_v = R_C/R_E = R_1/R_2$ and $R_1 \simeq R_2$. The problem with this circuit is that the two output impedances are unequal, since looking back into the emitter a low impedance is seen, while looking back into the collector the impedance equals approximately R_1. This difference can be compensated for by adding some resistance in series with C_2.

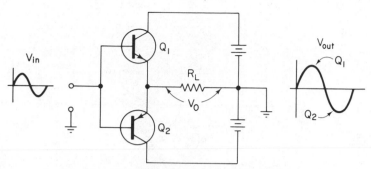

FIGURE 7-14 *npn-pnp class B push-pull amplifier.*

An even better solution to phase-splitting problems is to use a circuit that does not require the two equal but 180° out-of-phase input signals. The use of one *npn* and one *pnp* transistor makes such a design possible. Almost all today's PAs utilize this technique. Figure 7-14 shows a possible design for this technique, and this circuit arrangement is known as *complementary symmetry,* since *npn–pnp* transistor combinations are called *complementary pairs.* Both Q_1 and Q_2 are operating as emitter followers. Q_1 will conduct only during the positive half-cycle of the input, whereas Q_2 is cut off by this same signal. The situation reverses itself during the input's negative excursion, with the resulting combination of signals through R_L, as shown in Fig. 7-14. The use of identical or matched *npn* and *pnp* transistors (matched complementary pairs) will provide equal positive and negative output signals and thereby minimize distortion.

Crossover Distortion

Unfortunately, class B operation introduces severe distortion at the very low signal levels due to the turn-on base–emitter voltage of a transistor (≈ 0.7 V) and the nonlinearity in the low-signal area. The distortion so introduced is termed *crossover distortion,* because it occurs during the time that operation is crossing over from one transistor to the other in the push-pull amplifier. Figure 7-15 provides a dynamic

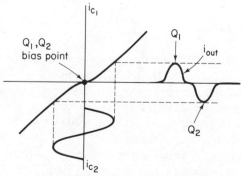

FIGURE 7-15 *Crossover distortion.*

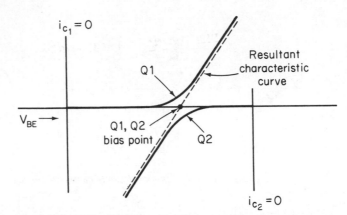

FIGURE 7-16 *Resultant characteristic curve for class AB operation.*

transfer characteristic for a class B amplifier with a slightly exaggerated amount of crossover distortion to clearly illustrate the effect. The bias point for Q_1 and Q_2 is at the origin, as shown.

Most applications cannot tolerate this amount of distortion, and to eliminate it, class AB operation is used. Class AB operation gives each transistor in the push-pull amplifier a small forward bias. It is defined as transistor conduction for more than 180° but less than 360°. Thus, instead of operating for exactly 180° as in class B, each transistor may operate for perhaps 200° out of a full 360° cycle, which results in a small overlap or period of time when both transistors are providing output current. A typical transfer characteristic for a class AB push-pull amplifier is shown in Fig. 7-16. The dashed line shows the composite curve that is the result of both stages. The resultant is a linear relationship, as shown, which should now faithfully reproduce the input signal. The small forward bias causes the class AB amplifier to be somewhat less efficient than class B, but still much more efficient than class A. Recall that the maximum theoretical efficiency for class A is 25%. Class B is 78%, and for class AB 70% would be a typical figure, but if, of course, depends on just how much forward bias is used.

FET Power Amps

Up until the late 1970s, production difficulties prevented large-scale manufacture of power field-effect devices. However, new construction techniques have overcome these problems, and the use of these devices becomes increasingly attractive as their price become competitive with BJT power devices.

The use of power FETs provides a major advantage over the BJT power amp studied so far. Because of the extremely high input impedance of field-effect devices, the amplifying stage prior to the high-power output stage can be very low power. This contrasts with BJT power amps, where "medium"-power driver stages must

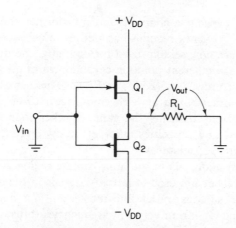

FIGURE 7-17 *FET push-pull power amplifier.*

precede the output stage. Thus, the use of FET power amps can minimize the number of required driver stages and consumes very little power except in the output stage.

The FET power amp shown in Fig. 7-17 is very similar in appearance to the BJT varieties. The use of FETs in power amplifiers offers some advantages with respect to signal distortion and transistor reliability over BJT designs. This is in addition to the low-power drive circuit advantage.

7-4 FREQUENCY RESPONSE

Our study of amplifiers up to this point has mainly concerned itself with only one range of frequencies. That is the midfrequency range, where changes in ac input frequency had little or no effect on our analysis. Figure 7-18 shows a typical frequency-response curve for an amplifier. The input signal is assumed to be of constant amplitude and is varied from 0 Hz (dc) up to a very high frequency where the amplifier no longer has any appreciable output. The result is an area of relatively flat response, known as the *midband,* and two areas of reduced response: one at low frequencies and one at high frequencies.

The low-frequency response is determined by emitter or source bypass capacitors and the coupling method between the signal source, amplifier stages, and load. If direct coupling is used throughout and no bypass capacitors are used, the amplifier should respond down to 0 Hz and is then called a *dc amplifier.* The other possible forms of coupling cause a reduction in low-frequency gain. The most often used

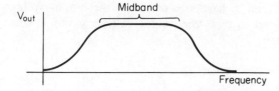

FIGURE 7-18 *Typical amplifier frequency-response curve.*

methods of coupling are capacitive coupling and transformer coupling. Transformer coupling limits the high-frequency response as well as low frequency response.

Since transformers are now seldom used for coupling, the major cause of high-frequency loss of gain is the inherent junction capacitances of the active amplification devices—bipolar junction transistors (BJTs) or FETs. These small capacitances tend to develop a low enough reactance at high frequencies to effectively short out a portion of the signal, thereby reducing the gain. Another source of high-frequency attenuation is the stray wiring capacities inherent in any circuit, and they have the same effect as the junction capacitances.

Amplifiers that have midbands in the approximate region 20 Hz to 20 kHz are termed *audio amplifiers*. They are used to amplify signals in the range of frequencies heard by the human ear, such as speech and music. A *video amplifier* must have a very wide frequency response—from very low frequencies (often down to direct current) on up to several hundred kilohertz or more. A *radio-frequency amplifier* (RF amplifier) is operated at only high frequencies. Typical frequencies range from 100 kHz up to 10,000 MHz. Those amplifiers that are operated in only a very small frequency range in the frequency spectrum are termed *tuned amplifiers*. They have special properties that reject all frequencies except a narrow desired range. This is usually accomplished through the use of *LC* tuned circuits or active-device solid-state filters.

RC Filters

Virtually all the high- and low-frequency effects in today's amplifiers are caused by capacitance. A quick review of the basics is therefore in order. It would also be wise for you to review the *RC* filter sections of Chapter 3 at this time. Consider the circuit shown in Fig. 7-19(a). At very high input frequencies the capacitor *C*

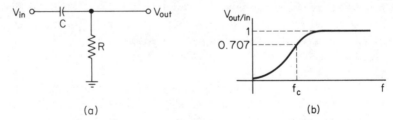

(a)   (b)

FIGURE 7-19   *High-pass filter and response.*

appears as a short circuit. Since $X_c = 1/2\pi fC$, increases in f cause a decrease in X_c. As the frequency goes down, however, X_c becomes larger, and eventually the voltage divider action of C and R begins to attenuate the output voltage [see Fig. 7-19(b)]. Of particular interest to us is the output voltage across R when $X_c = R$. By the voltage divider rule,

$$v_o = v_i \times \frac{R}{Z}$$

where

$$Z = \sqrt{R^2 + X_c^2}$$ (3-7)

If $R = X_c$, then

$$v_o = v_i \frac{R}{\sqrt{R^2 + R^2}} = v_i \frac{R}{\sqrt{2R^2}}$$

$$= 0.707 v_i$$

We see that when $R = X_c$, the output voltage is 0.707 of the input voltage, and by definition we shall call this the *low-frequency cutoff*, f_c. Figure 7-19(b) shows a graph of v_o/v_i versus frequency for a high-pass filter. Since by definition f_c is the frequency where $X_c = R$, we can say that

$$R = X_c = \frac{1}{2\pi f_c C}$$

or

$$f_c = \frac{1}{2\pi RC}$$ (3-9)

In many cases, the response of an RC circuit or total amplifier is expressed in terms of decibels (dB) instead of just an ordinary gain or attenuation ratio. In terms of the ratio of two powers, a *decibel* is defined as

$$dB = decibel = 10 \log_{10}\left(\frac{P_2}{P_1}\right)$$ (7-3)

where P_2 and P_1 are the two powers being compared. The logarithm in Eq. (7-3) is to the base 10, as indicated. Since power is related to the square of voltage and current, it follows that

$$dB = 20 \log_{10}\left(\frac{v_2}{v_1}\right)$$ (7-4)

since the log $v^2 = 2 \log v$ where v_2 and v_1 are measured across the same resistance. The following relationship is also useful:

$$\log\frac{1}{X} = -\log X$$ (7-5)

We could therefore calculate decibels down at the low-frequency cutoff f_c (where $v_o/v_i = 0.707$) as

$$\text{decibel} = 20 \log \frac{V_o}{V_i} = 20 \log .7071 = 20(-.1505)$$

$$\cancel{\text{decibel} = 20 \log \frac{1}{0.707}} = -20 \log 1.414$$

$$\approx 3 \text{ dB reduction or } -3 \text{ dB}$$

Thus, our low-frequency cutoff frequency is also the 3-dB "down" frequency as well as being the point where power to the load is one-half of its maximum value. Since $P = V^2/R$, then $(0.707 \text{ V})^2$ is half the original power, since $(0.707)^2 = 0.5$. Thus, at the low-frequency cutoff, f_c, the power is reduced to one-half of its original value, and the voltage is reduced to 0.707 of its original value.

EXAMPLE 7-2

Determine f_c for the circuit shown in Fig. 7-19(a) if $C = 1 \ \mu\text{F}$ and $R = 1 \text{ k}\Omega$. How many decibels down will the output be at one-tenth of f_c?

Solution:

$$f_c = \frac{1}{2\pi RC} = \frac{0.159}{10^3 \times 10^{-6}} = 0.159 \times 10^3$$

$$= 159 \text{ Hz}$$

At f_c or 159 Hz the value of X_c is equal to R, or 1 kΩ. Therefore, at $\frac{1}{10}f_c$, X_c will be $10R$ or 10 kΩ, since X_c is proportional to the inverse of f. Therefore, the output voltage at $f_c/10$ is

$$v_{\text{out}} = v_{\text{in}} \frac{R}{\sqrt{R^2 + X_c^2}}$$

The ratio $v_{\text{out}}/v_{\text{in}}$ is

$$\frac{v_{\text{out}}}{v_{\text{in}}} = \frac{1 \text{ k}\Omega}{\sqrt{1 \text{ k}\Omega^2 + 10 \text{ k}\Omega^2}} \cong \frac{1 \text{ k}\Omega}{10 \text{ k}\Omega} = 0.1$$

and thus, from Eq. (7-5),

$$\text{decibel} = 20 \log 0.1 = -20 \log \frac{1}{0.1}$$

$$= -20 \log 10 = \underline{-20 \text{ dB}}$$

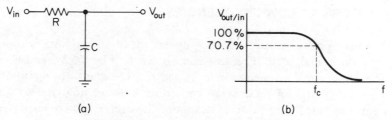

FIGURE 7-20   *Low-pass filter and response.*

The low-pass filter shown in Fig. 7-20(a) has the characteristics shown in part (b). It has a high-frequency cutoff, f_c, as shown at 0.707 of its maximum. Since the capacitor's reactance keeps decreasing as the frequency goes up, the output will drop by voltage divider action. The high cutoff frequency, f_c, also termed the 3-dB down frequency, half-power point, or break frequency, will occur when $X_c = R$ just as for the high-pass filter, so that f_c is still equal to $1/2\pi RC$.

EXAMPLE 7-3

Determine f_c for the circuit of Fig. 7-20 if $R = 1 \text{ k}\Omega$ and $C = 0.01 \ \mu\text{F}$.

Solution:

$$f_c = \frac{1}{2\pi RC} = \frac{0.159}{10^3 \times 10^{-8}}$$

$$= 0.159 \times 10^5$$

$$= \underline{15.9 \text{ kHz}}$$

If the circuits of Examples 7-2 and 7-3 were cascaded together, a band of frequencies would be passed, with the very low and high frequencies attenuated. The band of frequencies passed between the low- and high-frequency cutoffs is the *bandwidth* (BW) of the circuit, which is appropriately called a bandpass filter. The circuit and its response are shown in Fig. 7-21.

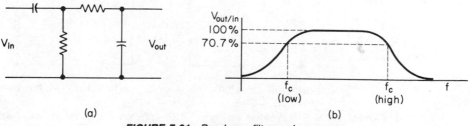

FIGURE 7-21   *Bandpass filter and response.*

7-5 AMPLIFIERS AT LOW FREQUENCIES

The major cause of low-frequency limitations in an amplifier is capacitance in series with the signal path. The amplifier shown in Fig. 7-22 has two such capacitors. The input signal is coupled into Q_1 via C_1. The transistor's output is coupled into R_L via C_2. Both of these series capacitors must be analyzed to ascertain the low-frequency cutoff they cause to determine the overall low-frequency response. Ideally, the capacitors serve to block dc levels and look like a short to ac signals. At low frequencies, however, they start developing appreciable impedance ($X_c = 1/2\pi fC$) and start to block even the ac signal. To calculate the low-frequency cutoff, f_c, caused by a series capacitance, the resistance "seen" by the capacitance must be determined. Then application of the familiar formula, $f_c = 1/2\pi RC$, will yield the 3-db down (cutoff) frequency, f_c.

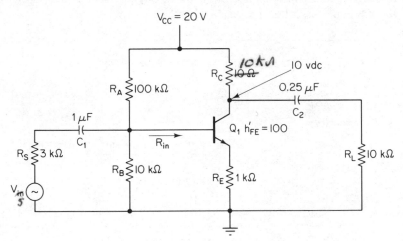

FIGURE 7-22 *Amplifier for low frequency analysis.*

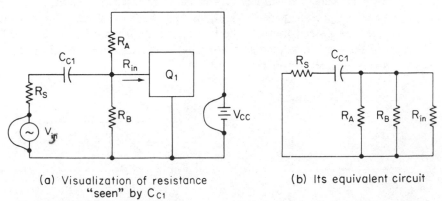

(a) Visualization of resistance "seen" by C_{C1}

(b) Its equivalent circuit

FIGURE 7-23 *Simplification of Figure 7-22 with respect to resistance "seen" by C_1.*

Let us do that for the amplifier of Fig. 7-22. First, we will calculate f_c due to C_1, the input coupling capacitor. The resistance "seen" by C_1 is not immediately apparent. To ease the process, refer to the simplifications in Fig. 7-23. The sources of voltage (dc and ac) have nearly zero resistance and thus are shown shorted out in Fig. 7-23(a). The resistance seen looking into the transistor's base, R_{in}, is added at Fig. 7-23(b) and the voltage sources are removed (shorted out). It can now be determined that the resistance R, seen by C_1, is

$$R = R_S + R_A \| R_B \| R_{in}$$

Since $R_{in} \simeq h'_{fe}R_E$ [Eq. (6-10)] or $100 \times 1 \text{ k}\Omega = 100 \text{ k}\Omega$,

$$\begin{aligned} R &= 3 \text{ k}\Omega + 100 \text{ k}\Omega \| 10 \text{ k}\Omega \| 100 \text{ k}\Omega \\ &= 3 \text{ k}\Omega + 8.33 \text{ k}\Omega \\ &= 11.33 \text{ k}\Omega \end{aligned}$$

Therefore, f_c due to C_1 is $1/2\pi R C_1$, or

$$f_c = \frac{1}{2\pi \times 11.33 \text{ k}\Omega \times 1 \text{ }\mu\text{F}} = 14 \text{ Hz}$$

To calculate f_c due to the output coupling capacitor, C_2, we must determine the resistance C_2 "sees." Looking out one side it "sees" the 10-kΩ load resistor to ground. Looking out its other side, it sees the collector resistor to ground (through the shorted battery) in parallel with the resistance seen looking into the transistor's collector. That resistance is very high, since the collector–base junction is reverse biased. The resistances seen by C_2 are summarized in Fig. 7-24. Since the resistance

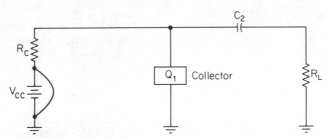

FIGURE 7-24 *Simplification of Figure 7-22 with respect to resistance "seen" by C_2.*

looking into the transistor's collector is so large, the value of it in parallel with R_c will be about equal to just R_c. Thus, the total resistance "seen" by $C_2 \simeq R_L + R_c$, or 20 kΩ. There, f_c due to C_2 is

$$\frac{1}{2\pi \times 20 \text{ k}\Omega \times 0.25 \text{ }\mu\text{F}} = 32 \text{ Hz}$$

Now since the cutoff frequency due to C_2 occurs at 32 Hz while the effects of C_1 do not cause cutoff until you are down to 14 Hz, C_2 determines the amplifier's cutoff frequency (32 Hz).

EXAMPLE 7-4

Determine f_c for the amplifier shown in Fig. 7-25.

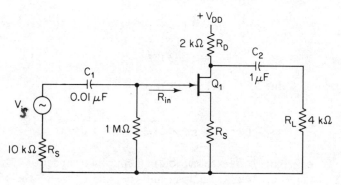

FIGURE 7-25   Circuit for Example 7-4.

Solution:

The input coupling capacitor C_1 sees 10 kΩ + (1 MΩ in parallel with R_{in} of the FET). Since R_{in} is very large, the resistance seen by the capacitor is just $R = (10 \text{ k}\Omega + 1 \text{ M}\Omega) \simeq 1 \text{ M}\Omega$, and then f_c for C_1 is

$$f_c = \frac{1}{2\pi CR}$$

$$= \frac{1}{2\pi(0.01 \times 10^{-6})(10 \text{ k}\Omega + 1 \text{ M}\Omega)}$$

$$\simeq \frac{10^8}{2\pi \times 10^6}$$

$$= \frac{100}{2\pi} \simeq \underline{16 \text{ Hz}} \qquad \text{due to } C_1$$

The output coupling capacitor C_2 "sees" 2 kΩ + 4 kΩ in series. The resistance seen looking into the FET's drain (r_{ds}) is very high and can be neglected with respect to the 2-kΩ drain resistor it is in parallel with. Thus, f_c is

---

$$f_c = \frac{1}{2\pi \times 10^{-6} \times 6 \times 10^3}$$

$$= \frac{0.159 \times 10^6}{6 \times 10^3}$$

$$= \underline{26.5 \text{ Hz}} \qquad \text{due to } C_2$$

Since this is almost double the 16-Hz cutoff, it is safe to say that the overall low frequency cutoff is around 26 Hz.

7-6   AMPLIFIERS AT HIGH FREQUENCIES

In Section 7-5 we saw how series capacitance limits the low-frequency response of an amplifier. Generally, these series capacitors are purposely put into a circuit to provide a block to dc and a short to ac. The high-frequency response of an amplifier is *not* normally caused by capacitance purposely put in a circuit. It is the result of small capacitances between the *pn* junctions of transistors and due to wiring capacitance in the circuit. Two wires that are close together form a small capacitance, since the wires act as plates and the insulation and/or air between them form the dielectric. Even though this capacitance is very small at very high frequencies, it can start shorting out the desired signal. Since X_c is inversely proportional to frequency, when f gets high enough, X_c can get small enough to start "shorting" a portion of the signal.

The CE amplifier stage in Fig. 7-26 shows the capacitances that affect high-frequency response. They are shown with dashed connecting lines since they exist

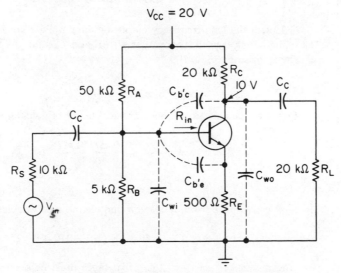

FIGURE 7-26   *High-frequency amplifier analysis.*

without physically being wired into the circuit. The capacitances C_{wi} and C_{wo} represent the wiring effects on the input and output, respectively, caused by the circuits physical layout. The junction capacitance of the transistor from base to collector is termed $C_{b'c}$ and is usually critical to an amplifier's high-frequency performance. The base–emitter junction capacitance, $C_{b'e}$, is usually less an effect on the high-frequency response. In cases where good high frequency response is necessary, a transistor with low values of junction capacitance is selected. These high-frequency transistors are specially fabricated to minimize junction capacitance and are more costly than standard transistors.

An interesting effect occurs with respect to junction capacitance that exists from input to output of a phase-inverting amplifier. The base–collector capacitance $C_{b'c}$ in Fig. 7-26 is the input to output capacitance for that amplifier. The value of $C_{b'c}$ is effectively multiplied by 1 plus the voltage gain of the amplifier and the capacitance actually appears from the amplifier's input lead to ground. This is termed the *Miller effect* (I take no credit) and is a dominant factor in amplifier high-frequency performance.

The Miller effect means that $C_{b'c}$ in Fig. 7-26 actually "appears" from base-to-ground (i.e., in parallel with C_{wi}) and has a value of $(1 + G_v)C_{b'c}$. The high-frequency cutoff, f_c, of most transistors can be approximated by combining $(1 + G_v)C_{b'c}$ in parallel with C_{wi} and calculating f_c as $1/2\pi RC$. The effects of $C_{b'e}$ and C_{wo} can usually be ignored. The transistor junction capacitances are usually supplied in transistor data sheets from the manufacturer.

Refer to supplemental notes

EXAMPLE 7-5

Calculate the high-frequency cutoff for the amplifier shown in Fig. 7-26. The input wiring capacitance, C_{wi}, is 40 pF and $C_{b'c}$ for the transistor is 10 pF. The transistor's h_{fe}' is 100.

Solution:

First the voltage gain of this amplifier should be calculated.

$$G_v \simeq \frac{R_c \| R_L}{R_E} = \frac{20 \text{ k}\Omega \| 20 \text{ k}\Omega}{500 \ \Omega}$$

$$= 20$$

The total capacitance from base to ground can now be calculated as

$$C_{wi} + (1 + G_v)C_{b'c} = 40 \text{ pF} + (1 + 20)10 \text{ pF} = 250 \text{ pF}$$

Now the resistance "seen" by this capacitance must be determined. Looking to the right, R_{in} ($\simeq h_{fe}R_E$) is seen, while to the left $R_A \| R_B \| R_S$ is seen. The total resistance R is then

$$R = R_{in} \| R_A \| R_B \| R_S$$
$$= h_{fe} R_E \| R_A \| R_B \| R_S$$
$$= 50 \text{ k}\Omega \| 50 \text{ k}\Omega \| 5 \text{ k}\Omega \| 10 \text{ k}\Omega$$
$$= 2.94 \text{ k}\Omega$$

Thus,

$$f_c = \frac{1}{2\pi RC}$$

$$= \frac{1}{2\pi \times 2.94 \text{ k}\Omega \times 250 \text{ pF}}$$

$$= \frac{1}{2\pi \times 2.94 \times 10^3 \times 250 \times 10^{-12}}$$

$$= \underline{217 \text{ kHz}}$$

It is interesting to note that amplifier high-frequency response is dependent upon gain. This is due to the Miller effect multiplication of input/output junction capacitance. The higher the gain, the greater the capacitance, and thus poorer high-frequency performance results. This leads to a term *gain-bandwidth product,* which tends to be a constant for a given transistor. A transistor with a voltage gain of 10 and a high-frequency cutoff of 100 kHz has a gain-bandwidth product of 10×100 kHz, or 1 MHz. This concept indicates that if its gain were increased to 20, its f_c would be 50 kHz, so that their product still equals 1 MHz.

7-7 MULTISTAGE AMPLIFIERS

Multistage amplifiers are nothing more than a combination of two or more active amplifying devices connected to give more gain than one of them could. In a subtle way, you have already been introduced to some simple multistage amplifiers in the previous work. This was done to facilitate explanation of effects and applications. It should also serve to ease the transition into this study. The major goal here is to provide enough experience with some common multistage amplifiers to enable you to grapple with whatever configurations you may run into.

Darlington Compound

A *Darlington compound* is a connection of two transistors that act as a single transistor with an effective h'_{fe} equal to the product of the h'_{fe} of each individual transistor. The most common form is shown in Fig. 7-27. The input current to the base of Q_1 is amplified by a factor of h'_{fe_1} in the emitter of Q_1. This emitter current of the first stage is the base current of the second stage. It gets amplified by a factor of h'_{fe_2} in the load resistor. Hence, the total current gain $G_i = h'_{fe_1} h'_{fe_2}$ and if identical transistors are used will simply equal $(h'_{fe})^2$.

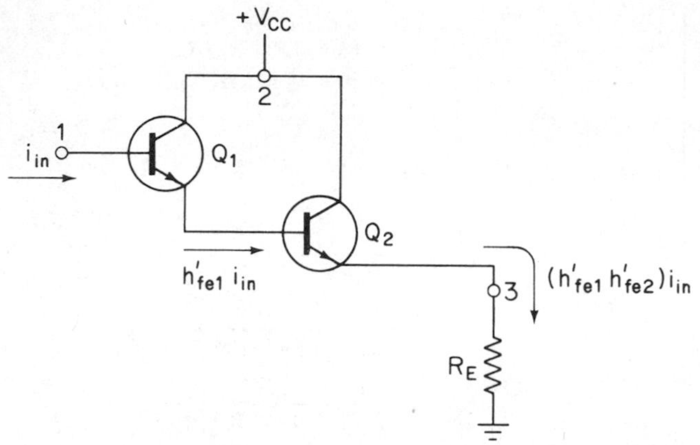

FIGURE 7-27 *Darlington compound.*

The Darlington pair acts as a single transistor with points 1, 2, and 3 of Fig. 7-27 equal to the base, collector, and emitter, respectively. They are commercially available in a single package with only those three leads brought out. This multistage package makes it possible to offer a much higher effective h_{fe} than is normally available with a single transistor. The input impedance is also of interest to us. Looking into the base of Q_2 we "see" $h_{fe_2} \times R_E$ ohms. But this input impedance is the load impedance of stage 1, so looking into the base of Q_1 we "see" $h_{fe_1}(h_{fe_2}R_E)$, or simply $h'^2_{fe}R_E$ if identical transistors are used. The Darlington compound is certainly capable of high input impedances. The impedance transformation capabilities of this circuit are enormous, as illustrated by the following example.

EXAMPLE 7-6

A CE amplifier stage is to drive a 100-Ω load to a 10-v p-p level. A 1-v p-p input signal is available. First, determine the resultant output if the 100-Ω load is capacitively coupled directly to the CE stage, as shown in Fig. 7-28(a), and then redetermine the output when a Darlington pair is used as a "buffer" between Q_1 and the load, as in Fig. 7-28(b). Assume that h'_{fe} is 50 for Q_2 and Q_3.

Solution:

The voltage gain of the CE amplifier in Fig. 7-28(a) is given by the approximation $(R_C \| R_L)/R_E$. Thus,

$$G_v \simeq \frac{100\ \Omega \| 5\ k\Omega}{470\ \Omega} \simeq \frac{1}{5}$$

It is seen that attenuation has occurred due to overloading the stage, and, by itself, it does not have enough gain to drive that heavy a load. The circuit of

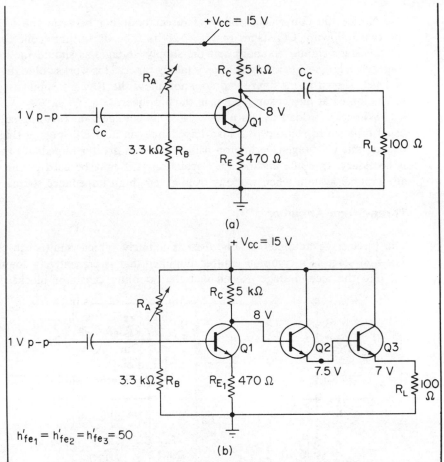

$h'_{fe_1} = h'_{fe_2} = h'_{fe_3} = 50$

(b)

FIGURE 7-28 (a) CE amplifier; (b) CE and Darlington compound.

Fig. 7-28(b) will provide a load to stage 1 equal to 5 kΩ with the resistance seen looking into the Darlington $(h'_{fe} \times R_L)$.

$$R_{L_1} = R_{C_1} \| (h'^{2}_{fe} \times R_L)$$
$$= 5 \text{ k}\Omega \| (50^2 \times 100 \ \Omega)$$
$$= 5 \text{ k}\Omega \| 250 \text{ k}\Omega$$
$$\simeq 5 \text{ k}\Omega$$

Hence, G_{v_1} is

$$G_{v_1} \simeq \frac{5 \text{ k}\Omega}{470 \ \Omega} \simeq 10$$

and the voltage gain of the Darlington is about 1, since each CC stage has a G_v of about 1. Therefore, the overall voltage gain is 10 and the design goal of a 10 v p-p output has been met.

Notice the convenient ability of direct coupling between the CE output and the two following CC stages in Fig. 7-28(b). The dc output (collector voltage) of the CE stage should be about half the supply voltage, as should the input of a CC stage when large-signal amplification is taking place. This works out to the approximate dc levels shown in the figure and does not allow the 10-v p-p signal at Q_1's collector to be clipped at any ensuing point in the amplifier.

Whenever a load causes a severe loss in voltage gain (loading effect), it is the usual practice to step up that load impedance via an FET stage, a single CC stage, or a double CC stage (Darlington pair) if an even greater impedance transformation is necessary. In some instances a Darlington pair may be used as the first stage of an amplifier system when working from a very high impedance source.

Three-Stage Amplifier

The three-stage amplifier in Fig. 7-29 is a fairly typical multistage configuration. The first stage is a common emitter amplifier that is capacitively coupled (through C_2) into the second stage. Recall that the coupling capacitor blocks dc levels and

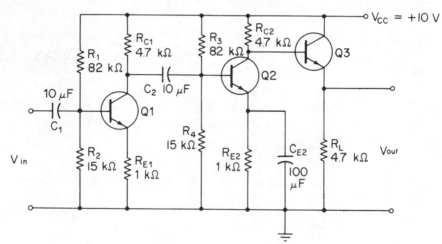

FIGURE 7-29   *Three-stage amplifier.*

acts as a short to the ac signal. Stage 2 (Q_2) is also common emitter, but notice the 100-μF capacitor across its emitter resistor. This *emitter bypass* capacitor acts as an open circuit to dc, and thus has no effect on the dc (bias) levels. However, it looks like a short to the ac signal and thus greatly increases the voltage gain of stage 2. The CE voltage gain changes from R_C/R_E to approximately R_C/h_{ib} when the emitter bypass is used. The value of h_{ib} is dependent on dc current levels in the transistor and is approximated as

$$h_{ib} \simeq \frac{0.026 \text{ V}}{I_E} \qquad\qquad (7\text{-}6)$$

Additionally, the impedance seen looking into the base of a bypassed CE stage is

$$R_{\text{in}} \simeq h'_{fe} h_{ib} \qquad (7\text{-}7)$$

The output of Q_2 in Fig. 7-29 is directly fed to an emitter follower (Q_3) that acts as a buffer between Q_2 and the output. The biasing of all three transistors is such that about 1 mA of dc collector and emitter current exists. Assuming an h'_{fe} of 250 allows calculation of this amplifier's voltage gain. The gain of the first stage, G_{v_1}, is simply $R_{L_1} \| R_E$. The only catch here is to realize that R_L for Q_1 is the 4.7-kΩ collector resistor (R_{C_1}) in parallel with the input impedance of the second stage. Since the emitter of Q_2 is bypassed,

$$Z_{\text{in}_2} = 15\ \text{k}\Omega \| 82\ \text{k}\Omega \| (h'_{fe_2} \times h_{ib_2})$$

since

$$h_{ib} \simeq \frac{0.026\ \text{V}}{I_E} \qquad (7\text{-}6)$$

$$= \frac{0.026\ \text{V}}{1\ \text{mA}} = 26\ \Omega$$

and

$$R_{\text{in}} \simeq h'_{fe} \times h_{ib} \qquad (7\text{-}7)$$

$$\begin{aligned} Z_{\text{in}_2} &= 15\ \text{k}\Omega \| 82\ \text{k}\Omega \| (250 \times 26\ \Omega) \\ &= 15\ \text{k}\Omega \| 82\ \text{k}\Omega \| 6.25\ \text{k}\Omega \\ &= 4.18\ \text{k}\Omega \end{aligned}$$

Therefore, $R_{L_1} = 4.7\ \text{k}\Omega \| 4.18\ \text{k}\Omega = 2.22\ \text{k}\Omega$, and we can calculate G_{v_1} as

$$G_{v_1} \simeq \frac{R_L}{R_E} = \frac{2.22\ \text{k}\Omega}{1\ \text{k}\Omega} = 2.22$$

Since the second stage is bypassed, G_{v_2} will equal R_{L_2}/h_{ib_2}. The load R_{L_2} is 4.7 k$\Omega \| Z_{\text{in}_3}$. The input impedance of the third stage is $h'_{fe_3} \times 4.7$ kΩ, or 250×4.7 kΩ, or 1.175 MΩ. Therefore, R_{L_2} is 4.7 k$\Omega \| 1.175$ M$\Omega \simeq 4.7$ kΩ and h_{ib_2} is 0.026 V/1 mA = 26 Ω. Thus,

$$G_{v_2} = \frac{4.7\ \text{k}\Omega}{26\ \Omega} = 181$$

Notice how the emitter bypass capacitor is able to provide extremely high voltage gains. The voltage gain of such a stage is R_L/h_{ib} instead of R_L/R_E since R_E is shorted

TABLE 7-1 *Amplifier analysis comparison.*

	Approximate Analysis	Computer Analysis	Measured
$G_{v_{oa}}$	360	353	340
Z_{in} (kΩ)	12.1	12.0	13

out by the bypass capacitor. Additionally, the impedance seen looking into the base of a bypassed stage is $h'_{fe}h_{ib}$ instead of $h'_{fe}R_E$ for the same reason. Therefore, the amplifier's *overall voltage gain*, $G_{v_{oa}}$, can be calculated, assuming a G_v of 0.9 for the CC stage, as

$$G_{v_{oa}} = G_{v_1}G_{v_2}G_{v_3}$$
$$= 2.22 \times 181 \times 0.9$$
$$= 360$$

The input impedance of the three-stage amplifier in Fig. 7-29 is $R_1 \| R_2 \| R_{in_1}$, or 82 k$\Omega \| 15$ k$\Omega \| h'_{fe_1} R_{e_1}$, or 82 k$\Omega \| 15$ k$\Omega \| (250 \times 1$ k$\Omega)$, or 12.1 kΩ. This amplifier was constructed using 2N3565 transistors (h_{fe} is specified as 120 to 750) and resistors and capacitors with $\pm 10\%$ tolerances. The voltage gain and input impedance were measured from this test circuit with the results shown in Table 7-1. Additionally, the table shows the result of a very detailed computer analysis of this circuit. Comparing these results with our approximate analysis shows rather close agreement.

QUESTIONS AND PROBLEMS

7–1–1. Describe the function of each block in the diagram for a regulated power supply shown in Fig. 7-1. Include the description of a feedback control system in your discussion.

7–1–2. Calculate the appropriate value for R_S in Fig. 7-2 given the following information:

$$\text{unregulated input} = 14 \text{ to } 22 \text{ V}$$

$$V_z = 9.5 \text{ V}$$

$$I_{ZK} = 0.1 \text{ mA}$$

$$h'_{fe} \text{ for } Q_1 = 80$$

7–1–3. Determine the worst-case power dissipation for R_S, Q_1, and D_1 in Problem 7–1–2. What is the output voltage of this power supply?

7–1–4. Explain how the circuit shown in Fig. 7-3 is able to provide a regulated dc output voltage.

7-1-5. How does a voltage regulator also serve as a filter in terms of reducing ripple?

7-2-6. Define and compare small-signal, large-signal, and power amplifiers.

7-2-7. Explain why a power amplifier is likely to introduce more distortion than is a small-signal amplifier.

7-2-8. Discuss some general considerations of power transistors with respect to heat dissipation. Include heat sinks, mica washes, and derating curves in your discussion.

7-2-9. Define the meaning of a class A amplifier. What is meant by amplifier efficiency, and what is typical for class A designs?

7-2-10. Describe the operation of the two-stage class A power amp shown in Fig. 7-7. If the emitter resistor of Q_2 is 12 Ω and has 5 V dc and 8 v p-p ac across it, calculate the amplifier's efficiency. The load resistance (R_L) is an 8-Ω speaker.

7-3-11. Describe class B operation and explain its major advantage over class A.

7-3-12. Draw a schematic of a class B, push-pull amplifier and explain its operation.

7-3-13. Provide a schematic of a paraphase amplifier. Explain its operation and application with respect to push-pull amplifiers.

7-3-14. Discuss all aspects of crossover distortion, including its cause and how class AB operation is able to remedy the effect.

7-3-15. Discuss several advantages of FET power amplifiers as compared to the more common BJT designs. Explain the reason for the greater use of BJTs in power amps.

7-4-16. In general, discuss the operation of an amplifier with respect to frequency. Include a definition of midband in your discussion.

7-4-17. Provide descriptions for audio, video, radio frequency, and tuned amplifiers.

7-4-18. Design a high-pass RC filter where $f_c = 30$ Hz. Provide its schematic and frequency-response curve.

7-4-19. Repeat Problem 7-4-18 for a low-pass design with $f_c = 18$ kHz.

7-4-20. The output power of a filter is 0.07 times the input. Calculate the attenuation in decibels. An amplifier has an output voltage 83 times the input. Calculate the gain in decibels.

7-4-21. Calculate the decibel attenuation for the filter from Problem 7-4-18 at 18 Hz.

7-4-22. Calculate the low-frequency cutoff for the circuit in Fig. 7-19 if $R = 3.3$ kΩ and $C = 10$ μF.

7-5-23. Calculate the cutoff frequency caused by C_{C_1} and C_{C_2} in Fig. 7-26. They are both 5 μF and the transistor's h_{fe} is 100. Which capacitor determines the overall low-frequency response, and why is this the case?

7-5-24. Calculate the low-frequency cutoff for the amplifier shown in Fig. 7-28(a). Assume that $R_A = 30$ kΩ, h_{fe} of 90 for Q_1, $C_{C_1} = 1$ μF, and $C_{C_2} = 10$ μF. What would be the most economical means of improving the amplifier's low-frequency response?

7–6–25. In general terms, discuss the factors involved in an amplifier's high-frequency response. Include the Miller effect, junction and wiring capacitance, and gain-bandwidth product in your discussion.

7–6–26. Calculate the high-frequency cutoff for the amplifier shown in Fig. 7-25. Assume a voltage gain of 5 and that the gate–drain capacitance of 8 pF is the only important cause of high-frequency cutoff.

7–6–27. Calculate the high-frequency response of the amplifier analyzed in Example 7-5. The value of $C_{b'c}$ is changed to 4 pF and R_S to 2.4 kΩ.

7–6–28. What do you suppose would be the dominant high-frequency cutoff factor for the amplifier in Fig. 7-29? Neglecting wiring capacitance, calculate f_c. Your answer should be close to the 81-kHz value that was measured for this circuit when tested in the laboratory.

7–7–29. Show a schematic for a Darlington compound and discuss its important characteristics and applications.

7–7–30. Explain the general characteristics of the multistage amplifier shown in Fig. 7-29.

7–7–31. What is the function of an emitter bypass capacitor? What effect does it have on the dc and ac analysis of an amplifier?

8

LINEAR
INTEGRATED CIRCUITS
AND
OPERATIONAL AMPLIFIERS

8-1 INTEGRATED-CIRCUIT BASICS

Integrated circuits (ICs) have had a profound effect on all phases of electronics. Integrated circuits became commercially available in the early 1960s and quickly snowballed into a giant industry in a few short years. A *monolithic integrated circuit* is a device in which a complete circuit (including all of its components) is formed upon or within a single piece of silicon crystalline material. A *thick-film hybrid integrated circuit* has the resistors, capacitors, and wire paths "screened" onto a ceramic substrate in paste form through a mesh mask. A high-temperature baking or curing cycle follows this process, and then the other components are externally added and interconnected by wire bonds. A *thin-film hybrid integrated circuit* has films deposited either by sputtering, evaporation, or chemical vapor deposition through a mask. Conductors, resistors, and capacitors are so deposited, and all other circuit elements must be added to the thin-film circuit as with the thick-film circuits. The essential difference between thick- and thin-film circuits is not their relative thickness but the *method* of depositing the films.

Integrated circuits offer a tremendous reduction in size over the standard printed circuits, which use standard discrete components. Since ICs lend themselves to high-volume, mass-production techniques, they also offer significant cost advantages over discrete circuits. This is especially true of monolithic ICs, with the hybrid ICs generally falling somewhere between the monolithic and discrete circuits in both size and cost considerations. In fact, hybrid circuits are, as their name implies, a blend of discrete and integrated techniques.

Thus far we have talked of integrated circuits only from the method of construction standpoint. They are also classified according to type of circuit—digital or linear. This chapter is concerned with linear circuits; however, it is important to realize that digital ICs represent perhaps 80% of the IC dollar market, with the great majority of these circuits being utilized in the computer industry. Digital ICs lend themselves to monolithic integration, because a computer uses large numbers of identical circuits, and monolithic integrated circuits become increasingly economical as the volume demand for one specific circuit increases. We shall study the digital ICs and their applications in Chapters 10, 11, and 12. Linear applications are smaller in volume and tend to require differences in performance from one system to another. Despite this difficulty, linear integrated circuits (LICs) are quickly displacing their discrete-circuit counterparts in many applications as their cost becomes competitive. They also demonstrate greater reliability, because so many external connections, a major source of circuit failure, are eliminated in ICs. Linear integrated circuits find wide application in military and industrial applications as well as in consumer products. They are used in many ways. The following list includes the bulk of these functions:

1. Operational amplifiers.

2. Small-signal amplifiers.

3. Power amplifiers.

4. Sense amplifiers.

5. RF and IF amplifiers.

6. Microwave amplifiers.

7. Multipliers.

8. Comparators.

9. Voltage regulators.

Operational amplifiers are by far the most versatile form for LICs since they can be manipulated to perform all the functions listed above except for the microwave amplifiers. Thus, they make up the bulk of LICs, and we shall study them in detail later in this chapter.

By now the variety of electronic circuits discussed has probably left you in a state of confusion. Figure 8-1 charts the possibilities for circuit construction. Notice the new terms MSI, LSI, and VLSI following the monolithic circuits. They stand for medium-scale integration (MSI), large-scale integration (LSI), and very large scale

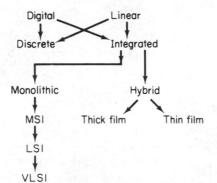

FIGURE 8-1 *Types of electronic circuits.*

integration (VLSI), respectively. They are really extensions of the monolithic technology whereby whole electronic systems rather than just a circuit are incorporated in one package. They are most suited to digital circuits, and the dividing line between MSI and LSI is somewhat arbitary. However, a good approximate rule of thumb is that MSI becomes LSI when over 100 separate circuits are incorporated and connected together in one package. The arbitrary dividing line between LSI and VLSI is 100,000 devices! VLSI is a major electronic system constructed on a single wafer of silicon.

8-2 MONOLITHIC IC FABRICATION

The remainder of this chapter will be concerned with monolithic integrated circuits. Even though industry makes widespread use of hybrid circuits for complete systems, it is the commercial availability of low-cost monolithic ICs that is currently having the most widespread effect in today's circuits. A photograph of the typical configurations of these devices is shown in Fig. 8-2. It is shown at approximately two times actual size. The manufacturer offers these devices in a number of circuit configurations that make their use on standard printed-circuit boards a common practice. Thus, these standard circuit configurations (in LIC form) in conjunction with any additionally required discrete components provide the circuit designer the high degree of flexibility required for most linear circuits. It also relieves the designer from the tedious design requirements of large numbers of single-stage amplifiers and allows, instead, designs of whole systems by piecing together the appropriate LIC building blocks. This is a much more efficient design approach than the 100% discrete circuit, and since it will be the most often encountered form of linear circuitry for years to come, it warrants this chapter of study.

    In order to skillfully deal with this monolithic-discrete component circuitry, it is helpful to first have some knowledge of how these LICs are fabricated and to learn of their special characteristics. The technology presently used for monolithic integrated circuits is based on the silicon planar techniques developed for discrete

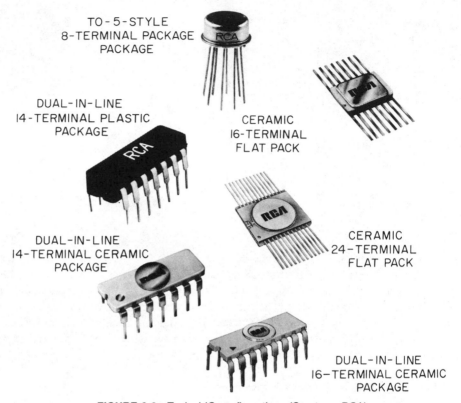

TO-5-STYLE
8-TERMINAL PACKAGE
PACKAGE

CERAMIC
16-TERMINAL
FLAT PACK

DUAL-IN-LINE
14-TERMINAL PLASTIC
PACKAGE

DUAL-IN-LINE
14-TERMINAL CERAMIC
PACKAGE

CERAMIC
24-TERMINAL
FLAT PACK

DUAL-IN-LINE
16-TERMINAL CERAMIC
PACKAGE

FIGURE 8-2 *Typical IC configurations (Courtesy, RCA).*

transistors. The starting point for this process is a uniform single crystal of silicon—usually p-type material, as shown in Fig. 8-3(a). In this explanation, a simple transistor, resistor, and capacitor circuit will be formed—those being the three possible circuit elements, but with a diode also possible by simply using either the base–emitter or base–collector junction of a transistor.

The silicon crystal has impurities introduced by a diffusion process. *Diffusion* is the introduction of controlled small quantities of material into the crystal structure. It modifies the electrical characteristics of the crystal in a tightly controlled high-temperature environment. Figure 8-3(b) shows the result of the diffusion of two n-type regions into the p-type substrate. The diffusion process introduces these regions to the desired depth and widths, with depth being controlled by the diffusion temperature and time, and the lateral dimensions controlled by silicon dioxide and photochemical masks. Diffusion of additional p-type and n-type regions then forms a transistor *(npn)* on the left; the element on the right omits the final n-type emitter diffusion, which then results in a p-type silicon resistor of controlled size (hence controlled resistance). The silicon wafer is then coated with an insulating oxide layer that is opened selectively to allow for interconnections, as shown in Fig. 8-3(c). The heavy,

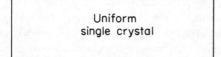

(a) Silicon wafer used as starting material for an integrated circuit.

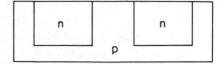

(b) Diffusion of n-type areas to provide isolated circuit nodes.

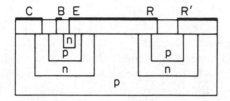

(c) Connection of contacts to p-type region to form integrated resistor.

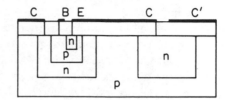

(d) Use of oxide as a dielectric to form integrated capacitor.

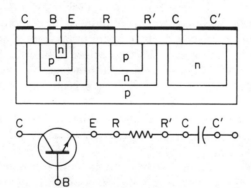

(e) Completed silicon chip containing transistor, resistor, and capacitor.

FIGURE 8-3   *Monolithic IC fabrication process.*

dark lines show the conductors that are formed by a metallization process. Metal (usually aluminum) is evaporated over the otherwise completed circuit and allowed to form a thin coating over the entire circuit. Then by a photosensitizing, masking, and etching process, the metal is selectively removed to leave the desired interconnect pattern.

Figure 8-3(d) shows a transistor–capacitor combination. A capacitor may be formed by using the oxide layer as the capacitor's dielectric with the metallization serving as the two plates. Figure 8-3(c) shows a transistor–resistor–capacitor combination and the schematic diagram for this circuit.

The major cost of a monolithic IC is the tightly controlled processing steps it is subjected to. Cost is thus directly proportional to size, and the smaller the circuit the lower will be the cost. This is true because the required processing ovens can only handle a limited silicon surface area at one time, and because all such ICs go through the same processes, regardless of the mix of transistors, capacitors, and resistors. Since resistors and capacitors occupy considerably more area than transistors or diodes (typically by a factor of 2 or $3:1$), their use is minimized. The ratio of active to nonactive devices is therefore as high as possible—in direct contrast with standard discrete circuits, in which the active devices are generally the most expensive components.

Another factor that changes the design rules for IC circuits as compared to discrete circuits is the inability to obtain close resistor value tolerances in monolithic circuits, since resistors are of silicon material and highly temperature-sensitive. These silicon resistors drift in a predictable fashion, which allows for very tight resistor ratio tolerances. This characteristic can often be used to compensate for the high resistance value drift with temperature. In addition, high-valued resistances (over about 10 kΩ) are expensive from the standpoint of requiring a lot of space. This is even more critical for even moderate-sized capacitors, with special capacitance multipliers used to generate capacitances of about 100 pF or greater.

Transistors in LICs have somewhat reduced high-frequency response, even though the same basic construction techniques are utilized for monolithic as well as discrete units. In addition, the ability to manufacture both *pnp* and *npn* units in the same IC is very difficult and not often done in practice. The *pnp* transistors are generally inferior in quality, even if they are the only polarity used, and we thus see a preponderance of *npn* units in monolithic circuits. However, since monolithic transistors are manufactured in very close proximity to each other and in the same processes, they generally offer very similar characteristics, and this fact is often taken advantage of.

Because of the aforementioned variations from discrete components, the monolithic circuit schematic usually takes on a strange appearance when compared to the schematic for a discrete circuit performing the same function. The inability to provide inductors in monolithic circuits is another factor that must be contended with. In summary, then, the following circuit design guidelines are applied to the design of monolithic circuits:

1. Maximize the number of active devices.

2. High-valued resistors and capacitors are impractical.

3. Use resistor ratios rather than absolute values.

4. Take advantage of the closely matched transistor characteristics.

5. Inductors are not available.

6. Designs incorporating only *npn* transistors are desirable.

8-3 SPECIAL DESIGN CONSIDERATIONS

In this section some of the circuit peculiarities made possible and/or necessary by monolithic circuits are discussed.

DC Level Shifting

Recall that isolation between the various stages of discrete amplifiers is most often done with a coupling capacitor. One side of the capacitor may be a transistor's collector at 6 V dc and the other side at the base of the next transistor at 2 V dc. The capacitor isolates between the dc levels while still allowing the ac signal through. These capacitors are too large to obtain in LICs, and therefore a method of dc level shifting is often employed. A level shift is possible with resistive dividers, but this of course causes undesirable signal attenuation. The circuit of Fig. 8-4 solves this problem by taking the input signal at some dc level and providing the output signal at 0 V dc. This system requires dual voltage supplies, which are usually used in circuits using LICs anyway. Transistor Q_2 in Fig. 8-4 acts as a dc constant current source since no signal is applied to its base. The value of its constant current is determined by the base bias resistors R4 and R5. Its constant current is used to provide a constant dc voltage drop across R_1. R_1's dc voltage drop added to the 0.5 V dc drops across Q_1 and Q_3's base–emitter junction constitutes the amount of dc level shift in the circuit. If the input signal to Q_1 were 100 mv p-p riding on a 10-V dc level, this circuit can be designed to remove the 10 V dc and provide nearly the full 100 mv ac at its output. If Q_2 delivered about 1 mA of constant dc current to R_1 and $R_1 = 900$ Ω, a 9-V drop occurs. Thus from input to output the following dc drops are encountered—0.5 V (V_{BE} of Q_1), 9 V (V_{R_1}), and 0.5 V (V_{BE} of Q_3). Thus, the 10-V drop causes the dc output to be 10 V (input) − 10 V (drop), or 0

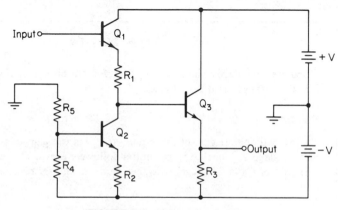

FIGURE 8-4 *Voltage-level shifter.*

231

V dc. The ac output is just slightly less than the 100-mv input after passing through two emitter-follower stages ($Q_1 + Q_2$) and R_1.

Multiplication of Capacitance

Whenever a larger capacitor is needed than is possible to fabricate in a LIC, a capacitance multiplier may be utilized. Values above about 0.01 μF are too bulky to include in ICs. Recall from our study of the Miller effect that the effective value of a capacitor C connected between the input and output of a phase-inverting amplifier is multiplied by 1 plus the amplifier's voltage gain. Hence, the circuit of Fig. 8-5 could be used to accomplish this goal. Notice that the voltage amplifier, Q_2, is buffered from the input by Q_1 to keep from having the resulting capacitance shunted by a low resistance. Thus, the resistance seen looking into Q_1's base is $h'_{fe_1} \times (h_{ib_1} + h'_{fe_2} h_{ib_2})$, and this should be high enough to provide a fairly high quality capacitance.

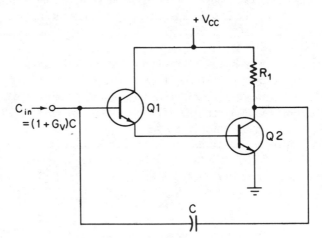

FIGURE 8-5 *Capacitance multiplier.*

EXAMPLE 8-1

Calculate the effective input capacitance for the circuit of Fig. 8-5, $R_1 = 10$ kΩ, $C = 100$ pF, and $I_E = 1$ mA.

Solution:

The overall voltage gain of this circuit $(G_{v_{oa}})$ is the gain of stage 1 times the gain of stage 2. Since Q_1 is an emitter follower, a gain of 1 can be assumed. The gain of Q_2 (a CE stage) is R_1/h_{ib_2} since no emitter resistor is used. Recall that

$$h_{ib} = \frac{0.026 \text{ V}}{I_E} = \frac{0.026 \text{ V}}{1 \text{ mA}} = 26 \text{ }\Omega$$

Thus

$$G_{v_{oa}} \simeq 1 \times \frac{10 \text{ k}\Omega}{26 \text{ }\Omega}$$

$$\simeq 400$$

Therefore, C_{in} is

$$C_{in} = (1 + G_{v_{oa}})C$$
$$\simeq 400 \times 100 \text{ pF}$$
$$= 40{,}000 \text{ pF} = \underline{0.04 \text{ }\mu\text{F}}$$

    Example 8-1 showed how a 100-pF capacitor can be made to look like one nearly 400 times as big. This technique is used in ICs requiring a high-valued capacitor.

Differential Amplifier

Common-emitter (CE) amplifiers offer high gain if no emitter resistance is used. For stability reasons, an emitter resistor is desirable, but is used at the expense of gain, however. The use of an ac short (a capacitor) across the emitter resistor (emitter bypass capacitor) restores that gain potential, but unfortunately a large capacitor is needed if good low-frequency response is required. A solution to this dilemma that is very practical with LICs is the *differential amplifier* shown in Fig. 8-6. Even though it requires three transistors, they are very cheap in LICs and the resulting high-

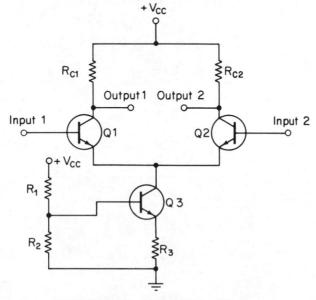

FIGURE 8-6 *Differential amplifier.*

gain amplifier works at low frequencies, even down to dc (0 Hz). The R_1–R_2 combination biases Q_3 such that it acts as a constant current source for the emitters of Q_1 and Q_2. Hence, the emitter-to-ground voltage of Q_1 and Q_2 will always remain constant, regardless of the ac signal applied to either input. Therefore, the entire ac input signal is dropped across the base–emitter junction to which it is applied, and no emitter resistor voltage drop occurs at any frequencies down to direct current. Thus, a highly temperature-stable, high-voltage gain amplifier results.

The output may be at either Q_1's or Q_2's collector. If a noninverting amplifier is needed, the output should be taken at the collector of the transistor opposite to where the input is applied. The voltage gain is one-half that of either of the transistors operating at the same dc current levels with no emitter resistance. This is because the total dc emitter current (supplied by Q_3) is split evenly between Q_1 and Q_2 if they have the desired matched characteristics. With one-half of the dc emitter current, the gain is cut in half, since the two are directly proportional. However, if the output is taken differentially across the two collectors, the gain is doubled and is thus back to the same level as with a single-transistor amplifier. If signals are applied to both inputs, the output between the two collectors is proportional to the difference between the input signals; hence the name differential amplifier. This circuit is the heart of most operational amplifiers, and we shall pursue it in more detail later. The importance of this circuit now, however, is that a high-voltage gain is now possible in a highly stable amplifier without requiring the emitter bypass capacitor.

EXAMPLE 8-2

Determine the single-ended and differential output for the circuit of Fig. 8-7.

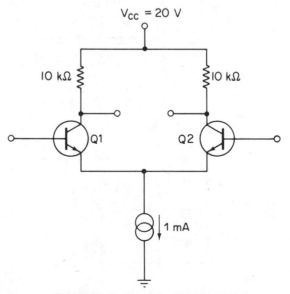

FIGURE 8-7 Circuit for Example 8-2.

Solution:

The current source supplies 1 mA of dc current, which is split evenly between Q_1 and Q_2. Thus, the single-ended gain will be R_c/h_{ib}, where $h_{ib} = 0.026/0.5$ mA $= 52\ \Omega$. Thus,

$$R_c/2h_{ib} = Gv_{se}$$

$$G_v \simeq \frac{10\ k\Omega}{52\ \Omega} \simeq \underline{200}$$

for the single-ended output, whereas taking the output differentially across the two collectors results in double that gain, or about ~~400~~. $200\ G_{v_{DIFF}} = \frac{R_c}{h_{ib}}$

Simulation of Inductance

As previously stated, inductors cannot be fabricated in ICs. However, by making use of the *gyrator* principle, inductance can be simulated. Figure 8-8 provides a simple

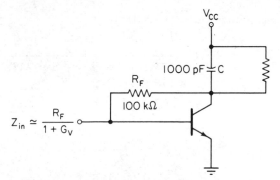

FIGURE 8-8   *Gyrator.*

circuit illustrating the gyrator effect. The biasing circuit is omitted. The amplifier's input impedance is approximately

$$Z_{IN} \simeq \frac{R_F}{1 + G_V} \simeq \frac{R_F}{G_V}$$

and its voltage gain is

$$G_V \simeq \frac{R_c}{h_{ib}} \simeq \frac{X_c}{h_{ib}}$$

so that

$$Z_{IN} \simeq \frac{R_F}{G_V} = R_F \times \frac{h_{ib}}{X_c} \qquad \textbf{(8-1)}$$

If we let the $R_F \times h_{ib}$ equal a constant, A, then

$$Z_{IN} = \frac{A}{X_c} = \frac{A}{X\underline{|-90°}} = AB\underline{|+90°}$$

where B is a constant equal to $1/X$. The capacitive load reactance therefore appears as an inductive reactance at the amplifier's input terminals. Recall from Chapter 3 that capacitive reactance has a $-90°$ phase angle, whereas an inductive reactance has a $+90°$ phase angle. A $-90°$ in the denominator of an expression becomes $+90°$ in the resultant.

EXAMPLE 8-3

Determine the inductance presented by the input terminals of the circuit shown in Fig. 8-8 at a frequency of 10 kHz. Assume that $h_{ib} = 26\ \Omega$.

Solution:

The capacitive reactance at 10 kHz is

$$X_c = \frac{1}{2\pi fC} = \frac{1}{2\pi \times 10\ \text{kHz} \times 1000\ \text{pF}}$$
$$= 15.9\ \Omega\underline{|-90°}$$

Thus,

$$Z_{IN} \approx \frac{R_F \times h_{ib}}{X_c}$$
$$= \frac{100\ \text{k}\Omega \times 26\ \Omega}{15.9\ \Omega\ \underline{|-90°}}$$
$$= 1.64 \times 10^5\ \Omega\underline{|+90°}$$

Thus, the input impedance is 164,000 Ω of inductive reactance. Solving for L at 10 kHz, we obtain

$$L = \frac{X_L}{2\pi f} = \frac{1.64 \times 10^5}{2\pi \times 10^4}$$
$$= 2.6\ \text{H}$$

8-4 THE OPERATIONAL AMPLIFIER

The recent availability of low-cost monolithic integrated circuits (ICs) has revolutionized the field of linear electronics. Whole circuits and subsystems are now available, predesigned, in extremely compact sizes, and with low cost and high reliability. As mentioned earlier, these circuits are replacing many of the discrete designs in today's electronics systems. The most common circuit configuration for LICs is the operational amplifier. An operational amplifier is so named because it originally was used to perform mathematical "operations" in analog computers. However, because of its versatility and the low cost made possible by monolithic manufacturing techniques, its use has spread into virtually every area of electronics. An operational amplifier (op amp) is a very high gain direct-coupled amplifier that uses external feedback to control its gain-impedance characteristics. The most common circuit form for operational amplifiers is two cascaded differential amplifier stages directly coupled into one or more emitter followers to provide a low output impedance. The differential amplifier is an ideal circuit for monolithic techniques, since it offers high gains without requiring capacitors or large-sized resistors, requires the closely matched transistor characteristics offered by LICs, and offers high gains determined by resistor ratios instead of absolute values. This circuit can be adapted to virtually any amplification situation, thus allowing high-volume operational amplifier production for the diverse linear circuit applications. This high-volume production, you may recall, is a requirement for making monolithic ICs economically attractive.

Ideal Operational Amplifier

In its simplest form, an operational amplifier is represented by a triangle with two inputs and one output, as shown in Fig. 8-9. It does not show any of the necessary connections such as for dc power, feedback connections, or phase compensation.

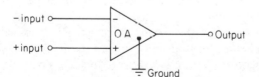

−input ○

+input ○

O A

○ Output

⏚ Ground

FIGURE 8-9 *Operational amplifier symbol.*

When operated without a resistor from output to one of the two inputs (feedback resistor), the op amp is said to be in the *open-loop* condition. The "(−) input" is termed the *inverting* input since signals applied between it and ground appear inverted at the output (with respect to ground). For similar reasons the "(+) input" is referred to as the *noninverting* input. The input may also be applied differentially (i.e., between the two inputs instead of to just one of them with respect to ground).

The *ideal* operational amplifier has the following characteristics:

1. Infinite gain (open loop).

2. Infinite input impedance (open loop).

3. Infinite bandwidth.

4. Zero output impedance.

These characteristics are obviously not possible, but in reality the first two are so closely attained that analysis can be made by using them as approximations. This greatly simplifies development of operational amplifier gain-impedance relationships without adversely affecting the results obtained. Therefore, we shall assume practical operational amplifiers to have infinite gain and infinite input impedance.

Let us first analyze the inverting operational amplifier shown in Fig. 8-10. Notice that the positive input is grounded and that resistor R_2 provides feedback from output

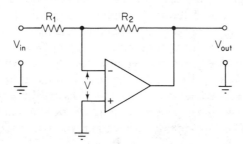

FIGURE 8-10 *Inverting operational amplifier.*

to input. If we let v be the voltage from the inverting input to ground, we can write the following relationship:

$$\frac{v_{in} - v}{R_1} = \frac{v - v_{out}}{R_2}$$

This equation is valid since the operational amplifier's impedance is ideally infinite looking into the inverting input, and thus no current flows into the amplifier. Hence, the current through R_1 must equal the current through R_2. Now since the open-loop gain of the operational amplifier is ideally infinite (and is extremely high, practically), it follows that the voltage v is zero; therefore, the equation reduces to

$$\frac{v_{out}}{v_{in}} = \frac{-R_2}{R_1} \tag{8-2}$$

This is the *closed-loop gain* of the amplifier (with feedback), and we shall hereafter refer to it as G_F. Thus,

$$G_F = \frac{-R_2}{R_1} \tag{8-3}$$

The negative sign simply indicates that the output is 180° out of phase (inverted) with respect to the input. We shall call the amplifier's *open-loop gain* G_O. We have assumed G_O to be infinite in this derivation and will throughout the chapter. The input impedance for the circuit of Fig. 8-10 will equal the input voltage divided by the input current. Thus,

$$\cancel{R_{IN}} = v_{in} \left/ \frac{v_{in} - v}{R_1} \right.$$

Since v is nearly zero, this will reduce to

$$R_{IN} = R_1 \tag{8-4}$$

   The noninverting version of the operational amplifier is shown in Fig. 8-11. Note that the (−) terminal is not grounded. Since the ideal gain of the operational amplifier itself is infinite, it seems logical that the voltage between the inverting and

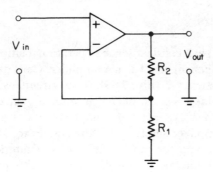

FIGURE 8-11 *Noninverting amplifier.*

noninverting terminals is negligible, and hence the voltage across R_1 should equal v_{in}. Also, since no current will ideally be drawn by the inverting input, the currents through R2 and R1 are equal. Thus,

$$\frac{v_{out} - v_{in}}{R_2} = \frac{v_{in}}{R_1}$$

and this can be manipulated to give a closed-loop gain of

$$G_F = \frac{v_{out}}{v_{in}} = 1 + \frac{R_2}{R_1} \tag{8-5}$$

Thus, the gain is positive (no phase inversion) and once again is determined by two resistors that are externally added to the operational amplifier package. The input impedance of the ideal noninverting circuit is infinite. In practice, however, some finite high value of input resistance is present. The operational amplifier manufacturer provides this as the open-loop input resistance, and it is usually given the symbol Z_{IN}. In the feedback mode of operation this impedance is multiplied by the ratio of G_O to G_F. Thus, the closed-loop circuit input impedance for the noninverting input is

$$R_{IN} = Z_{IN} \frac{G_O}{G_F} = \frac{Z_{IN} G_O}{1 + (R_2/R_1)} \tag{8-6}$$

8-5 OP-AMP APPLICATIONS

Even though we have considered the operational amplifier mainly from an ideal stand-point up to this point, it is useful to consider some of their applications before further study of the device. In fact, many scientific people, with little electronics background, have learned to apply these devices in their work—treating the operational amplifier as a "black box" that when properly connected can be set to useful work.

Small-Signal Amplifiers

The first application we shall consider for the operational amplifier is that of a replace-ment for a discrete-component small-signal amplifier. Consider the three-stage ampli-fier we analyzed in Section 7-7 (Fig. 7-29). That circuit, with its three transistors, three capacitors, and nine resistors, offered a voltage gain of 360 over a 40 Hz to 83 kHz bandwidth with a 12.1-kΩ input impedance and an output impedance of 40 Ω. An operational amplifier and two resistors could exceed all those specifications, and in much less space and for much less money. Commercial-grade operational amplifiers with specifications more than adequate for this purpose are available for well under \$1, which makes it ludicrous to use the discrete circuits in this case. Figure 8-12 shows an operational amplifier circuit providing a gain of 360 from direct current (0 Hz) on up with an input impedance of at least 100 kΩ. Depending on the operational amplifier used, the output impedance of 40 Ω may or may not be obtained. However, if not, the addition of a single emitter follower would clear up that problem. One glance at the circuit (Fig. 7-29) replaced by Fig. 8-12 (at probably one-fifth the cost) should provide the reader with an indication of the opera-tional amplifier's great value. Notice that the noninverting configuration was used

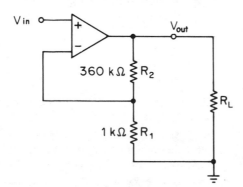

FIGURE 8-12 Small-signal amplifier using op-erational amplifier.

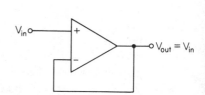

FIGURE 8-13 Operational amplifier follower circuit.

here, because the inverting circuit would have yielded an unacceptably low input impedance (1 kΩ).

Follower Circuit

An operational amplifier is also used as a follower circuit ($G_v = 1$), as in Fig. 8-13, whenever an extremely high input impedance is required. The input impedance is the circuit's open-loop gain G_O plus 1 multiplied by the unit's open-loop input impedance Z_{IN}.

$$Z_{in} \quad R_{in}$$
$$R_{IN} = Z_{IN}(1 + G_O) \tag{8-7}$$

Notice that it is the unit's input impedance that is multiplied, which then implies that the input capacity is divided, since $X_c = 1/2\pi fC$. The follower circuit therefore finds use in very high frequency applications, as well as for large impedance transformation applications.

Comparator

A comparator is, simply stated, a circuit that compares two signals. They are widely used for level detection. Shown in Fig. 8-14, the comparator is one of the simplest (along with the follower) operational amplifier circuits, with no additional external

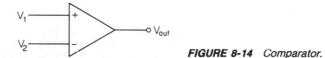

FIGURE 8-14 *Comparator.*

components necessary. The circuit works with full open-loop gain. If v_1 and v_2 are equal, then v_{out} should ideally be zero. If v_1 should change by even a very small amount from v_2's value, v_{out} should increase by a large amount because of the large value of open-loop gain, A_O. Because of the high gain involved, the output is usually "saturated" at the full positive or full negative level. Thus, a rectangular-type waveform is the normal output, regardless of the waveforms being compared. The circuit thus detects small level changes, which is another way of saying that it compares two signals.

Adder

The adder circuit provides an output either equal to or proportional to the sum of two or more signals. By prudently chosing resistor values, it can be used to provide an output equal to the average value of the inputs. Figure 8-15 shows a three-input adder circuit. All three are applied to the inverting input through separate resistors. Since the voltage at the noninverting input is virtually the same as at the inverting

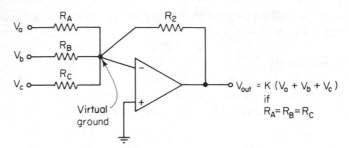

FIGURE 8-15 *Three-input adder circuit.*

input (due to the high open-loop gain), we can say that the inverting input is "almost" grounded. This effect is shown from Fig. 8-10, where the voltage between inputs, v, was nearly zero. This means that if one input is tied to ground, the other must be virtually grounded. It is known as a *virtual* ground and results in each source voltage "seeing" only its respective summing resistor as a load. The sum of the three input currents must equal the current through R_2, since our ideal operational amplifier draws no current itself. Therefore,

$$\frac{v_a}{R_A} + \frac{v_b}{R_B} + \frac{v_c}{R_C} = \frac{-v_{out}}{R_2} \tag{8-8}$$

and if we let $R_A = R_B = R_C = R$, Eq. (8-8) can be simplified to

$$v_{out} = \frac{-R_2}{R}(v_a + v_b + v_c) \tag{8-9}$$

Thus the output is proportional to the algebraic sum of the inputs. If $R_2 = R$, the output exactly equals the sum of the inputs. If $R_2 = \frac{1}{3}R$, the output would equal the sum of the inputs divided by 3, or an output equal to the average of the input signals.

Subtractor

As would be expected, the subtractor circuit shown in Fig. 8-16 provides an output equal to the difference of two signals. Since neither operational amplifier input is at ground potential we shall assume an input potential, with respect to ground, of v_3 for each one. Remember that the potentials at each input will be almost equal, since the operational amplifier itself has such a high loop gain. Now recalling that the operational amplifier inputs effectively draw zero current,

$$\frac{v_1 - v_3}{R_1} = \frac{v_3 - v_{out}}{R_2}$$

and

$$\frac{v_2 - v_3}{R_1} = \frac{v_3}{R_2}$$

By eliminating v_3 from these two equations and solving for v_{out}, we obtain

$$v_{out} = \frac{R_2}{R_1}(v_2 - v_1) \qquad\qquad (8\text{-}10)$$

If $R_2 = R_1$, the v_{out} simply equals the difference between v_2 and v_1.

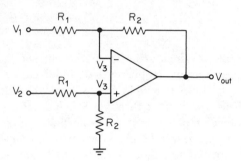

FIGURE 8-16   *Subtractor circuit.*

Integrator

There is considerable demand for a circuit that provides the mathematical integral of a given input signal. A simple example of integration is shown in Fig. 8-17, where the input signal is a dc level and the integral is a linearly increasing ramp output. Applying a dc level to an *RC* circuit, as shown in Fig. 8-18(a), results in an integrator-type output initially, but as the voltage charge across the capacitor builds up, its charging current decays exponentially because the voltage across *R* is decreasing. An operational amplifier circuit such as shown in Fig. 8-19 can supply the constant

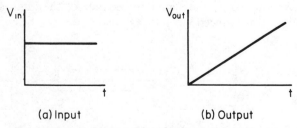

(a) Input   (b) Output

FIGURE 8-17   *Input and output signals for an integrator.*

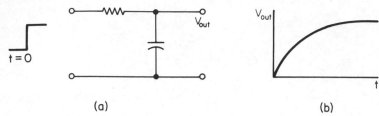

   (a)   (b)

FIGURE 8-18   *RC integrator and output signal.*

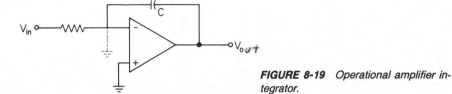

FIGURE 8-19   *Operational amplifier integrator.*

current necessary to allow a highly linear charge current to the capacitor, which results in a *highly linear* output. Applying a dc input causes the voltage across R to be constant, since the inverting input is a virtual ground (as shown in dashed lines in Fig. 8-19). Thus, the current through R is constant, and it is supplied only to C, since the operational amplifier itself draws negligible current. The voltage across C therefore increases at a linear rate and is equal to the negative value of v_0. The operational amplifier provides a method of generating highly accurate ramp voltages, as may be required to drive cathode-ray tubes, and also can provide highly accurate integrals of any other signal.

Differentiators

By interchanging the resistor and capacitor of the integrator circuit, a differentiator circuit is formed. Differentiation is a mathematical operation that provides the rate of change of a signal. Thus, feeding a linear ramp that has a constant rate of change into a differentiator yields a constant dc output, as shown in Fig. 8-20. Unfortunately, the operational amplifier differentiator circuit shown in Fig. 8-21 has a gain that increases with frequency, making it highly susceptible to high-frequency noise. A solution to this problem is to add some series resistance to the input capacitor (sometimes accomplished by the source impedance of the input signal) so that the high-frequency gain is reduced.

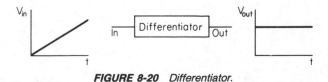

FIGURE 8-20   *Differentiator.*

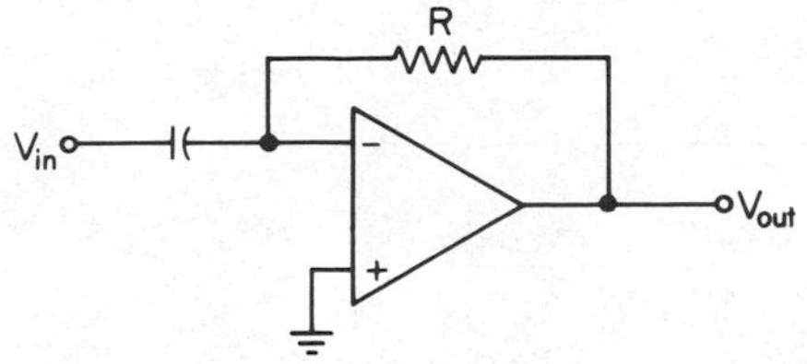

FIGURE 8-21 *Differentiator circuit.*

The preceding op-amp applications are but a few of the many possibilities. It should be apparent that this versatile circuit is a simple way to accomplish many useful functions.

8-6 INTERNAL CIRCUITRY

We have thus far studied the operational amplifier and some of its applications without any clear idea as to exactly what this device contains. Most operational amplifiers contain a high input impedance differential amplifier that provides high gain, another stage of gain that is usually another differential amplifier, and an impedance transformation output stage—often some emitter followers and/or a class AB push-pull stage. Figure 8-22 provides a schematic used by a whole series of RCA operational amplifiers.

Transistors Q_1 and Q_2 in Fig. 8-22 comprise the differential amplifier for the noninverting and inverting inputs. Transistor Q_6 provides a constant current to the emitters of Q_1 and Q_2 to allow for proper differential amplifier operation. The diode D1 is used to temperature-compensate Q_6 so that the current it supplies is constant over a wide temperature range. Resistors R_5 and R_6 form a voltage divider bias for Q_6 with pin 1 (we shall use the pin numbers enclosed by a square) usually grounded. It may, however, be connected to some variable voltage that is used to control the open-loop gain of this amount of dc emitter current to Q_1 and Q_2, and hence vary their voltage gain. Recall that $G_v = R_c/h_{ib}$, where $h_{ib} = 0.026/I_E$.

The outputs at the collectors of Q_1 and Q_2 drive the second differential stage made up of Q_3 and Q_4. Bias stabilization for this stage is provided by constant current transistor Q_7, which is compensated itself by diode D_2. The compensating diode, D_2, also provides thermal stabilization for Q_9. Transistor Q_5 is used to keep the output signal as close to zero as possible when the same signal is applied to both inputs of the operational amplifier. Under those conditions, the signal at Q_5's base should be zero. If it is not, it develops a signal across R_2 through Q_5's amplification in the proper phase to reduce the error. This same signal is reflected into Q_7's base from Q_5's emitter to further reduce the error. Transistor Q_5 also serves to reduce the error in output signal from zero when the two inputs are at the same level, because of drift by either the $+$ or $-$ power supply.

The output of the second differential amplifier stage is taken at Q_4's collector and then fed into an impedance-transforming dc level shifting network comprised of Q_8, Q_9, and Q_{10}. Since the output (pin 9) is to be at ground potential with no

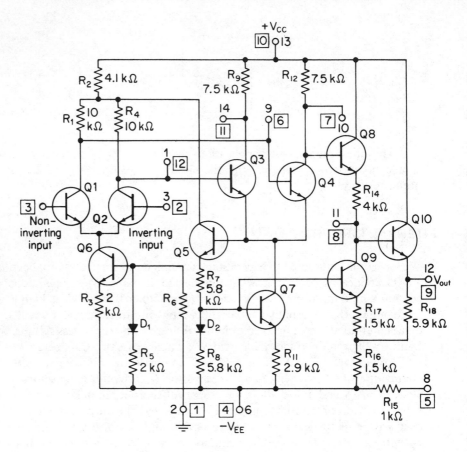

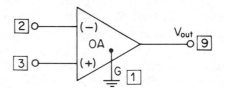

FIGURE 8-22 *Operational amplifier schematic (Courtesy of Radio Corporation of America.)*

input signals, it is the function of Q_9 to supply a constant current for biasing Q_8 such that the resulting dc level at Q_{10}'s emitter is as close to ground as possible, regardless of supply voltage variations and imbalances. Transistors Q_8 and Q_{10} also provide a low output impedance, being two cascaded emitter followers.

Although the internal circuitry shown in Fig. 8-22 is typical, there are many varieties of op amps available. Some varieties combine FETs and BJTs on the same chip and are termed *BIFETs*. The FETs are generally used as the input transistors to provide extremely high input impedance and low input leakage currents while BJTs are used throughout the remainder of the op amp. Power amps are available that can provide a few watts of ac output power that can be used to drive a small speaker or motor, for example.

8-7   SLEW RATE AND COMMON MODE EFFECTS

Slew Rate

An important op-amp characteristic of interest at high frequencies is known as its *slew rate*. The slew rate is the internally limited rate of change in output voltage with a large-amplitude step function applied to the input. In other words, limitations exist that will not allow the output to change as rapidly as the bandwidth information would lead one to believe for large-signal operation. The slewing rate then is the maximum rate at which the output can change in response to a step or changing input. Slew rate is usually specified in volts per microsecond. Although the slew rate and bandwidth both limit the usable frequency range of an operational amplifier, they produce effects that are different in a subtle way. The high-frequency cutoff or bandwidth is a frequency where the sinusoidal output is getting smaller, although it can still be purely sinusodial. Exceeding the slew rate, however, would not necessarily result in a reduction of the output level, but does result in distortion of the output wave.

Recall the equation of a sine wave as

$$v = V_P \sin (2\pi f t)$$

The maximum rate of change for a sine wave occurs as it passes through zero, and this rate of change is

$$\left. \frac{\Delta v}{\Delta t} \right|_{t=0} = 2\pi f V_p \qquad \text{(8-11)}$$

The term $2\pi f V_p$ in Eq. (8-11) should not exceed the manufacturer's published slew rate if an undistorted sinusodial output is desired.

EXAMPLE 8-4

The bandwidth of an operational amplifier is measured as 500 kHz. The slew rate is given as 1 V/μs. Determine the maximum amplitude output at its high-frequency cutoff.

Solution:

Setting the slew rate equal to $2\pi f V_p$ at 500 kHz yields

$$1 \text{ V}/\mu\text{s} = 2 \times 500 \text{ kHz} \times V_p = \frac{1 \text{ V}}{10^{-6} \text{ s}}$$

Rearranging and solving for V_p, we obtain

$$V_p = \frac{10^6}{2\pi \times 0.5 \times 10^6} = \frac{1}{\pi} = \underline{0.318 \text{ V}}$$

Thus, at 500 kHz a pure sine-wave output can be obtained only up to an output level of 0.318 V peak. At 250 kHz (one-half of the original frequency), we could obtain twice the undistorted output signal, and so on.

Common-Mode Effects

One very useful feature of an operational amplifier is the fact that ideally it will provide 0-V out for a zero input or inverting and noninverting inputs that are equal. Unfortunately, the practical operational amplifier cannot meet the ideal specification but typically has an output of 1 to 10 mV under these conditions. In addition, this *offset*, as it is termed, varies with temperature and supply voltage. Since we are talking about very small voltage levels (<10 mV), it is only in the most critical applications that this operational-amplifier characteristic becomes a factor. It is specified on manufacturer's data sheets in several fashions. *The common-mode voltage gain, A_c, is the ratio of the small offset signal to the change in common-mode (nonzero) input when that input signal changes to cause a zero output.* This gain is much less than 1. The *common-mode rejection ratio* (CMRR) is the ratio of the common-mode gain (A_c) to the open-loop voltage gain (A_O):

$$\text{CMRR} = \frac{A_c}{A_O} \qquad (8\text{-}12)$$

It is most frequently expressed in decibels as

$$\text{CMRR (dB)} = 20 \log \frac{A_c}{A_O} \qquad (8\text{-}13)$$

Typical operational amplifier values for the CMRR are 70 to 100 dB down. The CMRR is a measure of the ability of the differential amplifier to discriminate between differential and common-mode input signals.

Offset and Offset Null

Operational amplifiers are usually given an input offset voltage specification. The input *offset voltage* is the voltage that must be applied between the two input terminals to obtain zero output voltage. This voltage varies with temperature, but in critical applications an adjustable resistive divider is set up to null the operational amplifier for one specific temperature. Some operational amplifiers provide two specific terminals for offset nulling. A potentiometer is connected across these two terminals with the wiper usually connected to the negative supply. The *input offset current* is sometimes specified, and it is the difference in the two input currents necessary to provide 0-V output.

One last offset definition to consider is the supply-voltage rejection ratio. The *supply-voltage rejection ratio* is the ratio of the change in input offset voltage to the undesired change in the supply voltage producing it. This specification allows the designer to choose a power supply with adequate regulation and stability for a given application.

8-8   THE μA741 OPERATION AMPLIFIER

In this section the manufacturer's complete specifications for a modern operational amplifier are presented and discussed. The student will learn much about operational amplifiers by studying this information, and also will become familiar with one of the easiest-to-use and most versatile monolithic operational amplifiers developed. It has specifications that compare to some of the best hybrid operational amplifiers of 1965, which cost $100, yet this device is priced below $0.20 in very large quantities and at under $1 in single quantities. There can be little question why devices such as this have changed the character of the entire linear electronics industry.

Table 8-1 presents the complete specifications and a number of applications for the Fairchild μA741 operational amplifier. It is one of the second-generation mono-lithic operational amplifiers, because it offers considerable improvement over most characteristics of the initially offered monolithic units. The problems of the first-generation units, which the μA741 overcomes, include:

1. Complicated frequency stabilization networks.

2. Lack of output short-circuit protection.

3. Low allowable differential input voltages.

4. Lack of simple offset null method.

5. Instability with capacitive loads.

Referring to the equivalent schematic on page 251, several "different"-looking configu-rations are apparent. The unit is essentially a two-stage amplifier comprising a high-

TABLE 8-1 Frequency-compensated operational amplifier

FEATURES:
- **NO FREQUENCY COMPENSATION REQUIRED**
- **SHORT-CIRCUIT PROTECTION**
- **OFFSET VOLTAGE NULL CAPABILITY**
- **LARGE COMMON-MODE AND DIFFERENTIAL VOLTAGE RANGES**
- **LOW POWER CONSUMPTION**
- **NO LATCH UP**

GENERAL DESCRIPTION — The μA741 is a high performance monolithic operational amplifier constructed on a single silicon chip, using the Fairchild Planar* epitaxial process. It is intended for a wide range of analog applications. High common mode voltage range and absence of "latch-up" tendencies make the μA741 ideal for use as a voltage follower. The high gain and wide range of operating voltage provides superior performance in integrator, summing amplifier, and general feedback applications. The μA741 is short-circuit protected, has the same pin configuration as the popular μA709 operational amplifier, but requires no external components for frequency compensation. The internal 6dB/octave roll-off insures stability in closed loop applications.

ABSOLUTE MAXIMUM RATINGS
Supply Voltage	± 22 V
Internal Power Dissipation (Note 1)	500 mW
Differential Input Voltage	± 30 V
Input Voltage (Note 2)	± 15 V
Voltage between Offset Null and V−	± 0.5 V
Storage Temperature Range	−65°C to +150°C
Operating Temperature Range	−55°C to +125°C
Lead Temperature (Soldering, 60 sec)	300°C
Output Short-Circuit Duration (Note 3)	Indefinite

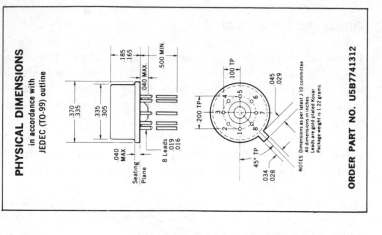

PHYSICAL DIMENSIONS
in accordance with
JEDEC (TO-99) outline

.185
.165

.040 MAX.

.500 MIN.

.370
.335

.335
.305

.040
MAX.

8 Leads
.019
.016

Seating
Plane

.100 TP

.200 TP

45° TP

.034
.028

.045
.029

NOTES: Dimensions as per latest J-10 committee
All dimensions in inches
Leads are gold plated Kovar
Package weight is 1.22 grams

ORDER PART NO. U5B7741312

TABLE 8-1 *(continued)*

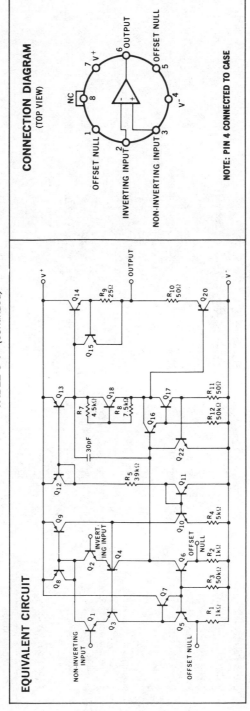

CONNECTION DIAGRAM
(TOP VIEW)

OFFSET NULL 1
INVERTING INPUT 2
NON-INVERTING INPUT 3
V⁻ 4
OFFSET NULL 5
OUTPUT 6
V⁺ 7
NC 8

NOTE: PIN 4 CONNECTED TO CASE

EQUIVALENT CIRCUIT

NOTES:

(1) Rating applies for case temperatures to 125°C; derate linearly at 6.5 mW/°C for ambient temperatures above +75°C.

(2) For supply voltages less than ±15 V, the absolute maximum input voltage is equal to the supply voltage.

(3) Short circuit may be to ground or either supply. Rating applies to +125°C case temperature or +75°C ambient temperature.

*Planar is a patented Fairchild process.

251

TABLE 8-1 (continued)

ELECTRICAL CHARACTERISTICS ($V_S = \pm 15$ V, $T_A = 25°C$ unless otherwise specified)

PARAMETERS (see definitions)	CONDITIONS	MIN.	TYP.	MAX.	UNITS
Input Offset Voltage	$R_S \le 10$ kΩ		1.0	5.0	mV
Input Offset Current			20	200	nA
Input Bias Current			80	500	nA
Input Resistance		0.3	2.0		MΩ
Input Capacitance			1.4		pF
Offset Voltage Adjustment Range			±15		mV
Large-Signal Voltage Gain	$R_L \ge 2$ kΩ, $V_{out} = \pm 10$ V	50,000	200,000		
Output Resistance			75		Ω
Output Short-Circuit Current			25		mA
Supply Current			1.7	2.8	mA
Power Consumption			50	85	mW
Transient Response (unity gain)	$V_{in} = 20$ mV, $R_L = 2$ kΩ, $C_L \le 100$ pF				
Risetime			0.3		μs
Overshoot			5.0		%
Slew Rate	$R_L \ge 2$ kΩ		0.5		V/μs
The following specifications apply for $-55°C \le T_A \le +125°C$:					
Input Offset Voltage	$R_S \le 10$ kΩ		1.0	6.0	mV
Input Offset Current	$T_A = +125°C$		7.0	200	nA
	$T_A = -55°C$		85	500	nA
Input Bias Current	$T_A = +125°C$		0.03	0.5	μA
	$T_A = -55°C$		0.3	1.5	μA
Input Voltage Range		±12	±13		V
Common Mode Rejection Ratio	$R_S \le 10$ kΩ	70	90		dB
Supply Voltage Rejection Ratio	$R_S \le 10$ kΩ		30	150	μV/V
Large-Signal Voltage Gain	$R_L \ge 2$ kΩ, $V_{out} = \pm 10$ V	25,000			
Output Voltage Swing	$R_L \ge 10$ kΩ	±12	±14		V
	$R_L \ge 2$ kΩ	±10	±13		V
Supply Current	$T_A = +125°C$		1.5	2.5	mA
	$T_A = -55°C$		2.0	3.3	mA
Power Consumption	$T_A = +125°C$		45	75	mW
	$T_A = -55°C$		60	100	mW

TABLE 8-1 *(continued)*
TYPICAL PERFORMANCE CURVES

OPEN LOOP VOLTAGE GAIN AS A FUNCTION OF SUPPLY VOLTAGE

OUTPUT VOLTAGE SWING AS A FUNCTION OF SUPPLY VOLTAGE

INPUT COMMON MODE VOLTAGE RANGE AS A FUNCTION OF SUPPLY VOLTAGE

TABLE 8-1 (continued)

TYPICAL PERFORMANCE CURVES

INPUT RESISTANCE
AS A FUNCTION OF
AMBIENT TEMPERATURE

INPUT BIAS CURRENT
AS A FUNCTION OF
AMBIENT TEMPERATURE

POWER CONSUMPTION
AS A FUNCTION OF
SUPPLY VOLTAGE

POWER CONSUMPTION
AS A FUNCTION OF
AMBIENT TEMPERATURE

INPUT OFFSET CURRENT
AS A FUNCTION OF
AMBIENT TEMPERATURE

INPUT OFFSET CURRENT
AS A FUNCTION OF
SUPPLY VOLTAGE

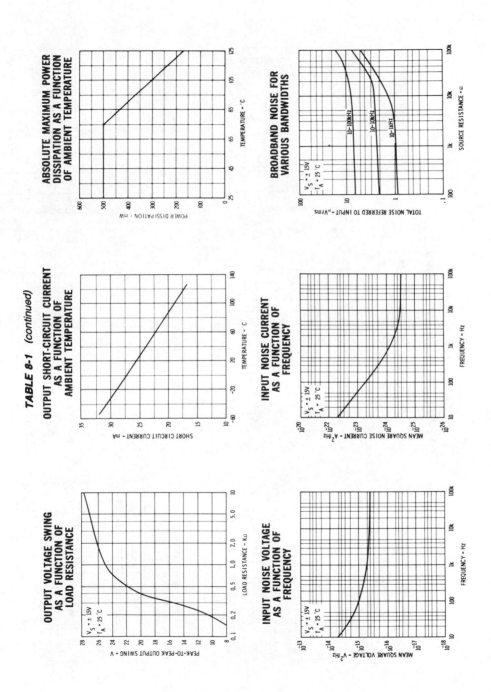

TABLE 8-1 *(continued)*

ABSOLUTE MAXIMUM POWER
DISSIPATION AS A FUNCTION
OF AMBIENT TEMPERATURE

OUTPUT SHORT-CIRCUIT CURRENT
AS A FUNCTION OF
AMBIENT TEMPERATURE

OUTPUT VOLTAGE SWING
AS A FUNCTION OF
LOAD RESISTANCE

BROADBAND NOISE FOR
VARIOUS BANDWIDTHS

INPUT NOISE CURRENT
AS A FUNCTION OF
FREQUENCY

INPUT NOISE VOLTAGE
AS A FUNCTION OF
FREQUENCY

255

TABLE 8-1 (continued)
TYPICAL PERFORMANCE CURVES

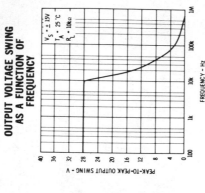

OUTPUT VOLTAGE SWING
AS A FUNCTION OF
FREQUENCY

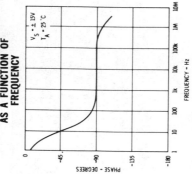

OPEN LOOP PHASE RESPONSE
AS A FUNCTION OF
FREQUENCY

OPEN LOOP VOLTAGE GAIN
AS A FUNCTION OF
FREQUENCY

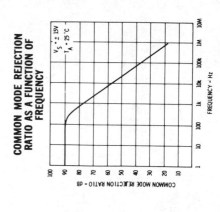

COMMON MODE REJECTION
RATIO AS A FUNCTION OF
FREQUENCY

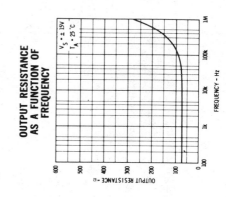

OUTPUT RESISTANCE
AS A FUNCTION OF
FREQUENCY

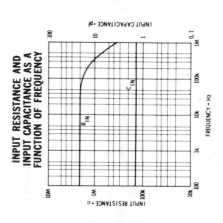

INPUT RESISTANCE AND
INPUT CAPACITANCE AS A
FUNCTION OF FREQUENCY

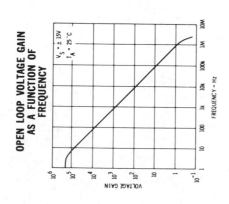

256

VOLTAGE FOLLOWER LARGE-SIGNAL PULSE RESPONSE

FREQUENCY CHARACTERISTICS AS A FUNCTION OF AMBIENT TEMPERATURE

TABLE 8-1 (continued)
TRANSIENT RESPONSE TEST CIRCUIT

VOLTAGE OFFSET NULL CIRCUIT

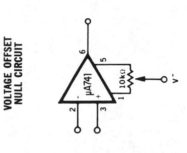

TRANSIENT RESPONSE

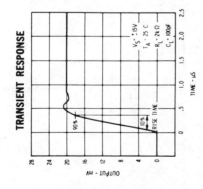

FREQUENCY CHARACTERISTICS AS A FUNCTION OF SUPPLY VOLTAGE

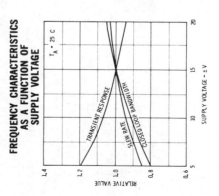

TABLE 8-1 (continued)

DEFINITION OF TERMS

INPUT OFFSET VOLTAGE — That voltage which must be applied between the input terminals to obtain zero output voltage. The input offset voltage may also be defined for the case where two equal resistances are inserted in series with the input leads.

INPUT OFFSET CURRENT — The difference in the currents into the two input terminals with the output at zero volts.

INPUT BIAS CURRENT — The average of the two input currents.

INPUT RESISTANCE — The resistance looking into either input terminal with the other grounded.

INPUT CAPACITANCE — The capacitance looking into either input terminal with the other grounded.

LARGE-SIGNAL VOLTAGE GAIN — The ratio of the maximum output voltage swing with load to the change in input voltage required to drive the output from zero to this voltage.

OUTPUT RESISTANCE — The resistance seen looking into the output terminal with the output at null. This parameter is defined only under small signal conditions at frequencies above a few hundred cycles to eliminate the influence of drift and thermal feedback.

OUTPUT SHORT-CIRCUIT CURRENT — The maximum output current available from the amplifier with the output shorted to ground or to either supply.

SUPPLY CURRENT — The DC current from the supplies required to operate the amplifier with the output at zero and with no load current.

POWER CONSUMPTION — The DC power required to operate the amplifier with the output at zero and with no load current.

TRANSIENT RESPONSE — The closed-loop step-function response of the amplifier under small-signal conditions.

INPUT VOLTAGE RANGE — The range of voltage which, if exceeded on either input terminal, could cause the amplifier to cease functioning properly.

INPUT COMMON MODE REJECTION RATIO — The ratio of the input voltage range to the maximum change in input offset voltage over this range.

SUPPLY VOLTAGE REJECTION RATIO — The ratio of the change in input offset voltage to the change in supply voltage producing it.

OUTPUT VOLTAGE SWING — The peak output swing, referred to zero, that can be obtained without clipping.

TYPICAL APPLICATIONS

UNITY-GAIN VOLTAGE FOLLOWER

$R_{IN} = 400 \ M\Omega$
$C_{IN} = 1 \ pF$
$R_{out} < 1 \ \Omega$
$B.W. = 1 \ MHz$

NON-INVERTING AMPLIFIER

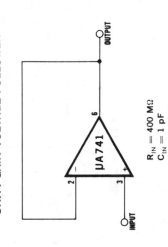

GAIN	R_1	R_2	B.W.	R_{IN}
10	1 kΩ	9 kΩ	100 kHz	400 MΩ
100	100 Ω	9.9 kΩ	10 kHz	280 MΩ
1000	100 Ω	99.9 kΩ	1 kHz	80 MΩ

258

CLIPPING AMPLIFIER

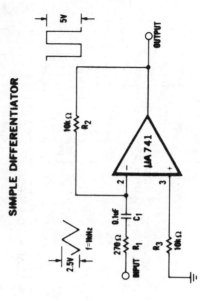

$$\frac{E_{out}}{E_{IN}} = \frac{R_2}{R_1} \quad \text{if } |E_{out}| \leq V_Z + 0.7 \text{ V}$$

where V_Z = Zener breakdown voltage

SIMPLE DIFFERENTIATOR

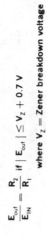

$$E_{out} = -R_2 C_1 \frac{dE_{IN}}{dt}$$

INVERTING AMPLIFIER

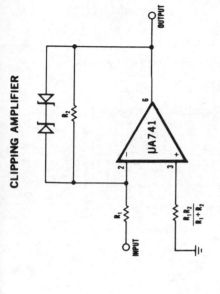

GAIN	R_1	R_2	B.W.	R_{IN}
1	10 kΩ	10 kΩ	1 MHz	10 kΩ
10	1 kΩ	10 kΩ	100 kHz	1 kΩ
100	1 kΩ	100 kΩ	10 kHz	1 kΩ
1000	100 Ω	100 kΩ	1 kHz	100 Ω

SIMPLE INTEGRATOR

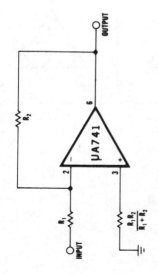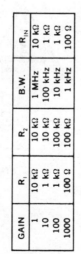

$$E_{out} = -\frac{1}{R_1 C_1} \int E_{IN} dt$$

TABLE 8-1 *(continued)*

HIGH SLEW RATE POWER AMPLIFIER

+15V

2N5005

180Ω

10pF

51kΩ

μA741

7

6

2

3

4

5.1kΩ

INPUT

5.1k

OUTPUT

.005uF

47Ω

2N5004

180Ω

-15V

R_L

15Ω

LOW DRIFT LOW NOISE AMPLIFIER

+15V

μA741

3

2

6

OUTPUT

R_{ADJ}
330kΩ

μA727

9

6

7

1

8

50kΩ

50kΩ

3

2

50Ω

50Ω

INPUT

Voltage Gain = 10^3
Input Offset Voltage Drift = 0.6 μV/°C
Input Offset Current Drift = 2.0 pA/°C

NOTCH FILTER USING THE μA741 AS A GYRATOR

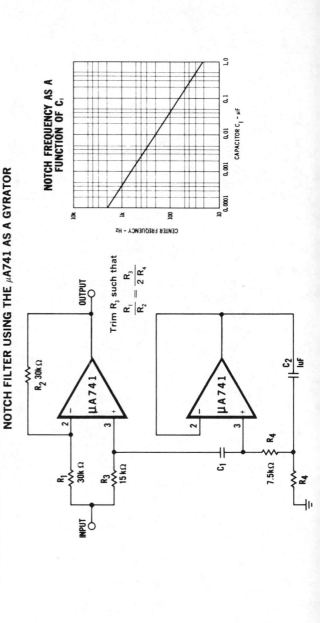

NOTCH FREQUENCY AS A FUNCTION OF C_i

CENTER FREQUENCY – Hz

CAPACITOR C_i – uF

R_2 30kΩ

μA741

2

3

OUTPUT

Trim R_3 such that
$$\frac{R_1}{R_2} = \frac{R_3}{2 R_4}$$

R_1
30kΩ

R_3
15kΩ

INPUT

μA741

2

3

C_1

R_4

7.5kΩ

R_4

C_2
1uF

260

gain differential input stage, followed by a high-gain driver with a class AB output. In the input stages (Q_1, Q_2, Q_3, Q_4), notice the use of *npn–pnp* combinations. Since high h_{fe} pnps involve costly additional processing steps, the input uses a combination of high-h_{fe} *npn*'s and low-h_{fe} *pnp*'s to achieve high input impedances and gain and low input bias currents. To obtain high gain the collectors of Q_5 and Q_6 are used as resistive loads for the differential stages, giving effective values of about 2 MΩ. The constant current source and bias stabilization are made up of Q_7, Q_8, Q_9, and Q_{10}.

In cases when input offset voltage control is required, an external 10-kΩ potentiometer should be connected between the emitters of Q_5 and Q_6 with the wiper connected to the negative supply. Using this technique, the input offset can be adjusted approximately ±25 mV from its initial value.

The output from the input differential stage is then taken from the collector of Q_4 and fed into a Darlington stage (Q_{16}, Q_{17}) to avoid loading. It then feeds a complementary symmetry network with Q_{14} and Q_{20} operating class AB with about 60 μA of quiescent bias current to eliminate crossover distortion.

The 30-pF capacitor between the bases of Q_{14} and Q_{16} is the basis of the internal frequency compensation. The inclusion of this high-voltage dielectric, high-valued (relatively speaking) capacitor in a monolithic circuit is the result of new, improved processing techniques. Its inclusion internal to the operational amplifier package results in increased reliability, more compact circuit boards, and lower assembly cost. If an application does require reduced high-frequency responses (for reasons other than instability), the connection of capacitance between the output and Q6's emitter (pins 5 and 6) will suffice.

The electrical characteristics of the μA741 are presented on page 252. Notice that they are provided in duplicate, once at room temperature, 25°C, and again for over the entire device operating range of $-55°$ to $+125°$C. This device has a very high open-loop gain *(A_O)* of 200,000, typically with a minimum value of 50,000.

Following the electrical characteristics are a rather complete set of performance curves on pages 253–257. The student will learn a great deal about operational amplifiers if each of these curves is analyzed and understood. With this information the μA741 can be successfully applied to virtually any application. The specification sheet for the μA741 is completed on pages 258–261 with some definitions and applications. Notice the additional resistor equal to the parallel combination of R_1 and R_2 in the inverting and noninverting amplifier circuits. It is added to minimize output errors due to input offset currents and is required in high-frequency applications.

QUESTIONS AND PROBLEMS

8–1–1. Define the following types of integrated circuits: monolithic, thick-film hybrid, and thin-film hybrid. Explain why monolithic ICs are the most widely used.

8-1-2. Provide some general guidelines as to when a monolithic IC becomes MSI, LSI, or VLSI. Explain the meaning of these abbreviations.

8-2-3. Briefly explain the process whereby a monolithic IC is fabricated.

8-2-4. Why do active devices (i.e., transistors) cost less than passive ones (i.e., resistors, capacitors) in a monolithic IC? Explain why resistors and capacitors with large values are not used in monolithic ICs.

8-2-5. Describe the different design criteria for a monolithic IC as compared to a discrete circuit.

8-3-6. Describe the function and application for a dc level shifting circuit. Provide a schematic and briefly explain its operation.

8-3-7. Why is a capacitance multiplier circuit desirable in monolithic ICs? Provide a schematic and describe how the circuit provides capacitance multiplication.

8-3-8. Provide a schematic for a differential amplifier. Why is this circuit particularly attractive for use in monolithic ICs? Calculate the single-ended and differential voltage gain for the circuit of Fig. 8-7 if the constant current source were changed to 2.2 mA.

8-3-9. What is a gyrater, and why is it needed for monolithic ICs? With the help of some equations, explain how it functions.

8-3-10. Calculate the inductance presented by the circuit shown in Fig. 8-8 if $h_{ib} = 40 \ \Omega$, $f = 7$ kHz, and C is changed to 470 pF.

8-4-11. Explain how the operational amplifier was named. Why is it so well suited to monolithic ICs? What type of circuitry does it generally include?

8-4-12. List the characteristics of the ideal operational amplifier. Which of these are nearly attained in practice?

8-4-13. Provide a schematic for the inverting operational amplifier and develop the expression for its voltage gain.

8-4-14. Develop the expression for the voltage gain of the noninverting operational amplifier. Show its schematic.

8-4-15. Provide a table that shows the voltage gain and R_{in} for the inverting and noninverting operational amplifier configurations. Explain the meaning of the terms G_0 and Z_{in}.

8-4-16. Figure 8-23 shows a current-to-voltage operational amplifier circuit. Develop an expression for v_{out} as a function of i_{in}.

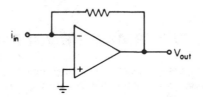

FIGURE 8-23   *Current-to-voltage circuit for Problem 8-4-16.*

8-4-17. Provide a schematic of an operational amplifier that can provide 180° of phase shift and increase a 20-mV signal up to 1 V. The input resistance of this circuit must be at least 5 kΩ.

8-5-18. What are the characteristics of the follower circuit? Provide a schematic and explain the good high-frequency performance of this circuit.

8-5-19. The waveform shown in Fig. 8-24 is applied to the inverting input of a comparator circuit. The noninverting input is grounded and ±12-V power supplies are used. Sketch v_{out} as a function of time directly below your sketch of the input signal.

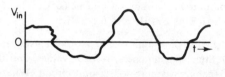

FIGURE 8-24 Problem 8-5-19.

8-5-20. Explain the concept of virtual ground as it relates to an operational amplifier. If one of the two inputs is grounded, why is the other virtually grounded?

8-5-21. Provide a schematic for a four-input adder that has an output equal to the average of the four inputs. Show resistor values.

8-5-22. Determine an expression for v_0 for the circuit shown in Fig. 8-25. Why is the circuit known as a scaling adder?

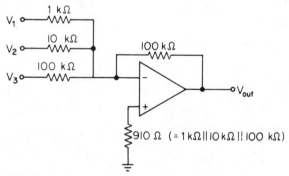

FIGURE 8-25 Scaling adder circuit for Problem 8-5-22.

8-5-23. Determine the output voltage for the circuit in Fig. 8-16 if $C_1 = C_2 = 3$ V, $R_1 = 3$ kΩ and $R_2 = 5$ kΩ.

8-5-24. Show a schematic and typical input and output signals for an integrator and differentiator circuit.

8-6-25. Explain the function of each transistor for the operational amplifier internal schematic shown in Fig. 8-22.

8-7-26. An operational amplifier has a bandwidth of 1 MHz and a slew rate of 2 V/μs. Determine the maximum sinusoidal output at 1 MHz. Calculate the highest frequency at which a 10-v p-p sinusoidal output is possible.

8–7–27. Explain the general aspects of common-mode effects. Specifically define CMRR and the input offset voltage.

8–8–28. Based on the typical performance curves for the μA741 operational amplifier circuit, determine:

(a) The open-loop voltage gain when the $\pm$V supply voltage is 14 V. Express the gain in dB and as a ratio.

(b) The input bias current at 60°C.

(c) The input resistance at 20°C.

(d) The input offset current at 25°C with $\pm$15-V supplies.

(e) The output short-circuit current at 60°C.

(f) The CMRR at 100 Hz and 0.1 MHz.

8–8–29. Show a schematic of a noninverting amplifier using the μA741 to provide a voltage gain of 10. Show pin numbers, resistor values, and include $\pm$15-V supplies in the schematic. Include a voltage offset null circuit. What is the purpose of the resistor from pin 3 to ground that equals $R_1 \| R_2$?

9

COMMUNICATION
SYSTEMS

9-1 INTRODUCTION

The ultimate goal of studying circuits is to apply them in a usable fashion. A major application of electronic circuits is in communication systems. These systems had their beginning with the discovery of various electrical, magnetic, and electrostatic phenomena prior to the twentieth century. Lee DeForest's investigations of the triode vacuum tube in the early 1900s allowed the first form of electronic amplification and opened the door to wireless communication. In 1948, another major discovery in the history of electronics occurred with the development of the transistor by Shockley, Brittain, and Bardeen.

The function of a communication system is to transfer information from one point to another via some communications link. The very first form of "information" transferred was the human speech in the form of a code, which was then converted back to words. Human beings have a natural desire to communicate rapidly between any points on the earth, and that initially was the major concern of these developments. As that goal became a reality with the evolution of new technology following the invention of the triode vacuum tube, new and less basic applications were also realized, such as entertainment, radar, and radio telescopes. The communications field is still a highly dynamic one, with new semiconductor devices and integrated circuits constantly making new equipment possible or allowing improvement of the old units. Communications was the basic origin of electronics, and no other major field developed until the transistor made modern digital computers a reality. We now have two major subcategories in the field of electronics—communications and digital systems.

Basic to the field of communications is the concept of modulation. *Modulation* is the process of impressing information onto a high-frequency carrier for transmission. It follows then that once this information is received the intelligence must be removed from the high-frequency carrier—a process known as *demodulation*. At this point you may be thinking why we bother to go through this modulation–demodulation process—why not just transmit the information directly and save the bother? The problem is that the frequency of the human voice ranges from about 20 to 4000 Hz. If everyone transmitted those frequencies directly as radio waves, interference between them would cause them all to be ineffective. As it turns out, it is virtually impossible to transmit such low frequencies anyway, since the required antennas for efficient propagation would have to be miles in length.

The answer to these problems then is modulation, which allows long-distance propagation of the low-frequency intelligence with a high-frequency carrier. The high-frequency carriers are chosen such that only one transmitter in an area operates at the same frequency to eliminate interference, and that frequency is high enough such that antenna sizes are physically small and manageable. There are three possible methods of impressing low-frequency information onto a high frequency carrier. Equation (9-1) is the mathematical representation of a sine wave, which we shall assume to be our high-frequency carrier:

$$v = V_P \sin (\omega t + \theta) \qquad (9\text{-}1)$$

where $v =$ instantaneous value

   $V_P =$ peak value

   $\omega =$ angular velocity $= 2\pi f$

   $\theta =$ phase angle

Any one of the last three terms could be varied in accordance with the low-frequency information signal so as to produce a modulated signal. If the amplitude term, V_P, is the parameter varied, it is termed *amplitude modulation* (AM). If the frequency

TABLE 9-1   *Radio-frequency spectrum.*

Frequency	Designation	Abbreviation
3–30 kHz	Very low frequency	VLF
30–300 kHz	Low frequency	LF
300 kHz–3 MHz	Medium frequency	MF
3–30 MHz	High frequency	HF
30–300 MHz	Very high frequency	VHF
300 MHz–3 GHz	Ultra high frequency	UHF
3–30 GHz	Super high frequency	SHF
30–300 GHz	Extra high frequency	EHF

is varied, it is termed *frequency modulation* (FM). Varying the phase angle, θ, results in *phase modulation* (PM).

Communication systems are often categorized by the frequency of the carrier. Table 9-1 provides the names for the various ranges of frequencies in the radio spectrum. Note the acronym abbreviations in the last column.

The extra high frequency range begins at the starting point of infrared frequencies, but the infrareds extend considerably beyond 300 GHz. After the infrareds in the electromagnetic spectrum (of which radio waves are a very small portion) come light waves, ultraviolet, X-rays, gamma rays, and cosmic rays. None of these frequencies (above 300 GHz) are classified as radio frequencies.

9-2 OSCILLATORS

The most basic building block in a communication system is an oscillator. An *oscillator* is a circuit capable of converting energy from a dc form to an ac form. In other words, an oscillator generates a waveform. The waveform can be of any type but occurs at some repetitive frequency.

One basic application for an oscillator is in a communication transmitter, where the audio information is combined with a high-frequency signal that is generated by an oscillator. The purpose of this is to allow transmission of low-frequency information at a higher carrier frequency, since transmission of low frequencies directly is impractical for a number of reasons, as explained in the preceding section. Modern communication receivers also require an oscillator to allow a stepdown to the intermediate frequency (IF).

Another important oscillator application is in electronic test equipment. To successfully check out many circuits, it is necessary to apply an oscillator output of correct waveform, frequency, and voltage level. Test generators (oscillators) whose output level and frequency can be varied over a wide range are commonly used for this purpose. They can often supply several types of waveforms, such as sine waves, square waves, triangular, and sawtooth waveforms.

A number of different forms of sine-wave oscillators are available for use in electronic circuits. They fall into the following broad classifications:

1. *LC* feedback oscillators.

2. *RC* phase-selective oscillators.

3. Negative-resistance oscillators.

The choice among these forms, as well as among individual varieties of these categories, is based upon the following criteria:

1. Output frequency required.

2. Frequency stability required.

3. Is the frequency to be variable, and if so, over what range?

4. Allowable waveform distortion.

5. Power output requirement.

These performance considerations, combined with economic factors, will then dictate the form of oscillator to be used in a given application.

Ringing in an *LC* Tank Circuit

A very close analogy exists between a swinging pendulum and a tank circuit. It is often easier for the student to visualize an electrical phenomenon by first examining its mechanical counterpart. With reference to Fig. 9-1(a), moving the pendulum from

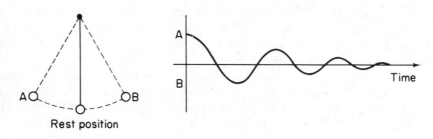

(a) Pendulum action   (b) Pendulum displacement versus time

FIGURE 9-1   *Pendulum: (a) Pendulum motion; (b) Pendulum displacement versus time.*

its rest position to point *A* provides the pendulum with *potential energy*—the energy of position. Release of the pendulum from point *A* starts a conversion process of potential energy into the energy of motion—*kinetic energy.* The pendulum now builds up speed and then passes right on through the rest position on to point *B*, where the kinetic energy has now been reconverted into potential energy. Since losses are introduced by friction and windage, the pendulum displacement at *B* will be somewhat less than the initial displacement at *A*. If the pendulum is allowed to continue to swing back and forth, each succeeding swing will be smaller than the preceding one until the motion ceases, as shown in Fig. 9-1(b). This is known as a *damped sinusoidal oscillation* and is a sinusoidal waveform. In a clock, however, the motion is not allowed to stop but is maintained by supplying a small force to the pendulum, at the right time, to compensate for the losses. The frequency of the pendulum's swing is a constant determined by its mass and length, and it regulates the speed of the clock's hands quite accurately.

The effect of charging the capacitor in Fig. 9-2(a) to some voltage potential and then closing the switch results in the waveform shown in Fig. 9-2(b), which is identical to the one shown in Fig. 9-1(b) for the pendulum. The switch closure (which

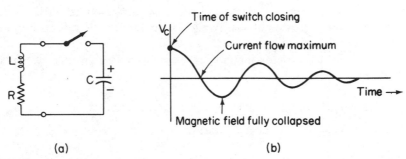

FIGURE 9-2 *Tank circuit "flywheel" effect: (a) Tank circuit; (b) "Flywheel" effect.*

corresponds to releasing the pendulum) starts a current flow as the capacitor begins to discharge through the inductor. The inductor, which resists a change in current flow, causes a gradual sinusoidal current buildup that reaches a maximum when the capacitor is fully discharged. (This corresponds to the pendulum just reaching its rest position after its initial release.) At this point the potential energy is zero, but since current flow is maximum, the magnetic field energy around the inductor is maximum (this corresponds to the kinetic energy of the pendulum). The magnetic field no longer maintained by capacitor voltage then starts to collapse, and its counter EMF will keep current flowing in the same direction, thus charging the capacitor to the opposite polarity of its original charge. The circuit losses (mainly the dc winding resistance of the coil) cause the output to become gradually smaller as this process repeats itself after the complete collapse of the magnetic field. The energy of the magnetic field has been converted into the energy of the capacitor's electric field, and vice versa. The process repeats itself at the natural or resonant frequency, f_0, as predicted by Eq. (3-18):

$$f_0 = \frac{1}{2\pi \sqrt{LC}} \tag{3-18}$$

For an *LC* tank circuit to function as an oscillator, an amplifier is utilized to restore the lost energy to provide a constant-amplitude sine-wave output. The resulting "undamped" waveform is known as a *continuous wave* in radio work. In the next section the most straightforward method of electronically restoring this lost energy is examined, and the general conditions required for oscillation are introduced.

9-3 BASIC *LC* OSCILLATORS

The *LC* oscillators are basically feedback amplifiers with the feedback serving to increase or sustain the self-generated output. This is called *positive feedback,* and it occurs when the fed-back signal is in phase with (reinforces) the input signal. It would seem, then, that the regenerative effects of this positive feedback would cause the output to continually increase with each cycle of fed-back signal. However, in practice, component nonlinearity and power-supply limitations limit the theoretically infinite gain.

The criteria for oscillation are formally stated by the *Barkhausen criteria* as follows:

1. The loop gain must be slightly greater than 1.

2. The loop phase shift must be n (360°), where $n = 1, 2, 3, \ldots$.

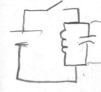

An oscillating amplifier adjusts itself to meet both of these criteria. The initial surge of power to such a circuit creates a sinusoidal voltage in the tank circuit at its resonant frequency, and it is fed back to the input and amplified repeatedly until the amplifier works into the saturation and cutoff regions. At this time, the flywheel effect of the tank is effective in maintaining a sinusoidal output. This process shows us that too much gain would cause excessive impurity (distortion) of the waveform and hence the gain should be limited to a level that is just greater than 1. This is necessary to maintain oscillations under all possible conditions.

The circuit shown in Fig. 9-3 will now be analyzed to show the basics of oscillator action. Oscillators of this form are known as *Franklin oscillators*. Note that two CE stages are utilized, which result in two 180° phase shifts or a total of 360° from input to output. Thus, the phase-shift criteria for an oscillator is satisfied. The circuit's resistor values must be set to cause a total loop gain of just greater than 1 so that the other Barkhausen criteria is met. When power is first applied to this

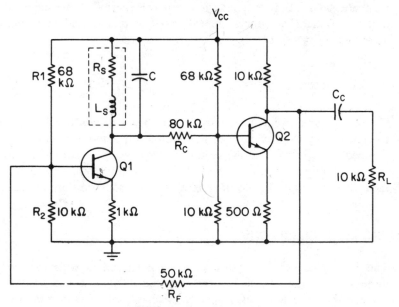

FIGURE 9-3 *Franklin oscillator.*

circuit, the *LC* tank circuit in Q_1's collector circuit starts a damped sinusoidal oscillation. That signal is fed to the base of Q_2 through the 80-kΩ coupling resistor. It is amplified by Q_2 and fed to the base of Q_1 through the 50-kΩ feedback resistor. The output of Q_1 (its collector) then starts this process over again.

A sine wave is created and the output is taken through the coupling capacitor (C_c) shown in Fig. 9-3. The 80-kΩ and 50-kΩ coupling resistors attenuate the signal while Q_1 and Q_2 provide gain. The total product of gain and attenuation should be just greater than 1 so that a continuous sine-wave output results.

The *LC* oscillators discussed in the rest of this section function on the same principles as the Franklin oscillator, but their action is somewhat more difficult to visualize. Their names are derived from some of the early pioneers in the field of radio, who developed these oscillators.

Hartley Oscillator

Figure 9-4 shows the basic *Hartley oscillator* in simplified form. The inductors L_1 and L_2 are a single center-tapped inductor. Positive feedback is obtained by mutual inductance effects between L_1 and L_2, with L_1 in the transistor output circuit and

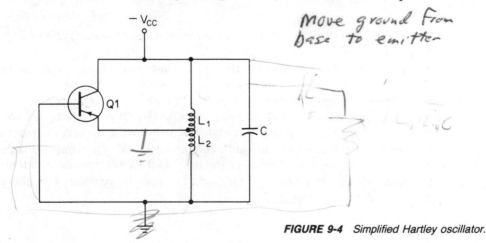

FIGURE 9-4   Simplified Hartley oscillator.

L_2 across the base emitter or input circuit. A portion of the amplifier signal in the collector circuit (L_1) is returned to the base circuit by means of inductive coupling from L_1 to L_2. As always in a CE circuit, the collector and base voltages are 180° out of phase. Another 180° phase reversal between these two voltages occurs because they are taken from opposite ends of an inductor (the L_1–L_2 combination) with respect to the inductor tap that is tied to the common transistor terminal—the emitter. Thus, the in-phase feedback requirement is fulfilled and loop gain is of course provided by Q_1. The frequency of oscillation is approximately given by

$$f_0 = \frac{1}{2\pi \sqrt{(L_1 + L_2)C}} \tag{9-2}$$

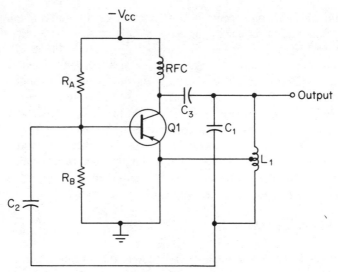

FIGURE 9-5 *Practical Hartley oscillator.*

and is influenced slightly by the transistor parameters and amount of coupling between L_1 and L_2.

Figure 9-5 shows a practical Hartley oscillator. A number of additional circuit elements are necessary to make a workable oscillator over the simplified one we used for explanatory purposes in Fig. 9-4. Naturally, the resistors R_A and R_B are for biasing purposes. The radio-frequency choke (RFC) is effectively an open circuit to the resonant frequency and thus allows a path for the bias (dc) current but does not allow the power supply to short out the ac signal. The coupling capacitor C_3 prevents dc current from flowing in the tank, and C_2 provides dc isolation between the base and the tank circuit. Both C_2 and C_3 can be considered as short circuits to the oscillator's frequency.

Colpitts Oscillator

Figure 9-6 shows a *Colpitts oscillator.* It is similar to the Hartley oscillator except that the tank circuit elements have interchanged their roles. The capacitor is now split, so to speak, and the inductor is single-valued with no tap. The details of circuit operation are identical with the Hartley oscillator and therefore will not be explained further. The frequency of oscillation is given approximately by the resonant frequency of L_1 and C_1 in series with the C_2 tank circuit:

$$f_0 = \frac{1}{2\pi \sqrt{[C_1 C_2/(C_1 + C_2)]L_1}}$$ (9-3)

$$C_1 = \frac{1}{\frac{1}{C_1} + \frac{1}{C_2}} = \frac{C_1 C_2}{C_1 + C_2}$$

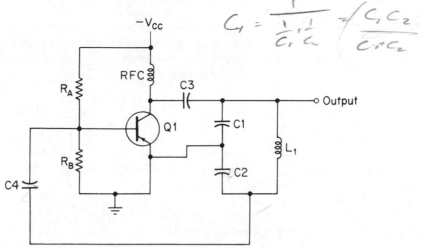

FIGURE 9-6 Colpitts oscillator.

The performance differences between these two oscillator forms are minor, and the choice between them is usually made on the basis of convenience or economics. They may both provide variable oscillator output frequencies by making one of the tank circuit elements variable.

Clapp Oscillator

A variation of the Colpitts oscillator is shown in Fig. 9-7. The *Clapp oscillator* has a capacitor C_3 in series with the tank circuit inductor. If C_1 and C_2 are made large enough, they will "swamp" out the transistor's inherent junction capacitances, thereby

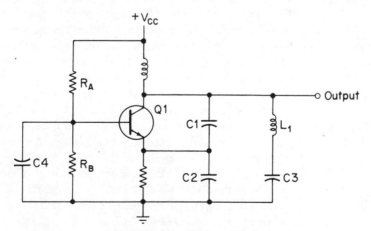

FIGURE 9-7 Clapp oscillator.

negating transistor variations and junction capacitance changes with temperature. The signal frequency is determined by the series resonant frequency of L_1 and C_3 when the series capacitance combination of C_1 and C_2 is much greater than C_3. Thus, the resonant frequency is

$$f_0 \simeq \frac{1}{2\pi \sqrt{L_1 C_3}} \tag{9-4}$$

and an oscillator with better frequency stability than the Hartley or Colpitts versions results. The possible range of frequency adjustment is not as large, however, with the Clapp oscillator.

9-4 OTHER OSCILLATOR CIRCUITS

The last three LC oscillators presented in this section are the ones most commonly used. However, many different forms and variations exist and are used for special applications.

Crystal Oscillator

When greater frequency stability than that provided by LC oscillators is required, a crystal-controlled oscillator is often utilized. A *crystal oscillator* is one that uses a piezoelectric crystal as the inductive element of an LC circuit. The crystal, usually quartz, also has a resonant frequency of its own, but optimum performance is obtained when it is coupled with an external capacitance. Applying an alternating electric potential across the two faces of the crystal results in mechanical vibrations that have maximum amplitude at the natural resonant frequency of the crystal.

The electrical equivalent circuit of a crystal is shown in Fig. 9-8. It represents the crystal by a series resonant circuit (with resistive losses) in parallel with a capacitance C_P. The resonant frequencies of these two resonant circuits (series and parallel) are quite close together (within 1%), and hence the impedance of the crystal varies sharply within a narrow frequency range. This is equivalent to a very high Q circuit, and in fact crystals with a Q-factor of 20,000 are common; a Q of up to 10^6 is possible. This compares to a maximum Q of about 1000 with high-quality inductors and capacitors. For this reason, and because of the good time and temperature stability characteristics of quartz, crystals are capable of maintaining a frequency to $\pm 0.001\%$ over a fairly wide temperature range. The $\pm 0.001\%$ term is equivalent to saying ± 10 parts per million (ppm), and this is a preferred way of expressing such very small percentages. Note that $0.001\% = 0.00001 = 1/100,000 = 10/1,000,000 = 10$ ppm. Over very narrow temperature ranges or by maintaining the crystal in a small temperature-controlled oven, stabilities of ± 1 ppm are possible.

Crystals are fabricated by "cutting" the crude quartz in a very exacting fashion. The method of "cut" is a science in itself and determines the crystal's natural resonant

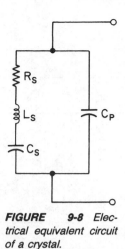

FIGURE 9-8 *Electrical equivalent circuit of a crystal.*

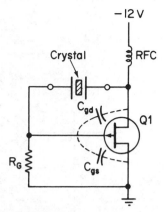

FIGURE 9-9 *Pierce oscillator.*

frequency as well as its temperature characteristics. Crystals are available at frequencies of about 15 kHz and up, with the higher frequencies providing the best frequency stability. However, at frequencies above 100 MHz, they become so small that handling becomes a problem.

Crystals may be used in place of the inductors in any of the previously discussed *LC* oscillators. A circuit especially adapted for crystal oscillators is the *Pierce oscillator,* shown in Fig. 9-9. The use of an FET is desirable, since its high impedance results in light loading of the crystal, provides for good stability, and does not lower the *Q*. This circuit is essentially a Colpitts oscillator with the crystal replacing the inductor and the inherent FET junction capacitances functioning as the split capacitor. Because these junction capacities are generally low, this oscillator is effective only at high frequencies.

A special advantage of the Pierce oscillator is the simple means of changing frequency. Since there are no tuned circuits, as such, the frequency can be changed by plugging a different crystal into the circuit at points *X* and *Y* in Fig. 9-9. This is a useful feature for many radio transmitters that are required to work at a number of different specific frequencies.

The high accuracy of the digital wristwatch can be attributed to the extreme accuracy of a crystal oscillator. A tiny quartz crystal provides the time base, which yields accuracy to a few seconds per month.

RC Phase-Shift Oscillator

If a sinusoidal oscillator is necessary for frequencies below about 10 kHz, *LC* oscillators become impractical. This is due to the bulk and expense of the high-valued inductors required. An alternative to the creation of sine waves via tank circuits is the use of *RC* selective filters. The *RC* phase-shift and Wien bridge oscillators are the two most common examples of *RC* selective filters.

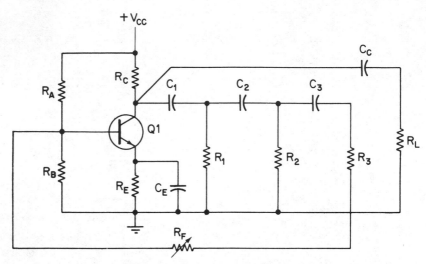

FIGURE 9-10   *RC phase-shift oscillator.*

Initially, one may think that the formation of a sine wave is impossible with the circuit of Fig. 9-10. The ability of this circuit to oscillate sinusoidally is based upon the fact that only one frequency can pass through the RC phase-shifting network with 180° of phase shift. This phase shift, coupled with the transistor's 180° input/output phase shift, means that only one sinusoidal frequency can successfully fulfill the Barkhausen phase-shift requirements, and hence regenerative effects occur for one frequency only and a sinusoidal output is possible.

Since the phase-shifting network must supply 180° phase shift, at least three RC sections must be used because a maximum of 90° phase shift can only be approached per section. Often four sections are used, since the attenuation introduced is actually less with four than with three sections.

The attenuation of the RC circuits must be compensated for by the gain of the transistor to allow a total loop gain greater than 1. Unfortunately, a gain much greater than 1 results in poor stability for this circuit, and it is often necessary to adjust the circuit's gain to the proper level to obtain satisfactory results. This can be accomplished by inserting a variable resistor in the feedback path, as shown in Fig. 9-10. In any event, a fairly high h_{fe} transistor must be used to overcome the RC network's losses.

The frequency of oscillation for the three-section RC oscillator when the three R and C components are equal is roughly approximated as

$$f_0 \simeq \frac{1}{18RC} \qquad\qquad\qquad (9\text{-}5)$$

The RC phase-shift oscillators do not lend themselves to frequency adjustment over a wide range because of the large number of resistors or capacitors that would have to be varied. In addition, a gain adjustment would be necessary because of the different attenuation that such adjustments would cause. Without gain adjustment the loop gain would subsequently drop below 1, causing oscillation to cease or become so much greater than 1 so as to cause instability. Distortion levels of 5% in the output signal are typical for this circuit.

Wien Bridge Oscillator

Whenever a wide range of frequencies is to be generated and a low distortion level is required, the *Wien bridge* oscillator shown in Fig. 9-11 is most often used. It is the circuit found in many laboratory sine-wave generators, which typically have a frequency range of 5 Hz to 500 kHz. The two capacitors in Fig. 9-11 are generally varied as a fine frequency adjust and a range switch may be used to switch the value of all components in the series RC and parallel RC circuits to allow large frequency changes. The resistors and capacitors are usually equal in value. The dashed lines between C_1 and C_2 indicate that these variable capacitors are "ganged" together (i.e., they are mechanically attached such that one control knob adjusts them both simultaneously). An operational amplifier is often used to provide the necessary gain.

     The basis of operation in this circuit is the fact that there is only one frequency that causes the fed-back signal to have zero phase shift, and it just happens to be the frequency that has the highest amplitude fed into the amplifier's input. For very low frequencies fed back into the series R_1–C_1 circuit, C_1 appears as an open circuit, thus making the voltage, v_{in}, very small. As the fed-back signal's frequency is gradually increased, v_{in} will increase until the shunt capacitor C_2 starts appearing as a short circuit, causing v_{in} to decrease. Thus, we see that the voltage, v_{in}, goes through a peak value, and this frequency happens to be the only one with zero phase shift from v_{out} to v_{in}. Thus, if the amplifier is noninverting, as with two CE stages cascaded

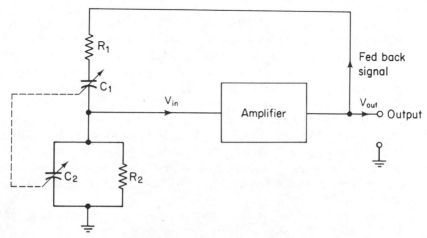

FIGURE 9-11 *Wien bridge oscillator.*

together or an op amp, the Barkhausen criteria are fulfilled for one frequency only and a stable oscillation results.

A mathematical solution of the series/parallel RC circuit shows that for equal values of R and C, the frequency of maximum fed-back signal and zero phase shift (and thus frequency of oscillation) is

$$f_0 = \frac{1}{2\pi RC} \tag{9-6}$$

9-5 COMMUNICATION SYSTEM NOISE EFFECTS

A communication system can be very simple, but it can also assume very complex proportions. Figure 9-12 represents a simple system in block-diagram form. Notice that the modulator accepts two inputs, the carrier and the information signal. It produces the modulated signal, which is subsequently amplified before transmission. Transmission can take place by any one of three means—antennas, waveguides, or transmission lines. The receiving unit of the system then picks up the transmitted signal but must reamplify it to compensate for attenuation that occurred during its transmission. Once suitably amplified it is fed to the demodulator (often referred to as the detector), where the information signal is extracted from the high-frequency carrier and fed to the power amplifier. The signal is brought to a suitably high level by the power amplifier to drive a speaker or any other load (output transducer).

A very important aspect of the field of electronic communications is the study of *electrical noise*. It may be defined as any form of energy tending to interfere with the reception and reproduction of electronic communication signals. It can be broadly categorized into two subgroups: that which is introduced in the transmitting medium (usually the atmosphere), and internal noise created within the transmitter and receiver.

The *external noise* affecting the reception of radio signals is caused by several different factors. *Atmospheric noise* is caused by lightning discharges and other naturally occurring electrical disturbances (such as cosmic radiation) in the atmosphere. It is generally heard as *static* in a radio receiver. Its frequency content is spread rather uniformly throughout the frequency spectrum up to about 30 MHz, above which it begins tapering off. Another source of external noise is the sun, and its emissions are termed *solar noise*. The sun is a variable source of noise, which reaches a peak of activity every 11 years. It affects some forms of communications drastically during these periods. The next peak in its emissions is due in 1990. Other stars besides our sun emit noise, and this is referred to as *cosmic noise*. Space noise is not significant below a frequency of 20 MHz or above 1.43 GHz (1 GHz = 10^9 Hz).

One final form of external noise is *man-made noise*. It is significant between the frequencies of 1 MHz and 600 MHz in most industrial areas. It is the dominant noise factor (considering both external and internal effects) in that range and is caused

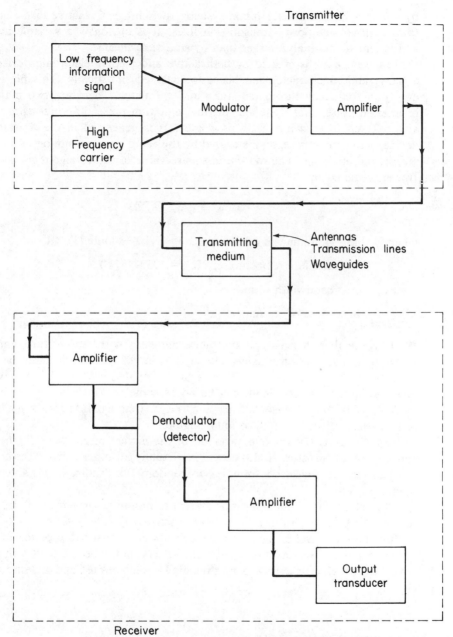

FIGURE 9-12 *Communication system block diagram.*

by automotive and other ignition systems, switching of reactive loads, and leakage from high-voltage power transmission lines. It is obviously a variable noise source and is difficult to analyze other than by statistical means.

Internal noise is created by both active and passive electronic devices. It is of a truly random character and hence is analyzed statistically. This noise is spread evenly in frequency throughout the radio spectrum. There are two major sources of internal noise, shot noise and thermal agitation noise. *Thermal agitation noise* is also referred to as *white* or *Johnson noise.* It is generated in the resistance of any device, active or passive, and is caused by the rapid random motion of its molecules, atoms, and electrons. The root-mean-square value of the voltage this noise produces (v_N) is predicted by

$$v_N = \sqrt{4kT\delta f R} \qquad \text{(9-7)}$$

where   k = Boltzmann's constant = 1.38×10^{-23} joule (J)/°K

            T = absolute temperature, °K (= 273° + °C)

            δf = bandwidth of interest

            R = value of resistance

We thus see that thermal agitation noise increases with temperature, bandwidth of the system, and resistance value. The peak value of this noise cannot be accurately predicted but seldom exceeds 10 times the calculated root-mean-square value. This noise, although usually small, can be bothersome at the front end of a high-gain receiver. If the input noise is 1 μv and the desired signal is also 1 μv, the output of this sensitive receiver will be unintelligible.

Shot noise is the random variation in the arrival of electrons (or holes) at the collector of a transistor. It also occurs in virtually all other active electronic devices. The paths taken by electrons (or holes) are random, thus leading to unwanted variations in current flow.

The ability of a communication system to minimize the effects of electrical noise to a great extent determines the receiver's sensitivity. The *sensitivity* may be defined as the minimum level of signal at the receiver's input that will produce an acceptable output signal at the speaker or headphones. This is not to be confused with a receiver's *selectivity,* which is its ability to discriminate between desired and undesirable frequencies.

9-6   AMPLITUDE MODULATION

Amplitude modulation (AM) is the process of varying the amplitude of a high-frequency sine wave (the carrier) in relation to the intelligence (information) signal. The amplitude of the carrier is made proportional to the instantaneous value of the modulating

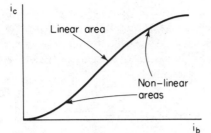

FIGURE 9-13 *Nonlinear transistor characteristics.*

voltage. It is produced by passing the carrier and information signal through a non-linear device. A transistor operating with very low bias or high bias has a nonlinear characteristic, as shown in Fig. 9-13. The result of passing two sine waves through a nonlinear device is shown in Fig. 9-14, and the output signal is an AM wave. Notice that its amplitude is varying at a frequency equal to the frequency of the information signal input. The AM waveform is made up of three sinusoidal components, one of them at the carrier frequency and also two sidebands. The upper sideband is at the carrier plus the modulating frequency, and the lower sideband is at the carrier minus the modulating frequency.

An AM waveform can be described mathematically as

$$v = V_c \sin \omega_c t + \frac{mV_c}{2} \cos (\omega_c - \omega_m)t - \frac{mV_c}{2} \cos (\omega_c + \omega_m)t \qquad \textbf{(9-8)}$$

where V_c = peak value of carrier

         ω_c = carrier's angular velocity $(2\pi f_c)$

         ω_m = information signal's angular velocity $(2\pi f_m)$

         m = modulation index

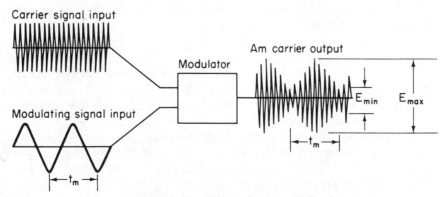

FIGURE 9-14 *AM waveform generation.*

The *modulation index, m,* is always ≤ 1 for AM. It is also expressed as a percentage and is then known as the percentage of modulation. It expresses a ratio of information signal level to carrier signal, and at 100% modulation (maximum) the V_{min} level shown in Fig. 9-14 is zero. The value of m can be calculated from the max and min values shown in Fig. 9-14 as follows:

$$m = \frac{V_{max} - V_{min}}{V_{max} + V_{min}} \tag{9-9}$$

It is often of interest to be able to calculate the amount of power contained in the sidebands as compared to the carrier. The following relationship allows for that calculation:

$$\frac{P_T}{P_C} = 1 + \frac{m^2}{2} \tag{9-10}$$

where P_T is the total transmitted power and P_C the carrier power. Once this ratio is calculated, the power in the carrier is

$$P_C = P_T - P_{SB} \tag{9-11}$$

and the power in the upper sideband equals the power in the lower sideband because they are equal. Hence,

$$P_{USB} = P_{LSB} = \frac{P_{SB}}{2} \tag{9-12}$$

The lower and upper sidebands are the second and third expressions in Eq. (9-8). The upper sideband is at the carrier frequency plus the information signal frequency $(f_c + f_m)$; the lower sideband is the difference frequency $(f_c - f_m)$.

Amplitude-modulated waveforms are usually generated by operating transistors in a nonlinear area. Figure 9-15 shows a very simple, yet effective means of accomplishing this. By making R_1 extremely large, the transistor is at very low bias (close to cutoff) and thus nonlinear. Other possibilities for AM generation include injecting one of the two signals at the base and the other at the emitter, and vice versa. Since many AM transmitters must generate high power outputs, they must use tubes in at least the output stages, since solid-state technology has not yet advanced very far into the high-power, high-frequency field.

Once the AM signal is generated, transmitted, received, and amplified, it becomes necessary to somehow extract the intelligence information from the signal. The circuit used almost exclusively to perform this function is the diode detector circuit shown in Fig. 9-16. The diode rectifies the incoming AM signal, leaving only the positive portion of the waveform, while the RC circuit effectively filters out the waveform's

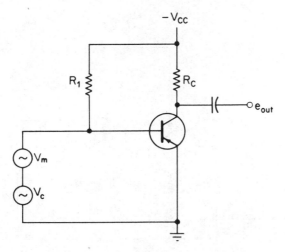

FIGURE 9-15 Simple AM modulator.

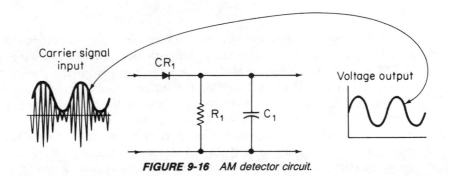

FIGURE 9-16 AM detector circuit.

high-frequency content. This then leaves only the outline of the positive portion of the AM wave, which is exactly the original modulating signal, as is desired. This detected signal is then suitably amplified to drive the final output transducer.

9-7 FREQUENCY MODULATION

Since phase modulation (PM) is seldom used and is so similar to frequency modulation (FM), we shall study FM only in this introduction to communications. In an FM system it is the frequency of the carrier that is made to vary in accordance with the modulating signal. This seems simple enough, but several subtleties often cause students to confuse the issue. The *rate* at which the FM carrier signal is made to vary is the rate (frequency) of the information (modulating) signal, and the *amount* of deviation from the original carrier frequency is made proportional to the strength (amplitude) of the modulating signal. Now carefully reread the previous sentence and then consider the simple FM transmitter shown in Fig. 9-17.

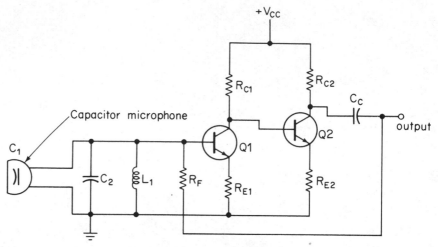

FIGURE 9-17 *Simple FM generation.*

The transmitter shown in Fig. 9-17 utilizes a capacitor microphone to generate an FM signal and is very helpful in aiding the student to understand the concept of FM generation. Let us say that you whistle into the microphone at a frequency of 1000 Hz. The sound waves striking the surface plate of the microphone cause its effective capacitance to vary at the frequency of the information signal, your whistle. Since the Q_1–Q_2 amplifier combination forms a simple Franklin-type oscillator circuit, it is seen that the frequency of oscillation is being changed at a 1000-Hz rate and will be rising and falling around the rest (carrier) frequency, since the whistle will cause the microphone's capacitance to rise and fall around its rest capacitance. The microphone's capacitance is part of the C_2–L_1 tank circuit that determines the frequency of oscillation.

The loudness of your whistle will also have an effect on this circuit. The loudness will affect the *amount* of capacitance change and therefore determine the amount of deviation from the center, rest, or carrier frequency, depending on how you wish to refer to it. Figure 9-18 provides a visualization of the FM-generated waveform. Notice that the FM signal's amplitude is nearly constant (ideally, it is perfectly constant). Its frequency, however, is varying at the frequency of the modulating signal.

Other more complex methods are usually used to generate FM than the capacitor microphone method just explained because of greater efficiency (greater frequency shift for a given amplitude-modulating signal). Regardless of the method of generation, the signal is ultimately amplified, transmitted, received, reamplified, and then must be detected. The FM slope detector is the most simple method available, and it involves converting the FM signal into an AM signal and then using the simple AM diode detector explained in Section 9-6. The conversion from FM to AM is accomplished by feeding the FM signal through a tank circuit tuned to a frequency somewhat above (or below) the FM carrier frequency. The tank output signal will

Generated carrier signal

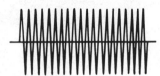

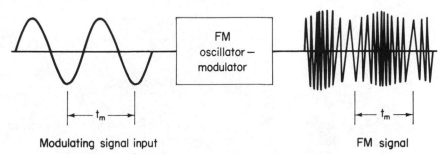

FIGURE 9-18 *FM signal generation.*

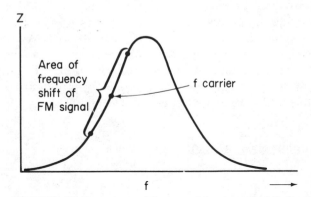

FIGURE 9-19 *Impedance (Z) versus f for a parallel LC circuit.*

then vary in amplitude according to the FM signal's frequency. This occurs because the tank presents a variable impedance to the FM signal, as shown in Fig. 9-19, and thus the output varies with frequency. This allows the original sinusoidal modulating signal to be reproduced at the detector's output. Figure 9-20 provides a circuit that performs this function with the waveforms shown to help you visualize the role of the three functions performed by the slope detector circuit.

Transmission of radio signals via FM techniques offers an advantage over AM systems with respect to minimization of noise effects. Since the receiver usually has a circuit to limit the signal to a constant amplitude, the effect of spurious noise pulses above that level is minimized. The FM system does not have information

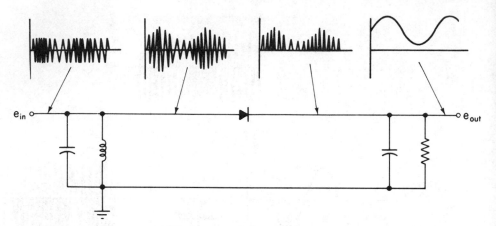

FIGURE 9-20   FM slope detector.

contained in amplitude changes as in AM. Thus, the amplitude changes caused by noise can be eliminated in FM systems without affecting your information. A more subtle source of noise reduction in FM receivers also occurs, but is beyond the scope of this book. It is, however, even more effective than the limiter action (just explained) in reducing noise effects.

The fidelity (quality) of reproduction on the standard FM stations (88 to 108 MHz) is better than standard AM stations (540 to 1600 kHz), primarily because of the greater bandwidth allocated for the FM stations. The FM stations operate within a 200-kHz bandwidth; AM stations use only 10 kHz. FM offers no great advantage over AM from an inherent standpoint, with respect to fidelity, but does perform better due to the increased bandwidth it is allocated.

9-8 TRANSMISSION METHODS

Antennas

The *wavelength* of any given frequency may be defined as the distance its wave travels in the time it takes to complete one cycle. Since radio waves travel at nearly the speed of light, c (3×10^8 m/s), the wavelength of any radio wave can be found by the following mathematical relationship:

$$\text{wavelength} = \lambda = \frac{c}{f} \qquad \textbf{(9-11)}$$

For an antenna to effectively radiate or receive a radio wave, it is usually necessary for it to be at least $\lambda/4$ in length or greater. An exception to this rule is the ferrite-core loopstick antenna common to most standard AM radio receivers. Since the standard AM broadcast band is such a low frequency, by radio standards, its wavelengths are quite long. For instance, at 1000 kHz we can calculate λ/4 as

$$\frac{\lambda}{4} = \frac{c}{4f} = \frac{3 \times 10^8 \text{ m/s}}{4 \times 1 \times 10^6 \text{ Hz}} = 75 \text{ m}$$

This means that the transmitting antenna for this station would be around 75 m tall (over 200 ft), which of course is out of the question for the receiver.

The transmitting antenna's function is to take the electrical energy from the transmitter and convert it into electromagnetic energy, which is capable of traveling through the atmosphere and most other dielectrics (nonconductors). The receiving antenna's function is to reverse that process—to change the electromagnetic energy into electrical energy. Generally, transmitting and receiving antennas are so similar that a *principle of reciprocity* exists. It states that all characteristics of antennas are identical whether they are being used for transmission or reception.

If an alternating current is passed through a conductor, a magnetic field *(H)* is set up around the wire and an electric field *(E)* is set up from one end to the other, as shown in Fig. 9-21. These two mutually perpendicular fields alternately receive energy from the electrical ac source as they build up, and return energy to the electrical source as they collapse. Whenever the electrical length of the wire is an appreciable fraction of the wavelength, a strange phenomenon takes place. The electromagnetic field does not collapse fully, but instead becomes propagated into space. The reciprocity

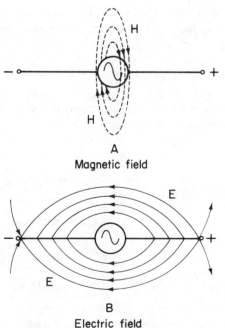

A
Magnetic field

B
Electric field

FIGURE 9-21 *Magnetic and electric fields around a wire.*

principle then tells us the reverse of this process occurs if a similar antenna is placed in the path of this radio wave. This receiving antenna then has a voltage and current induced in it, which is suitably amplified by the receiver circuitry.

The usual method by which these electromagnetic waves get from point *A* to point *B* is any one of the following:

1. Ground (surface) waves.

2. Sky-wave propagation—the ionosphere.

3. Space waves.

4. Satellite sky wave.

The waves from an antenna travel into space in all directions. Those which travel along the ground's surface are generally affected by the terrain features, and are called *ground waves*. These waves are attenuated by the earth's surface in varying degrees, depending upon the earth's conductivity and the frequency of the wave. Thus, the effective usable distance of transmission depends on the transmitted power level, the terrain, the frequency, and, of course, the receiver's sensitivity. This mode of propagation is normally useful at very low frequencies, owing to the otherwise high attenuation.

Sky waves depend upon the refractive effects on the transmitted wave by the various gases in the earth's upper atmosphere (the ionosphere). They serve to bend some radio waves around back toward the earth. The various layers of the ionosphere are dependent upon radiation from the sun, and hence this form of propagation varies between day and night and with variations in the sun's radiation. They are labeled the D, E, F_1, F_2, and F layers, as shown in Fig. 9-22. The figure also shows the elimination of the D and E layers and the combining of the F_1 and F_2 layers into the F layer at night. Their refractive abilities are reduced as the transmitter's frequency is increased, until a point is reached where the wave passes on through the ionosphere into free space and is no longer sent back to earth.

Figure 9-23 shows that lower frequencies radiated from a transmitter are refracted

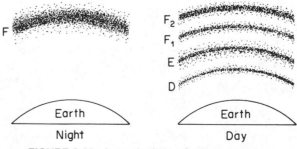

FIGURE 9-22 *Ionospheric layers about the earth.*

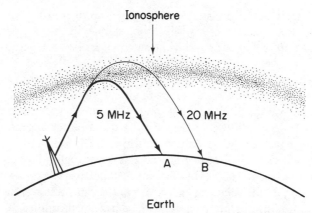

FIGURE 9-23 *Ionospheric refraction differences with frequency.*

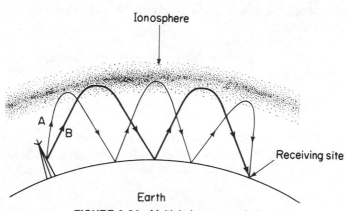

FIGURE 9-24 *Multiple-hop transmission.*

back to earth before higher ones. Reflections from the earth back to the ionosphere are shown in Fig. 9-24 and are known as *multiple-hop transmission*. The figure also shows the same signal "launched" from the antenna at different angles (as usually happens) and arriving at the same receiving site. If they arrive out of phase because of the different distances traveled, distortion will result in the receiver's output. Long-distance communication is possible with multihop transmissions. It is not uncommon for a link to occur halfway around the world.

Space waves behave very simply compared to sky waves. They travel directly (in relatively straight lines) between the transmitting and receiving antennas. They are therefore limited by the curvature of the earth and the height of the antennas. The following equation approximates this effective "line-of-sight" distance as

$$d \simeq \sqrt{2H_t} + \sqrt{2H_r} \qquad \text{(9-12)}$$

where d = useful distance of propagation, in miles
            H_t = height of transmitting antenna, in feet
            H_r = height of receiving antenna, in feet

Space-wave propagation is the usual mode for VHF frequencies and above, since they are too high to be refracted by the ionosphere and they are too greatly attenuated as ground waves. Standard FM and TV signals propagate via space waves. This explains the 50- to 70-mile line-of-sight distance for reception of these signals. On the other hand, standard AM signals are usually ground wave during the day and also sky wave at night. The combined F layer at night allows for skywave propagation at the 550-kHz to 1600-kHz AM band frequencies. This explains the long-distance reception that can occur on the AM band during the evening.

The use of *satellite communication* has mushroomed in recent years. A signal is beamed toward a satellite whose orbit is such that its position relative to the earth is constant. The satellite receives the signal, amplifies it, and retransmits back to earth. It uses solar-recharged batteries to maintain electrical power for its circuitry. Extremely reliable long-distance communication has been made possible by communication satellites.

As you know, antennas come in a wide variety of shapes and sizes. This is to provide the best characteristics for a given application. An analysis of the many varieties of transmitting and receiving antennas is beyond this book's intentions.

Transmission Lines

Although antennas are the most common method of transmitting radio energy, transmission lines and waveguides are also often used. The mode of energy transmission chosen for a given application would normally depend on the following factors:

1. Initial cost and long-term maintenance.

2. Frequency band and information-carrying capacity.

3. Selectivity or privacy offered.

4. Reliability and noise characteristics.

5. Power level and efficiency.

Naturally, any one mode of energy transmission will have only some of the desirable features. It therefore becomes a matter of sound technical judgment to choose the mode of energy transmission best suited for a particular application.

Transmission of energy down to zero frequency is practical with transmission lines, but waveguides and antennas inherently have a practical low-frequency limit. In the case of antennas, this limit is about 100 kHz, and for waveguides it is about 300 MHz. Theoretically, antennas and waveguides could be made to work at arbitrarily low frequencies, but the physical sizes required would become excessively large. How-

Dielectric material

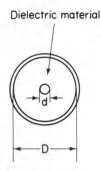

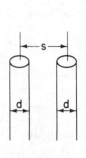

FIGURE 9-25 *Transmission-line types*
and geometry.

(a) Parallel-wire   (b) Coaxial

ever, with the low gravity and lack of atmosphere on the moon, it may be feasible
to have an antenna 10 miles high and 100 miles long for frequencies as low as a
few hundred hertz. As an indication of the sizes involved, it may be noted that for
either waveguides or antennas the important dimension is normally one half-wavelength. Thus, a waveguide for a 300-MHz signal would be about the size of a roadway
drainage culvert, and an antenna for 300 MHz would be about 1½ ft long.

A *transmission line* is simply two electrical conductors used to transmit electrical
energy. Thus, any two conductors can be technically considered to be a transmission
line. However, the consideration that separates transmission line study from simple
conductors is the effects of operation at high frequencies. At high frequencies some
new line characteristics are encountered other than just the resistive losses considered
at lower frequencies.

An important concept in the study of transmission lines is the characteristic
impedance of the line. The *characteristic impedance, Z_0,* is defined as the impedance
seen looking into an infinitely long line. This impedance can be calculated based
upon the physical dimensions of the line. Figure 9-25 shows the parallel-wire and
coaxial types of transmission line and their important dimensions. The characteristic
impedance of the parallel-wire transmission line shown in Fig. 9-25(a) is

$$Z_0 = 276 \log \frac{2s}{d} \ \Omega \qquad (9\text{-}13)$$

where s and d are as noted in Fig. 9-25(a).

The coaxial line shown in Fig. 9-25(b) consists of an inner conductor supported
by some dielectric material and an outer shield conductor that is usually grounded.
Its characteristic impedance is

$$Z_0 = \frac{138}{\sqrt{K}} \log \frac{D}{d} \ \Omega \qquad (9\text{-}14)$$

reletive

where K is the ᵛdielectric constant of the dielectric and D and d are as noted in
Fig. 9-25(b).

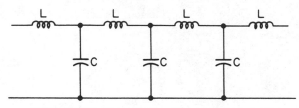

FIGURE 9-26   *Transmission-line equivalent circuit.*

A transmission line may be simulated at high frequencies by the *LC* circuit shown in Fig. 9-26. This is a somewhat idealized approach since no resistive losses are shown, but this is usually a justifiable simplification. The values of *L* and *C* for a given line are usually specified per unit length, because they occur continuously along the line. They cannot be assumed to be lumped at any one point. It can be mathematically shown that a line's characteristic impedance is given by

$$Z_0 = \sqrt{\frac{L}{C}} \ \Omega \qquad\qquad (9\text{-}15)$$

and thus still another means of determining Z_0 is available to us.

If a line is terminated in a resistance equal to the line's characteristic impedance, all the energy transmitted by a generator will be absorbed at the load. In addition, the line's input impedance will still be Z_0 in spite of the fact that the line is not infinite in length. Unfortunately, the load to which a line is delivering radio-frequency energy cannot often be made exactly equal to the purely resistive characteristic impedance. Under those conditions a portion of the energy delivered to the load is reflected back toward the generator. Methods of minimizing this inefficiency are a basic goal of transmission line work.

At low frequencies we are only concerned with the variations of voltage with time along a conductor. At high frequencies, however, the voltage along a conductor (transmission line) varies with distance also, since the line length is usually an appreciable fraction of a wavelength. Recall that a sine wave goes through one complete cycle in a distance of one wavelength. *Therefore, in transmission-line work we must be concerned with voltage variations with time at one specific point and also with voltage variations with location at one specific instant of time.*

Waveguides

A *waveguide* may be crudely though of as a coaxial transmission line with the center conductor removed. The energy is propagated in the electromagnetic mode as with antennas, but instead of radiating into space, the wave is contained by the metallic inner walls of the waveguide. Waveguides are seldom used below 3000 MHz because of size considerations. The most common shape is rectangular, and the longer of the two inner dimensions must be at least a half-wavelength. Circular waveguides are also commonly used. At high frequencies waveguides have less attenuation than transmission lines and can handle much higher power levels.

In recent years, the transmission of information using glass fibers has begun. The electrical information is converted to light energy via a light-emitting diode (LED) or a laser and then propagated in the glass fiber. The mode of propagation is identical to that in waveguides. The difference is simply the frequency of the electromagnetic energy. At the receiving end a photocell (see Chapter 13) can be used to convert from light to electrical energy. The fiber optics communication field is now expanding rapidly and is especially attractive in telephone communication links. A single glass fiber can handle many calls simultaneously and thus replace much copper wire.

QUESTIONS AND PROBLEMS

9–1–1.   Give a definition of the term "modulation." Describe the three fundamental characteristics of a sine wave that can be varied in the modulation process.

9–2–2.   What does an oscillator do? List the three broad classifications into which oscillators fall.

9–2–3.   Fully explain the analogy between a pendulum and an LC tank circuit.

9–3–4.   Explain the Barkhausen criteria for oscillation. Describe the relationship to positive feedback in your explanation.

9–3–5.   Draw a schematic for a Franklin oscillator and briefly describe the process whereby it generates and sustains a sine-wave output.

9–3–6.   Draw schematics for Hartley and Colpitts oscillators. Briefly explain their operation and their differences.

9–3–7.   Describe the reason that a Clapp oscillator has better frequency stability than the Hartley or Colpitts oscillators.

9–4–8.   List the major advantages of crystal oscillators over the LC varieties. Draw a schematic for a Pierce oscillator.

9–4–9.   The crystal oscillator time base for a digital wristwatch yields an accuracy of ±15 s/month. Express this accuracy in parts per million (ppm).

9–4–10.   Describe the process whereby an RC phase-shift oscillator is able to generate a sine wave and sustain it.

9–4–11.   Provide a schematic for a three-section RC phase-shift oscillator. Calculate the frequency of oscillation if the three resistors are 1.5 kΩ and the three capacitors are 0.1 μF.

9–4–12.   Draw a Wien bridge oscillator schematic using an operational amplifier. Label values that will provide an oscillation frequency of 120 kHz.

9–5–13. Briefly explain the function of each block in the communication system illustrated in Fig. 9-12.

9–5–14. Define electrical noise and briefly describe the major types of the external and internal forms.

9–5–15. Discuss the importance of noise considerations with respect to a highly sensitive communications receiver.

9–5–16. Calculate the noise voltage generated by a 1-MΩ resistor at 27°C in the audio-frequency range 20 Hz to 20 kHz.

9–6–17. Describe amplitude modulation and the way that it can be generated.

9–6–18. Sketch an AM waveform that illustrates about 50% modulation if the maximum p-p voltage is 4 V. Label the value of V_{min}. Calculate the carrier and sideband power if this AM signal is driven into an antenna that looks like a 50-Ω resistor.

9–6–19. Show how an AM signal can be demodulated at a receiver. Provide a block diagram of an AM receiver.

9–7–20. What is frequency modulation? Explain the effect of the amplitude and frequency of the information signal on the carrier in an FM modulating system.

9–7–21. Explain how the FM slope detector is able to convert an FM signal into an AM signal and then convert back to (detect) the information signal.

9–7–22. Describe why FM systems tend to be more immune to noise effects than are AM systems.

9–8–23. Explain the function of an antenna. Define wavelength (λ) and calculate the length of a $\lambda/4$ antenna operating at 88 MHz.

9–8–24. Explain the antenna principle of reciprocity.

9–8–25. List and briefly describe the four methods whereby electromagnetic waves get from one antenna to another.

9–8–26. Calculate the distance of useful propagation of space waves for a transmitting antenna height of 200 ft and receiving antenna height of 15 ft.

9–8–27. Explain why broadcast band FM and TV reception are limited to about 70 miles, whereas AM reception can be thousands of miles during the evening.

9–8–28. Define what is meant by a transmission line and its characteristic impedance.

9–8–29. A transmission line is 1 λ long. It is propagating a sine wave. Sketch the voltage versus time for a single point on the line and also the voltage versus *distance* on this line.

9–8–30. Describe a waveguide and explain when it might be used in preference over a transmission line.

9–8–31. Provide a simple block diagram for a fiber optics communication system.

10

DIGITAL CIRCUITS

10-1 BASIC CONCEPTS

Every individual has in some way been affected by recent advances in integrated-circuit technology. Specifically, the great strides made in the miniaturization of digital circuits has changed the life-styles of many of us. It is very likely that this morning you awoke to the beeping of a digital alarm clock, made breakfast in a digitally controlled radar range, and slipped on your digital watch (with alarm and calculator capabilities).

Today's student of electronics cannot afford to avoid the topic of digital electronics, since digital circuits and systems have found their way into areas that were traditionally considered linear or analog. Starting from the technician's education, where the hand-held calculator has replaced the slide rule all the way to where we now have major advances in microprocessor-based industrial control applications, an understanding of digital electronics is as important as an understanding of basic ac and dc circuits.

In this chapter basic digital circuits will be introduced and a comparison of digital and analog circuits made. In Chapter 11 digital systems will be briefly described, and Chapter 12 contains an introduction to the basics of microprocessors and microcomputers.

Most measured quantities in the real world are analog variables (linear); that is, they vary in a continuous manner. For example, a standard wristwatch has hands that are swept in continuous motion indicating the passage of time. On the other hand, a digital watch provides the time of day in the form of decimal digits which represent hours, minutes, and seconds. The digital watch does not change in a continuous manner, but in discrete steps of one per second (or one per minute). Because of this discrete representation, there is less likelihood of ambiguity in the reading of a value in a digital system as compared to an analog system.

Digital electronics has become very popular because of some distinct advantages it has over analog systems.

1. Digital systems can be *easier to design* based on two aspects of digital circuits. First, digital circuits function on either the presence or absence of a proper signal and are not concerned with the exact voltage (within certain limits) of that signal. Therefore, we are dealing with switching circuits and not linear amplifiers. Second, digital systems are made up by configuring (combining) a number of simple basic logic circuits.

2. *Storage* of information for relatively long periods of time is quite simple in digital systems, whereas in analog circuits it is very difficult.

3. The major advantage of digital systems is that they can be *fabricated into integrated-circuit chips.* Based on the fact that a complete system can be built up by interconnecting many basic digital devices, digital circuits become prime candidates for the integrated circuit process.

In digital systems, the information being operated on is usually present in binary form. Any device that has only two operating states can represent information in binary. For example, a switch can represent binary information by defining a binary 0 as an open switch condition and binary 1 as a closed switch position. Other devices, such as light-emitting diodes, relays, photocells, and regular diodes, can all be used to represent binary information.

In digital electronic systems, binary information is represented in the form of an electrical signal present at the inputs and outputs of various digital circuits. A binary 0 can be represented by a voltage range of, for example, 0.0 to 0.8 V, with binary 1 representing a voltage range 2.0 to 5.0 V. Since binary information is typically changing in digital systems, we end up dealing with pulse waveforms. Shown in Fig. 10-1 is binary information represented as two voltage levels; a logic 0 is shown

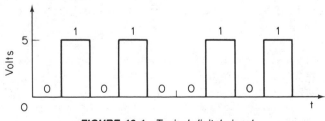

FIGURE 10-1 *Typical digital signal.*

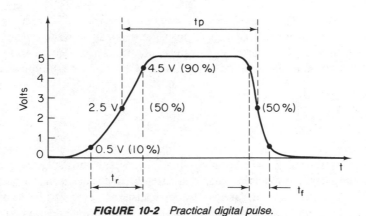

FIGURE 10-2 *Practical digital pulse.*

as 0 V and a logic one as 5 V. These are also referred to as a logic low (LO) and a logic high (HI), respectively.

The pulse waveform shown in Fig. 10-1 is an ideal pulse waveform since it has instantaneous changes in voltage levels. When dealing with real-world circuits it is important to become familiar with some of the characteristics of fast-changing practical digital signals. Figure 10-2 shows a single practical digital pulse waveform. The first characteristic that should be noted is that it takes a finite amount of time for voltage level changes and that the time to go from a logic 0 to a logic 1 need not be the same as the time to go from a logic 1 to a logic 0. The pulse's *rise time, t_r,* is defined as the time it takes for the signal to rise from 10% to 90% of its amplitude. Similarly, the pulse's *fall time, t_f,* is the time it takes for the signal to drop from 90% to 10% of its amplitude. In digital circuits t_r and t_f are typically in the range of microseconds or nanoseconds. The *pulse width, t_p,* is defined as the amount of time between the 50% points shown on the waveform.

Another important characteristic is shown in Fig. 10-3, where we see that there is a certain amount of delay in the response of a digital circuit to a change in the

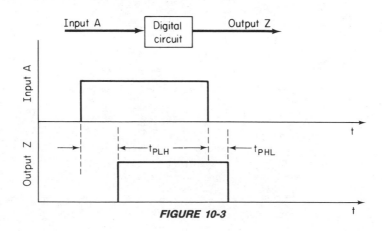

FIGURE 10-3

input signal. This delay is called *propagation delay, t_{pd}.* There are really two propagation delay specifications: t_{PLH} is propagation delay for the output going from a low to high, and t_{PHL} is propagation delay for the output going from a high to low. In digital circuits, propagation delay times of 1 to 30 ns are typical.

10-2 LOGIC GATES

A *logic gate* for our purposes is an electronic circuit that makes logic decisions. These gates will be the basic circuits from which most digital systems will be built. Logic gates are available today in integrated circuit form in various families. The most popular families are transistor-transistor logic (TTL), emitter-coupled logic (ECL), metal-oxide-semiconductor (MOS), and complementary metal-oxide-semiconductor (CMOS). We will look at the structure and performances of these families later; now, an introduction to the various basic gates and their function is in order.

OR gate: One way to construct an *OR gate* is shown in Fig. 10-4(a). In order for an output voltage to be present, diodes D_1 or D_2 (or both) must be forward-biased. The table shown in Fig. 10-4(b) lists the possible input combinations and the corresponding output voltages. This table is often referred to as a *truth table* and completely describes the functioning of the logic circuit. The more frequently used form of the truth table is shown in Fig. 10-4(c), where the actual voltages

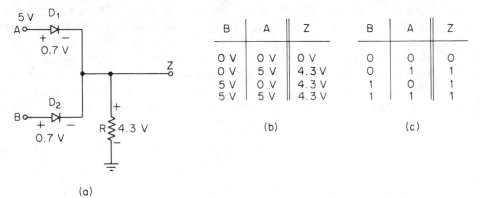

B	A	Z
0 V	0 V	0 V
0 V	5 V	4.3 V
5 V	0 V	4.3 V
5 V	5 V	4.3 V

(b)

B	A	Z
0	0	0
0	1	1
1	0	1
1	1	1

(c)

(a)

FIGURE 10-4 OR-gate.

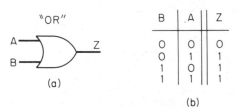

(a)

B	A	Z
0	0	0
0	1	1
1	0	1
1	1	1

(b)

FIGURE 10-5 OR-gate symbol and truth table.

have been replaced by logic 0's (LO) and logic 1's (HI): This table defines the logic OR operation since the output is 1 (HI) when inputs A or B, or both, are 1 (HI). The OR operation can be implemented by various electronic devices in certain circuit configurations. Rather than drawing the schematic of the logic gate, a symbol for the OR operation is used; it is shown in Fig. 10-5(a) and the corresponding truth table in Fig. 10-5(b).

AND gate: Another basic logic gate is shown in Fig. 10-6. By examining the circuit and its truth table, we see that the output is 1 only when inputs A and B are both 1. Therefore, the AND operation is defined as a circuit having two or more inputs whose output goes to the logic 1 state only when all the inputs are in the logic 1 state. Figure 10-7 shows the AND symbol and its associated truth table

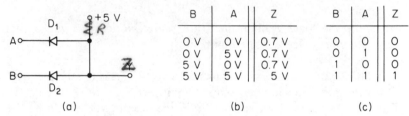

B	A	Z		B	A	Z
0 V	0 V	0.7 V		0	0	0
0 V	5 V	0.7 V		0	1	0
5 V	0 V	0.7 V		1	0	0
5 V	5 V	5 V		1	1	1

(a)   (b)   (c)

FIGURE 10-6   Diode AND-gate symbol and truth table.

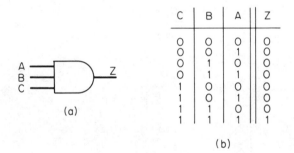

C	B	A	Z
0	0	0	0
0	0	1	0
0	1	0	0
0	1	1	0
1	0	0	0
1	0	1	0
1	1	0	0
1	1	1	1

(a)

(b)

FIGURE 10-7 Three-input AND-gate symbol and truth table.

for a three-input *AND gate*. Notice how the truth table shows all the possible input combinations and the outputs corresponding to those inputs. How do you go about arriving at all the possible input combinations? First, since there are three inputs, we have $2^3 = 8$ possible input combinations. Looking at the rightmost input column (A), you should see that the logic values alternate between 0 and 1 every other row, eight times, starting with a 0. The next input column to the left (B) alternates every two rows starting with a 0. The leftmost input column (C) alternates every four rows, starting with a 0.

NOT circuit: One of the most fundamental logic circuits is the *NOT* (inverter) *circuit* shown in Fig. 10-8(a). By observing the circuit and the truth table in (b), you can see that it is simply a common emitter inverter circuit. Figure 10-8(c) and

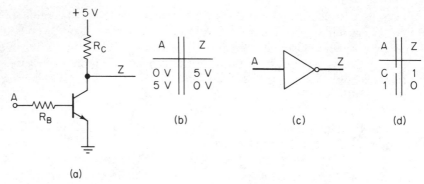

(a)

FIGURE 10-8 *NOT (inverter) circuit and truth table.*

(d) show the logic symbol and truth table. Notice that the output is always the opposite voltage level of the input.

NOR gate: One way to construct a *NOR gate* is to take the previously described OR gate and add onto the output an inverter or NOT circuit. This is shown in Fig. 10-9(a). If you look at point Y in the circuit, you would find that the voltages

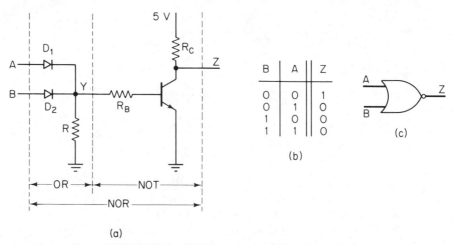

(a)

FIGURE 10-9 *NOR-gate, truth table and symbol.*

at this point conform to the OR gate outputs for various input combinations, as shown in Fig. 10-4. If these output values are inverted, the output for the NOR gate is obtained. The logic NOR operation is then defined as a circuit with two or more inputs whose output is a logic 1 (HI) only when all the inputs are logic 0 (LO). The symbol for the NOR operation is the same as the OR with the addition of a circle on the output, as shown in Fig. 10-9(c).

NAND gate: The construction of a *NAND gate* can once again be accomplished by the combining of the previously described AND gate in Fig. 10-6 with a NOT (inverter) circuit. Figure 10-10(a) and (b) show the circuit and the corresponding truth table. As can be seen from the truth table, the NAND operation can be defined

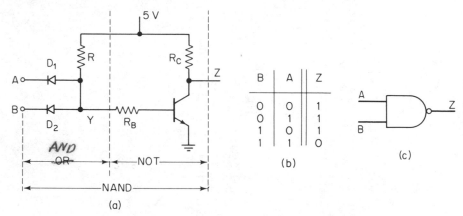

FIGURE 10-10   *NAND-gate, truth table and symbol.*

as a circuit with two or more inputs whose output is a logic 0 (LO) only when all the inputs are a logic 1. The symbol for the NAND operation is the same as that for the AND with the addition of a small circle on the output, as shown in Fig. 10-10(c).

10-3   BOOLEAN ALGEBRA

An Englishman by the name of George Boole, many years back, came up with a neat and simple mathematical way to represent complicated logical statements. Logic statements are defined as statements that can either be true or false. This system is referred to as *Boolean algebra* and has distinct application to the field of digital electronics, where variables that have only two states, either 1 (true) or 0 (false), are dealt with. Therefore, the output of each gate that has been described previously can be expressed as an algebraic expression. The basic Boolean algebra symbols (or operators) are the plus sign ($+$), which represents the OR operation; the multiplication sign ($\cdot$), which represents the AND operation; and the overbar ($-$), which represents the NOT (inversion) operation. The multiplication sign is sometimes omitted so that $A \cdot B$ and $A\,B$ are identical expressions. Figure 10-11 summarizes the basic logic symbols, truth tables, and Boolean expressions for their output. Notice how the Boolean expression for the NOR gate in Fig. 10-11(d) is derived and shows that the NOR gate output is the OR'ed output of A, B inverted, as indicated by the overbar over the expression $A + B$. The NAND gate shown in Fig. 10-11(e) is the AND'ed output of A, B inverted, as indicated by the overbar over the expression $A \cdot B$.

     Since digital systems are made up of interconnections of basic gates, and since each gate can be described using a Boolean expression, any system's output (or outputs)

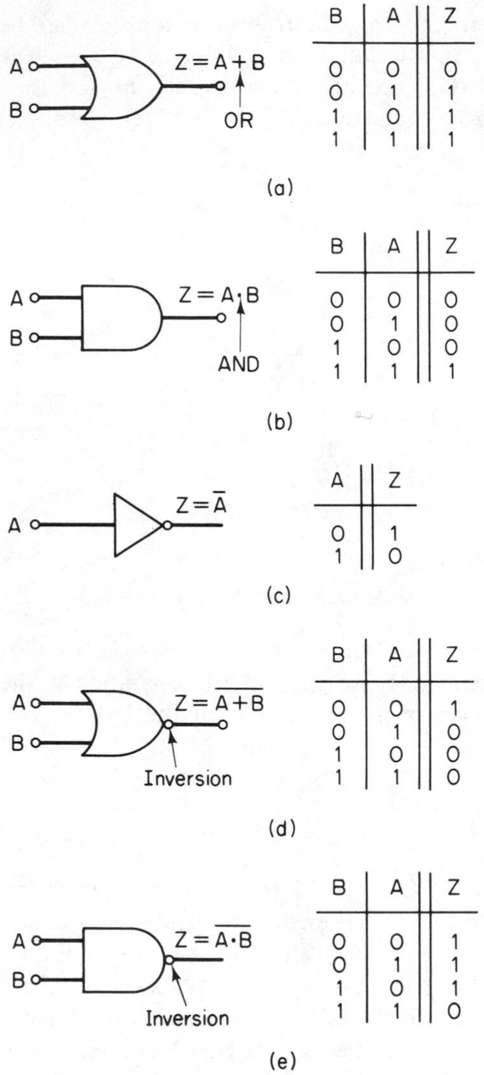

B	A	Z
0	0	0
0	1	1
1	0	1
1	1	1

$Z = A + B$

OR

(a)

B	A	Z
0	0	0
0	1	0
1	0	0
1	1	1

$Z = A \cdot B$

AND

(b)

A	Z
0	1
1	0

$Z = \overline{A}$

(c)

B	A	Z
0	0	1
0	1	0
1	0	0
1	1	0

$Z = \overline{A + B}$

Inversion

(d)

B	A	Z
0	0	1
0	1	1
1	0	1
1	1	0

$Z = \overline{A \cdot B}$

Inversion

(e)

FIGURE 10-11 *Summary of logic gates, Boolean expression for output and truth table.*

can be described (and evaluated) by a Boolean expression involving the basic Boolean operations described up to this point. Figure 10-12(a) shows a relatively simple digital circuit whose output expression is found by writing the Boolean expression for each gate as you progress through the circuit from left to right. If we were simply given the output expression, $Z = A + B \cdot C$, there *might* be two possible interpretations for this expression. First, A is OR'ed to the term $B \cdot C$, or that C is AND'ed to the term $A + B$. In a regular algebraic expression, we could have $X = 3 \times 2 + 5$, which could be interpreted as the term 3×2 summed to 5, which would make $X = 11$; or the term $2 + 5$ multiplied by 3, which makes $X = 21$. We should know that the correct interpretation is that multiplication precedes addition unless there

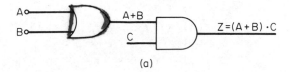

$$Z = (A + B) \cdot C$$

(a)

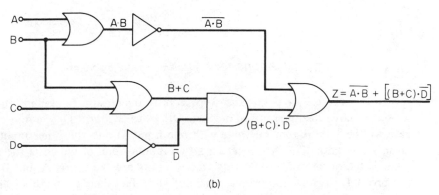

$$Z = \overline{A \cdot B} + \left[(B+C) \cdot \overline{D} \right]$$

(b)

FIGURE 10-12   *Logic circuits and their output expressions.*

are *parentheses* in the expression, in which case the operation inside the parentheses is performed first. The same holds true for the Boolean expression, in that AND'ing comes before OR'ing unless there are parentheses, in which case the terms inside the parentheses are evaluated first. Therefore, the correct output expression for the circuit of Fig. 10-12(a) must include parentheses around the A + B term. Figure 10-12(b) is a somewhat more complex digital circuit to evaluate. See if you can come up with the output expression shown.

Three things can be done with the Boolean expression for the output of a digital circuit or system. First, you can evaluate the system by generating a truth table for it. Remember: a truth table is a list of all the possible input combinations and the output expected for each variation. Therefore, all you have to do is plug the various combinations of 1's and 0's into the circuit's expression, get the output, and fill in the truth table. Second, given a Boolean expression, the logic circuit for that expression can be implemented. Figure 10-13 shows the digital circuit for the expression $Z = [(\overline{A} + B) \cdot C] + D$. In a sense, obtaining the circuit from an expression is the

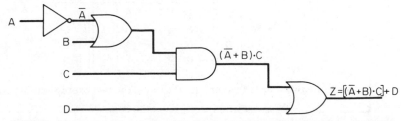

FIGURE 10-13   *Implementation of a logic circuit from the Boolean expression.*

(1) $Z + Z = Z$
(2) $Z + 0 = Z$
(3) $Z + 1 = 1$
(4) $Z + \bar{Z} = 1$
(5) $Z \cdot 0 = 0$
(6) $Z \cdot 1 = Z$
(7) $Z \cdot Z = Z$
(8) $Z \cdot \bar{Z} = 0$
(9) $Y + Z = Z + Y$

(10) $Y \cdot Z = Z \cdot Y$
(11) $X + (Y + Z) = (X + Y) + Z = X + Y + Z$
(12) $X(YZ) = (XY)Z = XYZ$
(13) $X(Y + Z) = XY + XZ$
(14) $Y + YZ = Y$
(15) $Y + \bar{Y}Z = Y + Z$
(16) $\overline{(Y + Z)} = \bar{Y} \cdot \bar{Z}$ ⎫ De Morgan's
(17) $\overline{(Y \cdot Z)} = \bar{Y} + \bar{Z}$ ⎭ theorems

FIGURE 10-14 *Boolean algebra theorems.*

reverse process used to obtain the expression from a circuit. Looking at the expression we see we have one quantity in brackets OR'ed to the term D. Therefore, the output gate will be a two input OR gate with one input D and the other input the output of an AND gate. The AND gate has two inputs—one of which is the term C, the other is the output of an OR gate whose inputs are the terms $\bar{A}$ and B. The third, and probably the most important, capability is the fact that, given a Boolean expression for a logic circuit, an attempt can be made to simplify that expression, reducing the number of terms and therefore the number of gates needed. This results in a more economical design.

The circuit expressions can often be simplified by use of Boolean theorems. Figure 10-14 lists the 17 common theorems of Boolean algebra. Any of the theorems can be shown to be correct by generating a truth table for the expression on each side of the equal sign. Once evaluated, the truth table outputs should be identical and, therefore, the expressions equal.

EXAMPLE 10-1

Show that theorem 16 in Fig. 10-14 is valid.

Solution:

Write a truth table for both sides of the equal sign and evaluate to compare the outputs, as shown below.

Y	Z	$\overline{Y + Z}$		Y	Z	$\bar{Y} \cdot \bar{Z}$
0	0	1		0	0	1
0	1	0		0	1	0
1	0	0		1	0	0
1	1	0		1	1	0

Since the two output columns are the same, the two expressions are equivalent.

EXAMPLE 10-2

Simplify the following expression:

$$R = XY + XYZ + \overline{X}Y + X\overline{Y}Z$$

Solution:

Looking at the expression, there are a number of ways to proceed for simplification, however, the correct use of theorems, whichever way started, should end up with the same final expression. If you start by using rule 13, you can factor out the XY from the first two terms, leaving

$$R = XY(1 + Z) + \overline{X}Y + X\overline{Y}Z$$

Using rule 3, the expression reduces to

$$R = XY(1) + \overline{X}Y + X\overline{Y}Z$$

Using rule 6 results in

$$R = XY + \overline{X}Y + X\overline{Y}Z$$

Looking at the partially reduced expression and using rule 9, terms can be rearranged, leaving

$$R = XY + X\overline{Y}Z + \overline{X}Y$$

from which, by using rule 13, you can factor an X from the first and second terms, leaving

$$R = X(Y + \overline{Y}Z) + \overline{X}Y$$

Using rule 15, this reduces to

$$R = X(Y + Z) + \overline{X}Y$$

Using rule 13 again leaves

$$R = XY + XZ + \overline{X}Y$$

Rearranging the terms by use of rule 9 leaves

$$R = XY + \overline{X}Y + XZ$$

Using rule 13 and factoring out a Y from the first two terms leaves

$$R = Y(X + \overline{X}) + XZ$$

Rule 4 reduces the term inside the parentheses to 1, leaving

$$R = Y(1) + XZ = Y + XZ$$

The final expression, then, is

$$R = Y + XZ$$

The reduced expression in Example 10-2 is obviously more economical than the original expression, because it will take many fewer gates to build. Additionally, the reduced circuit will operate at higher frequencies (more quickly), since there is less *propagation delay* involved in the circuit. Besides reducing the expression, the economy might be further increased by building the circuit with all one type of gate. This is accomplished by using what is referred to as the *universal gates,* which really refers to the NAND or NOR gate. Figure 10-15 shows how the NAND or NOR gate can be used to implement the basic logic operations. The economy comes from the fact that a manufacturer would only have to purchase and stock one type of circuit and would therefore probably get a better price.

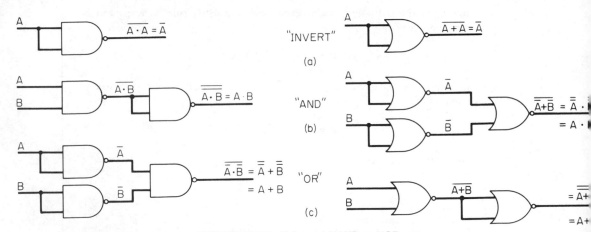

FIGURE 10-15 *Universal NAND or NOR gate.*

10-4 LOGIC CIRCUIT DESIGN

Section 10-3 showed how Boolean algebra is used to reduce circuit expressions to their simplest form. The question of how do we get from a verbal statement of a problem to a logic expression is going to be dealt with in this section.

B	A	Z
0	0	1
0	1	0
1	0	0
1	1	1

(a)

B	A	Z	
0	0	1	$\rightarrow \overline{A}\cdot\overline{B} = P1$
0	1	0	
1	0	0	
1	1	1	$\rightarrow A\cdot B = P2$

(b)

$$Z = \overline{A}\,\overline{B} + AB$$

(c)

FIGURE 10-16   *Sum-of-products.*

The *first* step is to clearly state the problem and define the input and output logic levels when necessary. For example, suppose that you need to determine the expression for a circuit that will compare two single-bit inputs and provide a logic 1 output if they are the same. The *second* step is to generate a truth table from the statement of the problem. Since there are two single-bit inputs, that means they only have two inputs and, therefore, four possible input combinations ($2^2 = 4$) with one output. The output will be a 1 whenever the inputs are identical. The truth table is shown in Fig. 10-16(a), and you should notice that there are only two cases where the output is high. The *third* step in the process is to look at the truth table and, wherever there is a logic 1 output, write an AND product (term) that will give a logic 1 output for the input conditions shown. Figure 10-16(b) shows the truth table and the AND products. Note the use of p_1 and p_2 to indicate product one and product two, respectively. The *fourth* step involves writing the expression that the output is a logic 1 whenever the p_1 term *or* p_2 term is a logic 1. Thus, $Z = p_1 + p_2$, which is the same as $Z = \overline{A} \cdot \overline{B} + A \cdot B$. This technique of obtaining an output expression is referred to as the *sum-of-products* form, since it is arrived at by summing (OR'ing) the individual product (AND) terms derived from the truth table. Once an expression is derived in this fashion, an attempt can be made to simplify using the Boolean theorems.

EXAMPLE 10-3

Determine an expression for a circuit that will compare two single-bit inputs and provide a logic 1 output if they are at opposite levels.

Solution:

First, generate the truth table for the expression of the problem.

B	A	Z
0	0	0
0	1	1
1	0	1
1	1	0

Next, develop product (AND) terms wherever the output is a logic 1.

B	A	Z	
0	0	0	
0	1	1 → $\overline{B} \cdot A$	
1	0	1 → $B \cdot \overline{A}$	
1	1	0	

Then, sum (OR) the product terms to get our expression for the output.

$$Z = \overline{B} \cdot A + B \cdot \overline{A}$$

If possible, the expression could be reduced using the Boolean theorems discussed previously.

The two previously stated problems of comparing bits to see if they are equal or opposite occurs so frequently in digital circuits that the expressions are given a special shorthand notation—the circuits are referred to by the special names of *exclusive-NOR* and *exclusive-OR,* respectively, and special symbols are used to represent these functions. Figure 10-17 summarizes these two new gates and symbols.

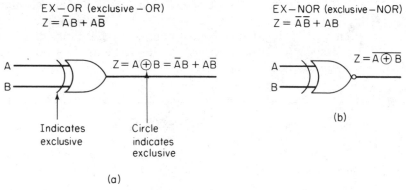

EX−OR (exclusive − OR)
$Z = \overline{A}B + A\overline{B}$

$Z = A \oplus B = \overline{A}B + A\overline{B}$

A
B

Indicates exclusive

Circle indicates exclusive

EX−NOR (exclusive−NOR)
$Z = \overline{A}\,\overline{B} + AB$

$Z = \overline{A \oplus B}$

A
B

(b)

(a)

FIGURE 10-17 *EX-OR and EX-NOR gates.*

10-5 FLIP-FLOPS

The logic circuits covered so far have outputs that represent the condition of the input signals at a specific instant in time. Any previous input-level conditions have no significance on the circuit or its output. In other words, these circuits have no *memory* and are referred to as *combinatorial* circuits.

This section introduces a group of circuits that have memory, that is, previous

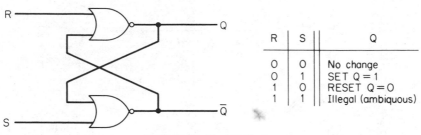

R	S	Q
0	0	No change
0	1	SET Q = 1
1	0	RESET Q = 0
1	1	Illegal (ambiquous)

FIGURE 10-18   *RS flip-flop.*

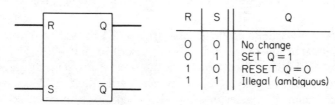

R	S	Q
0	0	No change
0	1	SET Q = 1
1	0	RESET Q = 0
1	1	Illegal (ambiquous)

FIGURE 10-19   *Flip-flop symbol.*

input conditions do influence the state of the output. The most popular memory element is the *flip-flop* (FF). The simplest FF circuit to construct is one that consists of two cross-coupled NOR gates, as shown in Fig. 10-18. This particular circuit arrangement is referred to as an R-S (Reset-Set) flip-flop. As can be seen from the truth table, the output of the FF can be controlled by the R and S inputs. This device can remember (store) information by making the R-S inputs both equal to zero. For example, with R = 0 and S = 1, the FF will be in the set state, which means that Q = 1, $\overline{Q}$ = 0. If S is now made zero, leaving R = 0 and S = 0, the FF will remain in the Q = 1 and $\overline{Q}$ = 0 state and, therefore, it remembers the previous input conditions. Notice, though, that R = 1 and S = 1 is an illegal condition, because it results in an ambiguous condition of both outputs being forced to the same logic level. This is improper for if one output is defined as Q and the other $\overline{Q}$, they cannot both be at the same logic levels and still be valid. The general symbol used to indicate the FF as a circuit element is shown in Fig. 10-19. Although the R-S FF just described is the most basic FF, the most popular FFs are the *synchronous* type. Since most digital systems perform a sequence of operations that are synchronized by a clock (pulse) signal, the most useful type of FFs are those that can synchronize their operation to this clock. These *synchronous* FFs have an additional input, labeled the clock input (CLK or CK), which will allow the FF to change states only on either the positive-going (rising) edge or negative-going (falling) edge of the clock signal. The R and S inputs now become preparatory inputs, in that they prepare the FF output to go to a particular state but the FF does not respond until the proper transition of the clock signal. Figure 10-20(a) shows a synchronous (edge-triggered) R-S FF that responds to positive transitions, and Fig. 10-20(b) shows a negative edge-triggered R-S FF.

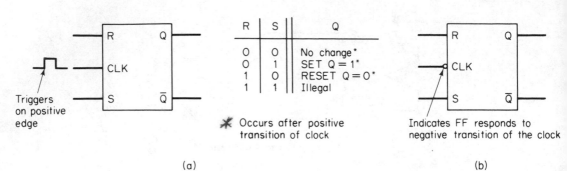

R	S	Q
0	0	No change*
0	1	SET Q = 1*
1	0	RESET Q = 0*
1	1	Illegal

Triggers on positive edge

✳ Occurs after positive transition of clock

Indicates FF responds to negative transition of the clock

(a)   (b)

FIGURE 10-20 *Synchronous (edge-triggered) flip-flop.*

The most popular of the edge-triggered FFs is the J-K FF. It operates in much the same fashion as the R-S FF just described, but when $J = 1$ and $K = 1$, instead of an illegal (ambiguous) condition, the J-K FF responds by toggling (changing states) after the occurrence of the proper clock transition. This occurs if $Q = 1$ and $\overline{Q} = 0$ and J and K are set equal to 1. When the proper clock transition occurs, the output will go to the opposite state of $Q = 0$, $\overline{Q} = 1$. Figure 10-21 shows the J-K FF symbol and truth table. The fact that the J-K FF does not have the illegal input condition makes it a much more versatile device.

Figure 10-22 shows the symbol and truth table for an edge-triggered D-type FF. The FF output will go to the state that is present on the D input after the occurrence of the proper clock transition. D FFs are usually used in systems involving

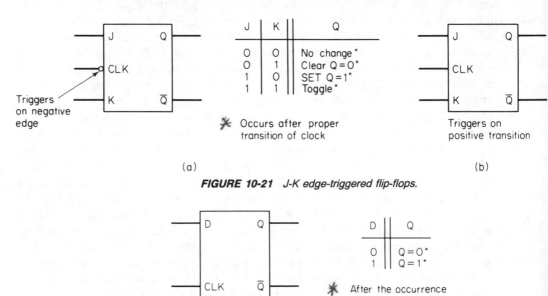

J	K	Q
0	0	No change*
0	1	Clear Q = 0*
1	0	SET Q = 1*
1	1	Toggle*

Triggers on negative edge

✳ Occurs after proper transition of clock

Triggers on positive transition

(a)   (b)

FIGURE 10-21 *J-K edge-triggered flip-flops.*

D	Q
0	Q = 0*
1	Q = 1*

✳ After the occurrence of the proper clock transition

FIGURE 10-22 *D-type edge-triggered FF.*

the transfer of information, since only one interconnecting line per FF is needed between the transmitter and receiver.

The edge-triggered FFs just described (J-K and D) are referred to as *synchronous* FFs and the preparatory inputs (R, S, J, K, D) are referred to as the synchronous inputs or control inputs. Most integrated circuit FFs also have asynchronous inputs. These additional inputs are referred to as *asynchronous* because they do not involve the use of the clock signal. As a matter of fact, they override the synchronous operation of the FF. For example, Fig. 10-23 shows an edge-triggered J-K FF with asynchronous inputs labeled *DC SET* and *DC CLEAR*. As long as these two inputs are kept HI, the FF behaves as a synchronous J-K FF. If, however, the DC SET is made a logic 0, the output will immediately be made Q = 1. It is important to realize that the asynchronous inputs are level-sensitive, meaning they respond to a logic LO signal for as long as it is present.

One important aspect of FF operation is a problem that occurs frequently in digital systems. Often, the outputs of one FF are connected to the preparatory inputs of another FF—both of which share a common clock signal. What results is that the preparatory input (for example, J-K) may be changing at the same time the clock transition is occurring. This could lead to unpredictable FF operation and therefore cause some problem in a digital system. There will be an example of where this problem occurs in counters that will be discussed in Chapter 11. To counteract the problem, a special circuit is used called a master-slave (M/S) FF. This special FF is really comprised of two FF's, one called the master and the other the slave, plus some control circuitry. Suppose that we have a negative edge-triggered J-K FF. During the positive portion of the clock signal, information on the J-K inputs is used to control the master FF. When the negative clock transition occurs, the J-K inputs are disabled and the information that was stored in the master FF is transferred to the slave FF, which then allows the outputs to change appropriately. With the use of M/S FF's, the problem of changing preparatory inputs at the time of the clock transition is eliminated since, even though the J-K inputs may have changing signals at the time of the negative clock transition, the slave FF will only respond to the outputs of the master FF, which is stable at this time.

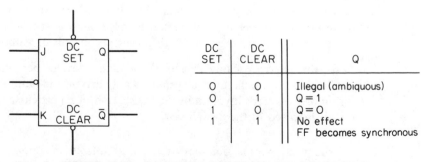

DC SET	DC CLEAR	Q
0	0	Illegal (ambiguous)
0	1	Q = 1
1	0	Q = 0
1	1	No effect
		FF becomes synchronous

FIGURE 10-23 *Edge-triggered JK FF with asynchronous inputs.*

10-6 LOGIC FAMILY CHARACTERISTICS

Most digital systems are made by combinations of the various logic functions we have discussed up to this point. All of the logic circuits discussed come in integrated-circuit form, of which the following are the most popular:

1. Transistor-transistor logic (TTL or T²L).

2. Emitter-coupled logic (ECL).

3. Metal-oxide semiconductor (MOS).

4. Complementary metal-oxide semiconductor (CMOS).

The scope of this book does not allow great detail concerning each family, but you should know their basic characteristics. With this knowledge in hand, you will have some idea, for example, why certain systems are built using CMOS as opposed to ECL.

There are many factors to consider in the choice of a particular IC family to use in a system design. Some of the most important factors are:

1. *Speed of operation*—this involves taking into account transition times (t_R and t_F) and propagation delays (t_{PLH}, t_{PHL}). Both of these times are increased as loading increases. The more inputs attached to an output, the more load that output has to handle.

2. *Noise immunity* (noise margin)—all of the circuits, to operate properly, require certain voltage ranges for logic 0's and 1's. Stray electric and magnetic fields can induce noise (spurious) signals in the connecting wires between the logic circuits and possibly cause unreliable operation. Therefore, the amount of noise voltage the various logic families can tolerate and still operate properly is of interest.

3. *Power dissipation*—the amount of power required by an integrated circuit is of interest in many systems since hundreds of circuits may be used. The resulting total power dissipation can be quite high—especially for the logic families that require relatively high power per circuit or gate.

4. *Cost*—things like packing density (how many devices per unit area), *fan-out* (how many inputs can one output reliably drive), and number of manufacturing process steps involved in the fabrication of a particular type of family are all involved in the overall cost.

One of the most popular families of logic circuits is *transistor-transistor logic* (TTL). Figure 10-24(a) shows the circuit diagram for a basic TTL NAND gate.

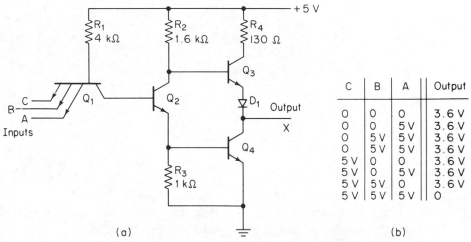

C	B	A	Output
0	0	0	3.6 V
0	0	5 V	3.6 V
0	5 V	5 V	3.6 V
0	5 V	5 V	3.6 V
5 V	0	0	3.6 V
5 V	0	5 V	3.6 V
5 V	5 V	0	3.6 V
5 V	5 V	5 V	0

(a)   (b)

FIGURE 10-24   *Three-input TL NAND gate and truth table.*

Notice that the input transistor, Q_1, has three emitters—it is referred to as a multiple emitter transistor. This is characteristic of all the standard series TTL device inputs. With all the inputs in the logic 1 state, Q_1 is turned off. The base–emitter junction of Q_1 is reverse-biased but the base–collector junction is forward-biased, turning Q_2 and Q_4 on. With Q_2 turned on, there is not enough base current to turn Q_3 on, so it is in the off state. With Q_3 turned off and Q_4 turned on, any load attached to the output will see a logic 0 state. Also, Q_4 has the capability of *sinking* (accepting) current from any input it is attached to. This is why TTL is referred to as *current sinking* logic. If any one of the inputs is made a logic 0 then Q_1 is turned on taking base current away from Q_2 turning it and Q_4 off. With Q_2 off, Q_3 will be turned on by +5 V and R_2. Q_3 will then act as an emitter follower, producing a high voltage (≈ 3.6 V $= 5$ V $- V_{BEQ_3} - V_{D_1}$) and good drive capability. As can be seen from the truth table in Fig. 10-24(b), the circuit functions as a NAND gate. The arrangement of Q_3 attached on top of Q_4, as shown in Fig. 10-24(a), is often referred to as *totem-pole output.*

The circuitry shown in Fig. 10-24 is of type found in the TTL series referred to as the 5400/7400 standard series. The difference between 5400 and 7400 is that the 5400 is meant for military use and can be operated over a wider temperature range and supply-voltage variation. The typical standard series gate will have characteristics as follows: average noise margin $\approx$ 400 mV, average power dissipation $\approx$ 10 mW, average propagation delay $\approx$ 9 ns, and a *fan-out* of 10 (one output can drive 10 other TTL inputs). Figure 10-25 shows a pin-out of a 7400 IC, referred to as a quad two-input NAND gate chip. Notice there are four separate two-input NAND gates, of which any or all may be used at any point in time.

A wide variety of logic devices can be found in the 7400 series of TTL logic.

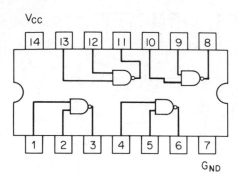

FIGURE 10-25 Pin-out for a 7400 IC chip.

There are a number of other TTL series that have been developed to allow for a wider choice of speed and power dissipation characteristics.

74L00 series—low-power TTL series, which has a typical average power dissipation of 1 mW per gate but results in an increased average propagation delay up to 33 ns.

74H00 series—high-speed TTL series, which has a typical average propagation delay of 6 ns but an increased power dissipation of 23 mW.

74S00 series—Schottky TTL series, has the highest-speed characteristics by using a Schottky barrier diode as a clamp from base to collector of each transistor in the circuit. A typical average propagation delay of 3 ns and an average power dissipation of 23 mW is common for this series.

74LS00 series—low-power Schottky series, which has a typical average propagation delay of 9.5 ns and an average power dissipation of 2 mW.

A relatively new development in the TTL series is referred to as *tristate TTL* logic. This circuitry allows for three possible output states. By the addition of a control input, the gate can operate as a normal TTL gate having two output states (0, 1) or by changing the control input the totem-pole output transistors (Q_3 and Q_4) can both be turned off, resulting in a high-impedance output for a third state. This effectively disconnects the output of the gate from whatever it is physically attached to. This will be very useful in bussing, (see Chapter 11) where we tie outputs of logic gates together so that information can be transferred using a common line.

Another series of logic devices is called *emitter-coupled logic* (ECL), also referred to as current-mode logic (CML). It has basic circuitry that is essentially a differential amplifier configuration. A fixed bias current (less than saturation) is made to switch between the two transistors, making up the differential amplifier, depending upon input logic levels. This arrangement allows for fast switching speed, since neither transistor is ever saturated. The power dissipation is increased since one transistor is always in the active region. Typical characteristics for an ECL circuit would show a worst-case noise margin of 250 mV, a fan-out of 25, a typical average propagation delay of 2 ns, and an average power dissipation of 25 mW.

Leaving the area of bipolar transistor digital integrated circuits, we come to digital circuits that are designed using *MOSFET* transistors. We can find logic devices being built with one of three types of MOSFET transistors: *p*-channel enhancement-mode MOSFETs (PMOS), *n*-channel enhancement-mode MOSFETs (NMOS), and complementary enhancement-mode MOSFETs (CMOS). Figure 10-26 shows a PMOS

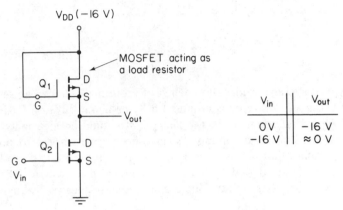

FIGURE 10-26   *PMOS inverter circuit.*

inverter circuit and a function table. Notice that transistor Q_1 has the gate tied to the supply V_{DD} (−16 V). This connection makes Q_1 behave as though it were a fixed (or load) resistor of ≈100 kΩ. Q_2 is a switching transistor. With 0 V applied to the gate of Q_2, the output voltage will be V_{DD} (−16 V), since Q_2 is OFF (≈10^{10} Ω) and Q_1 is acting as a resistor. With V_{DD} (−16 V) applied to the Q_2 gate, the MOSFET turns on (≈1 kΩ) and the output voltage drops to approximately 0 V.

PMOS and NMOS have greater packing density capabilities than CMOS, TTL, and ECL. NMOS is faster than PMOS since the more mobile electrons rather than holes are the current carriers. Typical characteristics for a PMOS device would be an average propagation delay of 100 ns, a noise margin of around 2 V, a fan-out of about 50, and an average power dissipation of around 1 mW.

CMOS logic circuits use both PMOS and NMOS devices in the same circuit. This has the advantage of decreasing power dissipation, increasing speed of operation, but increases the fabrication process cost and lowers the packing density. Figure 10-27 shows the basic CMOS inverter. When V_{DD} is applied to the input, Q_1 is turned OFF and Q_2 ON; therefore, the output is a logic LO ≈ 0 V. When 0V is applied to the input, Q_1 is turned ON and Q_2 OFF, producing a logic HI = +V_{DD}. Since one transistor is always OFF in the circuit, we get extremely low power dissipations in the CMOS family, typical average power dissipation of 12 nW per gate. When the output is a logic HI transistor, Q_1 is turned on and therefore can charge any load capacitance much faster than the load resistor of the PMOS inverter. This results

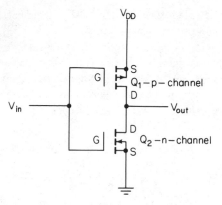

FIGURE 10-27 *CMOS inverter circuit.*

in typical average propagation delays of around 30 ns. The noise margin for a CMOS gate using a 5-V supply is around 1.5 V, or about 30% of V_{DD}.

As a final comment in the chapter, it should be mentioned that all the circuits discussed using logic 1's and 0's have been with respect to positive logic. *Positive logic* assumes that the most positive voltage is a logic 1 and the least positive voltage a logic 0. Some digital systems use *negative logic*, which defines a logic 1 as the most negative voltage and a logic 0 as the least negative voltage.

QUESTIONS AND PROBLEMS

10–1–1. What is the difference between digital and analog signals?

10–1–2. List three reasons why digital electronics has become very popular.

10–1–3. Consider the waveform in Fig. 10-28. From the waveform diagram, determine:
(a) The pulse width, t_p.
(b) The rise time, t_r.
(c) The fall time, t_f.

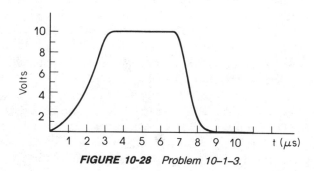

FIGURE 10-28 *Problem 10–1–3.*

10–1–4. Consider the waveforms in Fig. 10-29. From the waveforms, determine:
  (a)   The propagation delay, t_{PLH}.
  (b)   The propagation delay, t_{PHL}.

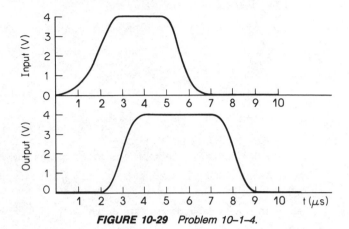

FIGURE 10-29 Problem 10–1–4.

10–2–5. Define the term "logic gate."

10–2–6. The two input signals in Fig. 10-30 are applied to the two-input OR gate of Fig. 10-5(a). Draw the output waveform.

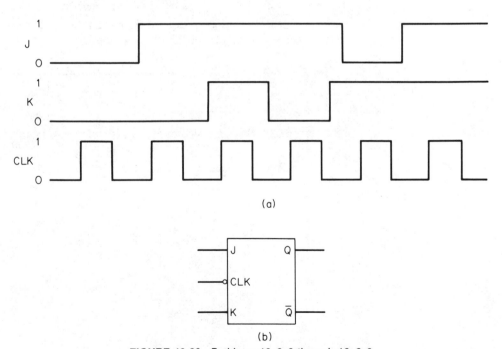

FIGURE 10-30 Problems 10–2–6 through 10–2–8.

10–2–7. The two input signals in Fig. 10-30 are applied to the two-input AND gate of Fig. 10-7. Draw the output waveform.

10–2–8. The two input signals in Fig. 10-30 are applied to the two-input NOR gate of Fig. 10-9(c). Draw the output waveform.

10–3–9. Generate a truth table and the Boolean expression for a four-input NOR gate.

10–3–10. Generate a truth table and the Boolean expression for a four-input NAND gate.

10–3–11. Simplify each of the following expressions:
(a) $Z = ABC + AB(\overline{C \cdot D})$
(b) $Z = ABC + AB\overline{C}$
(c) $Z = \overline{A}BC + A\overline{B}C + ABC + B\overline{C}$

10–3–12. (a) Construct the logic circuit for $Z = (A \cdot B) + (C \cdot D)$.
(b) Use all NAND gates to implement the circuit for the above expression.

10–4–13. Design a logic circuit whose output is high only when a majority of its three inputs are high.

10–5–14. What is the major difference between an R-S and a J-K flip-flop?

10–5–15. Explain the difference between synchronous and asynchronous inputs of a flip-flop.

10–5–16. Explain the difference between a J-K FF and a J-K master-slave FF. When are M/S J-K FF's most often used?

10–5–17. Apply the J, K, and CLK waveforms shown in Fig. 10-30(a) to the J-K FF shown in Fig. 10-30(b). Assume that $Q = 0$ initially and determine the Q waveform.

10–5–18. Apply the D and CLK waveforms shown in Fig. 10-31(a) to the D FF shown in Fig. 10-31(b). Assume that $Q = 0$ initially and determine the Q waveform.

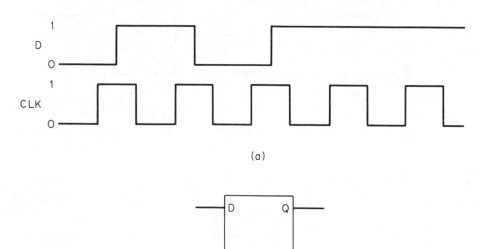

(a)

(b)

FIGURE 10-31 Problem 10–5–18.

10–5–19. Determine the Q waveform for the FF in Fig. 10-32. Assume that $Q = 0$ initially.

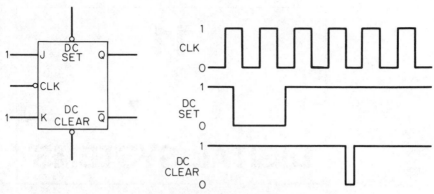

FIGURE 10-32 *Problem 10–5–19.*

10–6–20. List and explain four factors used in comparing logic families.

10–6–21. Explain what is meant when the statement "TTL is current sinking logic" is used.

10–6–22. In what state does the output of a TTL NAND gate act as a current sink?

10–6–23. Explain what is meant by the term "totem-pole output."

10–6–24. Explain the function of diode D_1 in Fig. 10-24.

10–6–25. Explain what tristate logic is as compared to standard logic circuits.

10–6–26. What is the basic configuration of circuitry for the input of emitter-coupled logic?

10–6–27. Why is the average power dissipation greater for ECL than for TTL?

10–6–28. Why is ECL transition time shorter than standard TTL?

10–6–29. Explain why NMOS logic is faster than PMOS logic.

10–6–30. Explain why CMOS has less power dissipation than PMOS or NMOS logic.

10–6–31. Explain why CMOS has faster transition times than NMOS or PMOS logic.

10–6–32. Which has greater packing density, PMOS or CMOS? Explain.

10–6–33. Explain the difference between positive and negative logic.

10–6–34. The truth table shown in Fig. 10-33 indicates:
(a) What type of gate using positive logic?
(b) What type of gate using negative logic?

B	A	Z
0 V	0 V	0 V
0 V	5 V	0 V
5 V	0 V	0 V
5 V	5 V	5 V

FIGURE 10-33 *Problem 10–6–34.*

11

DIGITAL SYSTEMS

In Chapter 10 we discussed basic digital circuits. Digital systems involve the putting together of various basic circuits so that they perform some useful function. Many of these functions are performed so often that manufacturers have put all of the involved components into one integrated circuit. To facilitate understanding the operation of these systems, we must remember that digital circuits and systems operate on *binary* information. We, therefore, must become familiar with how data are handled in a binary number system. To ease the process of learning the binary number system, we will first investigate some important characteristics of the decimal number system.

11-1 NUMBER SYSTEMS

Most often in digital systems, information is processed in the binary number system. There are times, however, because of certain advantages, when other systems are used. In this section we will begin by investigating the decimal number system and then progress to the binary, octal, and hexadecimal systems.

The decimal number system is the most familiar system because we deal with it every day. It is so familiar that we perform most operations intuitively, not really paying attention to some of the important characteristics. First, the decimal number system utilizes 10 symbols, 0 through 9, to represent any quantity. Since there are 10 symbols, we refer to this as a base-10 number system.

If we want to express a number greater than 9, we must increase the number

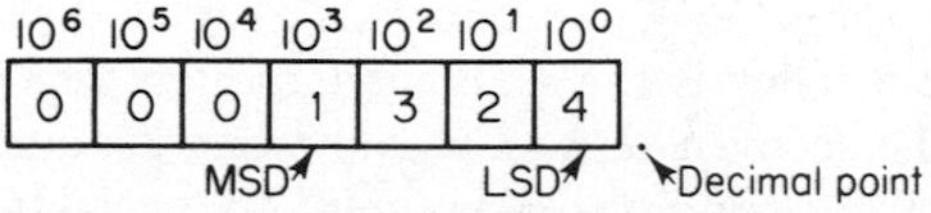

$$10^6 \quad 10^5 \quad 10^4 \quad 10^3 \quad 10^2 \quad 10^1 \quad 10^0$$

0	0	0	1	3	2	4

MSD ↗   LSD ↗   ↖ Decimal point

$1324_{10} = (1 \times 10^3) + (3 \times 10^2) + (2 \times 10^1) + (4 \times 10^0)$

FIGURE 11-1 *Decimal positional value system.*

of digits used to represent this quantity. Therefore, the decimal system is a *positional-value system,* meaning that the position of a digit in a number carries an associated weight. The weights are various powers of 10, as shown in Fig. 11-1. The 1 carries the most weight (1000's) and is referred to as the *most significant digit* (MSD); the 4 carries the least weight (1's) and is called the *least significant digit* (LSD).

The binary number system has many of the same characteristics as the decimal system except that it only uses two symbols—0 and 1. Since there are only two symbols, we refer to the binary number system as a base-2 system.

If we want to express a number greater than 1, we must increase the number of binary digits (referred to as *bits*) used to represent this quantity. The binary system is also a *positional-value system,* with the weight of each bit corresponding to some power of 2. The weights are shown in Fig. 11-2 together with a binary number.

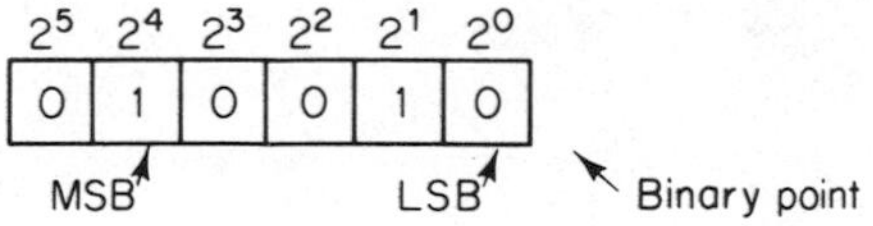

$$2^5 \quad 2^4 \quad 2^3 \quad 2^2 \quad 2^1 \quad 2^0$$

0	1	0	0	1	0

MSB ↗   LSB ↗   ↖ Binary point

$10010_2 = (1 \times 2^4) + (0 \times 2^3) + (0 \times 2^2) + (1 \times 2^1) + (0 \times 2^0)$

$\quad\quad = 16 \quad\quad + 0 \quad\quad + 0 \quad\quad + 2 \quad\quad + 0$

$\quad\quad = 18_{10}$

FIGURE 11-2 *Binary positional value system.*

Notice that the fifth bit to the left of the binary point carries the most weight (16's) and is referred to as the *most significant bit* (MSB); the bit just to the left of the binary point carries the least weight and is called the *least significant bit* (LSB). Throughout the rest of this chapter we will be using subscripts as in Figs. 11-1 and 11-2 next to numerical information to indicate the number system being used.

Most quantities outside the digital system are represented in the decimal number system, yet the digital system deals with binary information. We must therefore become familiar with how to convert back and forth between the two number systems.

To convert a decimal number to its binary equivalent, we can perform the following sequence of steps. Suppose that we want to convert 19_{10} to a binary equivalent:

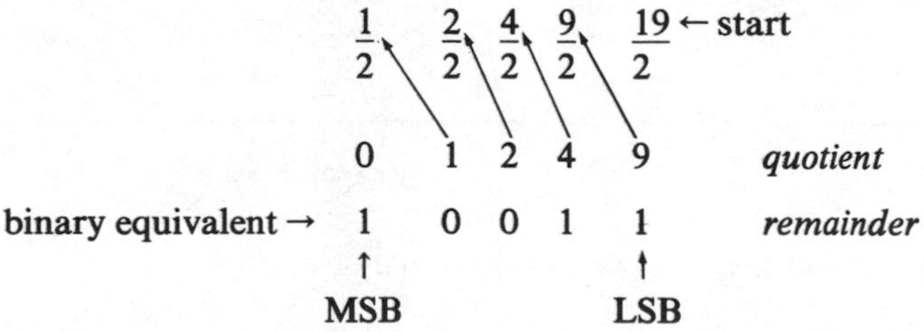

$$\frac{1}{2} \quad \frac{2}{2} \quad \frac{4}{2} \quad \frac{9}{2} \quad \frac{19}{2} \leftarrow \text{start}$$

$$\quad\quad 0 \quad\quad 1 \quad\quad 2 \quad\quad 4 \quad\quad 9 \quad\quad \textit{quotient}$$

binary equivalent →   1   0   0   1   1   *remainder*

           ↑   ↑

       MSB   LSB

To accomplish this, simply keep dividing the decimal number by 2, keeping track of the remainders. The first remainder is the LSB, and the last remainder is the MSB of the binary equivalent. The conversion is complete when the quotient is zero.

To convert from binary to decimal, we simply sum the weights of each bit that is a 1 in the binary number. For example,

$$101101_2 = ?_{10}$$

$$2^5 \quad 2^4 \quad 2^3 \quad 2^2 \quad 2^1 \quad 2^0 = (1 \times 32) + (0 \times 16) + (1 \times 8) + (1 \times 4) + (0 \times 2) + (1 \times 1)$$
$$= 32 + 8 + 4 + 1 = 45_{10}$$

As can be seen, the conversion from binary to decimal is fairly easy and straightforward.

Another popular number system used in digital system applications is the *octal number system*. As the name implies, this system uses eight symbols, 0 through 7, and is referred to as a base-8 system. If we want to express a number greater than 7, we must increase the number of digits used to represent this quantity. Each digit carries a weight as did the decimal and binary digits, but the weights are various powers of 8, as shown in Fig. 11-3(a).

$$8^5 \quad 8^4 \quad 8^3 \quad 8^2 \quad 8^1 \quad 8^0$$

0	0	2	1	7	5

(a)

$$2175_8 = ?_{10}$$
$$= (2 \times 8^3) + (1 \times 8^2) + (1 \times 8^1) + (5 \times 8^0)$$
$$= (2 \times 512) + (1 \times 64) + (7 \times 8) + (5 \times 1)$$
$$= 1024 + 64 + 56 + 5$$
$$= 1149_{10}$$

(b)

FIGURE 11-3 *Octal positional value system and conversion to decimal.*

When it is necessary to convert an octal number to a decimal equivalent, all that needs to be done is to sum the weights times the digit value for the octal number, as shown in Fig. 11-3(b). Notice that this is the same procedure as that used previously in the decimal and binary number systems.

In the process of converting a decimal number to an octal equivalent, simply repeat the repetitive division process as was done when converting from decimal to binary, but divide by eight.

EXAMPLE 11-1

Convert 459_{10} to an octal equivalent.

Solution:

$$
\begin{array}{ccc}
7 & 57 & 459_{10} \leftarrow \text{start} \\
8 & 8 & 8 \\
0 & 7 & 57 \qquad \textit{quotient} \\
7 & 1 & 3_8 \qquad \textit{remainder} \\
\uparrow & & \uparrow \\
\text{MSB} & & \text{LSD}
\end{array}
$$

The conversion is complete when the quotient is zero. The octal equivalent of 459_{10} is 713_8.

Notice that the conversion of a relatively large decimal number to an octal equivalent took only three division steps, whereas if we tried to convert directly to binary, many divisions-by-two steps would have to be taken. Also notice that to represent a large decimal quantity such as 459_{10} takes only three octal digits, whereas it would take nine binary bits. The chief advantages of the octal number system and reason for its use stem from two principal factors: (1) it is easier to handle data in the octal number system from the user's point of view, and (2) the conversion from octal to binary and binary to octal is so simple and straightforward that when it is necessary to look at what the digital system is doing (digital circuits work on binary information) the conversion is quick and simple.

To convert an octal number to a binary equivalent, we simply write a three-bit binary equivalent for each octal digit.

EXAMPLE 11-2

Find the binary equivalent of 713_8.

Solution:

$$
\begin{array}{ccc}
7 & 1 & 3 \;_8 \\
\underbrace{111} & \underbrace{001} & \underbrace{011}_2
\end{array}
$$

Therefore, the binary equivalent of 713_8 is 111001011_2.

The procedure for converting from binary to octal is as simple. Break the binary number into groups of three bits, starting from the LSB, and convert each group of bits to its octal equivalent.

EXAMPLE 11-3

Convert 101101111011_2 to an octal equivalent.

Solution:

$$
\begin{array}{cccc}
5 & 5 & 7 & 3_8 \\
\overline{101} & \overline{101} & \overline{111} & \overline{011}_2
\end{array}
$$

Therefore, the octal equivalent of binary 101101111011_2 is 5573_8.

Another frequently used number system in digital system applications is the hexadecimal number system. This is a base-16 number system which utilizes the symbols 0 through 9 plus A, B, C, D, E, and F as the 16 digit symbols. This is

Decimal	Binary	Hexadecimal
0 ⟶	0000 ⟶	0
1 ⟶	0001 ⟶	1
2 ⟶	0010 ⟶	2
3 ⟶	0011 ⟶	3
4 ⟶	0100 ⟶	4
5 ⟶	0101 ⟶	5
6 ⟶	0110 ⟶	6
7 ⟶	0111 ⟶	7
8 ⟶	1000 ⟶	8
9 ⟶	1001 ⟶	9
10 ⟶	1010 ⟶	A
11 ⟶	1011 ⟶	B
12 ⟶	1100 ⟶	C
13 ⟶	1101 ⟶	D
14 ⟶	1110 ⟶	E
15 ⟶	1111 ⟶	F

FIGURE 11-4 *Relationships among decimal, binary, and hexadecimal systems.*

also a weighted number system with each digit having some power of 16 as its weight. Figure 11-4 shows the relationship between decimal and hexadecimal.

To convert from decimal to hexadecimal, the decimal number is repetitively divided by 16, keeping track of the remainders.

EXAMPLE 11-4

Convert 412_{10} to a hexadecimal equivalent.

Solution:

$$1 \quad 25 \quad 412 \leftarrow \text{start}$$

$$16 \quad 16 \quad 16$$

$$0 \quad 1 \quad 25 \qquad \textit{quotient}$$

$$1 \quad 9 \quad 12 \qquad \textit{remainder}$$

$$\downarrow \quad \downarrow \quad \downarrow$$

$$1 \quad 9 \quad C_{16}$$

Therefore, the hexadecimal equivalent of 412_{10} is $19C_{16}$. Notice that the remainder of the first division (12) is changed to the hexadecimal equivalent.

When converting from a hexadecimal number to a decimal equivalent, simply take the sum of the weight times the digit symbol.

EXAMPLE 11-5

Convert $23A_{16}$ to a decimal equivalent.

Solution:

$$23A_{16} = (2 \times 16^2) + (3 \times 16^1) + (A \times 16^0)$$
$$= (2 \times 16^2) + (3 \times 16^1) + (10 \times 16^0)$$
$$= 256 + 48 + 10 \quad 512 + 48 + 10$$
$$= 314_{10} \quad 570_{10}$$

The advantages of the hexadecimal number system are apparent when handling large amounts of binary information. It is much more convenient and less prone to error if we represent the binary data in the hexadecimal system and convert back and forth whenever necessary. To convert a hexadecimal number to a binary equivalent, simply take each hexadecimal digit and replace it by a four-bit binary equivalent.

EXAMPLE 11-6

Convert $23A_{16}$ to a binary equivalent.

Solution:

$$
\begin{array}{ccc}
2 & 3 & A_{16} \\
\overbrace{0010} & \overbrace{0011} & \overbrace{1010_2}
\end{array}
$$

Therefore, $23A_{16} = 1000111010_2$. The leading zeros can be dropped.

To convert a binary number to a hexadecimal equivalent, take the binary number and break it down into four-bit groups, starting from the LSB. Then convert each four-bit group to a hexadecimal equivalent.

EXAMPLE 11-7

Convert 1011010111_2 to a hexadecimal equivalent.

Solution:

$$
\begin{array}{ccc}
 & & \text{LSB} \\
10 & 1101 & 0111_2 \\
\underbrace{} & \underbrace{} & \underbrace{} \\
2 & D & 7_{16}
\end{array}
$$

Therefore, the hexadecimal equivalent of 1011010111_2 is $2D7_{16}$.

EXAMPLE 11-8

Suppose that a computer is transmitting a group of eight-bit binary words, as shown below, to some output device. Convert this group to an octal and a hexadecimal equivalent listing.

0010	1101
0101	1011
1001	0101
0111	1011
1011	0110
0011	1010
0100	1111

Solution:

Binary	Octal			Hexadecimal	
00101101	00	101	101	0010	1101
	0	5	5_8	2	D_{16}
01011011	01	011	011	0101	1011
	2	3	3_8	5	B_{16}
10010101	10	010	101	1001	0101
	2	2	5_8	9	5_{16}
01111011	01	111	011	0111	1011
	2	7	3_8	7	B_{16}
10110110	10	110	110	1011	0110
	2	6	6_8	B	6_{16}
00111010	00	111	010	0011	1010
	0	7	2_8	3	A_{16}
00101111	00	101	111	0010	1111
	0	5	7_8	2	F_{16}

Therefore,

Binary	Octal	Hexadecimal
00101101	055	2D
01011011	233	5B
10010101	225	95
01111011	273	7B
10110110	266	B6
00111010	072	3A
00101111	057	2F

As can be seen from Example 11-8, if the user must deal with these data, it is easier and less prone to error to utilize the octal or hexadecimal number system as opposed to the binary number system. Also, whenever the need arises to be in binary, the user can easily and quickly convert from octal or hexadecimal to binary.

11-2 CODES

There are many reasons why codes are used in digital systems. These reasons usually revolve around ease of application (less circuitry involved) or more efficient operation (faster and less prone to error). First, we must realize what a code is. Your social security number is a code that identifies you from all other people in the work force. Another code is your driver's license number, which identifies you from all other drivers. In our case, we are interested in specifying characters (numeric or alphanumeric) in terms of the 1's and 0's that digital systems can deal with properly. In the preceding section we saw that it was possible to take decimal numbers and convert them to their *straight binary equivalent* or *octal equivalent* or *hexadecimal equivalent.* The circuitry necessary to convert a large decimal number to a true binary equivalent would be fairly complex, and yet digital systems are often involved in having to deal with the outside decimal world. Therefore, some means for converting decimal numbers to some form of binary information easily (in terms of circuits) would be most advantageous.

The BCD (binary-coded-decimal) code involves a simple process of taking any decimal number and replacing each digit of that number by a four-bit binary equivalent. To represent the decimal number 96_{10} in BCD, replace each digit by its four-bit binary equivalent.

$$9 \qquad 6_{10}$$

$$1001 \qquad 0110_{BCD}$$

EXAMPLE 11-9

Convert the following decimal numbers to BCD equivalents.

$$136_{10} = ?_{BCD}$$

$$57 \ = ?_{BCD}$$

$$1293 \ = ?_{BCD}$$

$$12 \ = ?_{BCD}$$

Solution:

Again, convert each decimal *digit* to a four-bit binary equivalent.

$$
\begin{array}{cccc}
& 1 & 3 & 6_{10} \\
136_{10} = & \overbrace{0001} & \overbrace{0011} & \overbrace{0110}_{BCD}
\end{array}
$$

$$
\begin{array}{ccc}
& 5 & 7_{10} \\
57_{10} = & \overbrace{0101} & \overbrace{0111}_{BCD}
\end{array}
$$

$$
\begin{array}{ccccc}
& 1 & 2 & 9 & 3_{10} \\
1293_{10} = & \overbrace{0001} & \overbrace{0010} & \overbrace{1001} & \overbrace{0011}_{BCD}
\end{array}
$$

$$
\begin{array}{ccc}
& 1 & 2_{10} \\
12_{10} = & \overbrace{0001} & \overbrace{0010}_{BCD}
\end{array}
$$

As you can see from Example 11-9, converting from decimal to BCD is very easy. Similarly, converting from BCD to decimal is simple. It is just the reverse process. The circuitry for converting from decimal to BCD is also simple because it is only necessary to convert one decimal digit to one four-bit binary code. In converting multidigit decimal numbers, the same circuitry can simply be cascaded. It is important to remember that there is a difference between BCD and straight binary coding. In Example 11-9, the last conversion was to change 12_{10} to its equivalent BCD, which was $12_{10} = 00010010_{BCD}$. To convert 12_{10} to straight binary would produce $12_{10} = 1100_2$. This leads to one of the disadvantages of BCD. BCD is not as efficient as true binary because only 10 of the 16 possible combinations of the four-bit code are used. Therefore, it took eight bits to represent 12_{10} in BCD and only four bits in straight binary. If storage space were a problem, straight binary would be more attractive. Another disadvantage to using BCD will appear when arithmetic operations are discussed later in the chapter.

Another commonly used code that is very similar to BCD is called the *excess-three code*. The excess-three code exhibits advantages in certain arithmetic operations. To convert a decimal number to an excess-three equivalent, first add three to each digit and then replace that digit by a four-bit binary equivalent.

EXAMPLE 11-10

Convert the following decimal numbers to an excess-three equivalent.

$$450_{10} = ?_{X-3}$$

$$126_{10} = ?_{X-3}$$

$$377_{10} = ?_{X-3}$$

$$629_{10} = ?_{X-3}$$

Solution:

Simply add three to each decimal digit, then write the four-bit binary equivalent for that digit.

$$450_{10} = \underbrace{7}_{4+3} \quad \underbrace{8}_{5+3} \quad \underbrace{3}_{0+3} \rightarrow 0111 \quad 1000 \quad 0011_{X-3}$$

$$126_{10} = \underbrace{4}_{1+3} \quad \underbrace{5}_{2+3} \quad \underbrace{9}_{6+3} \rightarrow 0100 \quad 0101 \quad 1001_{X-3}$$

$$377_{10} = \underbrace{6}_{3+3} \quad \underbrace{10}_{7+3} \quad \underbrace{10}_{7+3} \rightarrow 0110 \quad 1010 \quad 1010_{X-3}$$

$$629_{10} = \underbrace{9}_{6+3} \quad \underbrace{5}_{2+3} \quad \underbrace{12}_{9+3} \rightarrow 1001 \quad 0101 \quad 1100_{X-3}$$

TABLE 11-1 Binary, BCD, and excess-three code equivalents

Decimal	Binary	BCD	Excess-three
0	0	0000	0011
1	1	0001	0100
2	10	0010	0101
3	11	0011	0110
4	100	0100	0111
5	101	0101	1000
6	110	0110	1001
7	111	0111	1010
8	1000	1000	1011
9	1001	1001	1100

Table 11-1 lists the binary, BCD, and excess-three code representations for the decimal digit symbols.

There are many applications of digital systems that keep track of a rotating shaft's position. In other words, angular position is an important input to this digital system. Ideally, it would be nice to represent the rotating shaft position with a code that changes only one bit for each incremented change in shaft position. For example, suppose that the shaft were in a certain position and rotated one increment further. If the binary code is being used and the original shaft position was coded 0111, the new position would be 1000. This incremented change in position caused four bits of the code word to change. If slow-responding circuitry is used to sense this input code, some ambiguous code may occur and be detected and considered the input to the digital system, thus, causing an error. The *Gray code* is one of a group of codes that are designed to change only one bit in the code group when going from one step to the next. This code is an *unweighted code,* meaning that the bit positions in the coded words do not carry any weight. The advantage and, therefore, popularity of the Gray code is that it can be converted easily to a binary form, and since digital systems operate on binary information, this becomes very desirable. Table 11-2 presents a list of straight binary and Gray code equivalent for the decimal

TABLE 11-2 Binary and Gray code equivalents

Decimal	Binary Code	Gray Code
0	0000	0000
1	0001	0001
2	0010	0011
3	0011	0010
4	0100	0110
5	0101	0111
6	0110	0101
7	0111	0100
8	1000	1100
9	1001	1101
10	1010	1111

numbers 0 to 10. Notice that only one bit changes between any two successive Gray code words. The conversion from Gray to binary and binary to Gray is a cookbook-type process and can be found in most digital system texts.

All the codes dealt with up to this point involved numeric information. A large amount of information in digital systems involves not only numeric but also alphabetic and special characters. A standardized code that is widely used throughout industry is the *ASCII* code—*A*merican *S*tandard *C*ode for *I*nformation *I*nterchange. This is a seven-bit code which allows for $2^7 = 128$ different letters, numbers, and characters. The letter A is represented by the seven-bit code 1000001, B = 1000010, C = 1000011, and so forth. Table 11-3 is a partial listing of the ASCII code. See if you can use the table to decode the following message.

1000101 | 1001100 | 1000101 | 1000011 | 1010100 | 1010010 | 1001111 |

1001110 | 1001001 | 1000011 | 1010011 | 0100000 | 1001001 | 1010011 |

0100000 | 1000110 | 1010101 | 1001110 | 0100001

Whenever information is being transmitted from one location to another, there is a possibility for error, causing the receiver to read the wrong information. Many schemes have been developed to protect transmission of data against errors. One of the most popular is referred to as the *bit parity method* of error detection. Two parity schemes are available: even parity and odd parity. An extra bit is added to the code being transmitted and made a 1 or 0 (parity bit) to make the *total* number

TABLE 11-3 *ASCII code equivalents*

Character	ASCII Code	Character	ASCII Code
0	0110000	J	1001010
1	0110001	K	1001011
2	0110010	L	1001100
3	0110011	M	1001101
4	0110100	N	1001110
5	0110101	O	1001111
6	0110110	P	1010000
7	0110111	Q	1010001
8	0111000	R	1010010
9	0111001	S	1010011
A	1000001	T	1010100
B	1000010	U	1010101
C	1000011	V	1010110
D	1000100	W	1010111
E	1000101	X	1011000
F	1000110	Y	1011001
G	1000111	Z	1011010
H	1001000	Space	0100000
I	1001001	!	0100001

of 1's even or odd, depending on the parity scheme chosen. If the ASCII code for A = 1000001 is being transmitted and the system is designed for even-parity transmission with the parity bit being added to the MSB position, the actual data transmitted would be

$$01000001$$

parity bit

and this could be written in hexadecimal as 41_{16}. If the system is set up to transmit in odd parity, the data transmitted for the same letter would be

$$11000001$$

parity bit

or $C1_{16}$. The receiver for the system will be using the same parity as the transmitter and will have circuitry to detect for the proper number of 1's. If an error is detected, the system can provide appropriate action.

11-3 ARITHMETIC CIRCUITRY

To facilitate an understanding of arithmetic circuitry, it is necessary to look at the technique of binary arithmetic. *Binary addition* is performed on binary numbers just like decimal addition is performed on decimal numbers. In fact, binary addition is easier because, in adding two binary bits, there are only four possible combinations.

$$0 + 0 = 0$$

carry into position from previous bit addition

$$0 + 1 = 1$$

$$1 + 1 = 0 \quad \text{with a carry} = 1 \text{ into the next position}$$

$$1 + 1 + 1 = 1 \quad \text{with a carry} = 1 \text{ into the next position}$$

EXAMPLE 11-11

Add the following binary numbers:

$$\begin{array}{ll} 01010 \quad (10_{10}) & 01111 \quad (15_{10}) \\ \underline{00011} \quad (3_{10}) & \underline{00011} \quad (3_{10}) \end{array}$$

Solution:

Add each bit position starting from the LSB, taking any carry into account.

$$
\begin{array}{ll}
1 & 1111 \\
01010 \quad (10_{10}) & 01111 \quad (15_{10}) \\
\underline{00011} \quad (3_{10}) & \underline{00011} \quad (3_{10}) \\
01101 \quad (13_{10}) & 10010 \quad (18_{10})
\end{array}
$$

Addition is a very important procedure in digital systems because all the other operations, such as subtraction, multiplication, and division, are performed using addition as their only basic operation.

To be more efficient and save on circuitry, the operation of subtraction is basically performed by addition using what is called *2's complement*. The 2's complement of a number is formed by inverting each bit of the original binary number (1's complement) and adding 1 to the LSB position. For example, to take the 2's-complement of the binary number 01011011, invert each bit, leaving 10100100 (1's complement) and add 1 to the LSB, leaving 10100101 as the 2's-complement of the original binary number. Subtraction is achieved by 2's-complementing the *subtrahend* (the number to be subtracted) and adding it to the minuend (number being subtracted from). In effect, the subtrahend is being inverted and added to the minuend, which is the basic operation of a subtraction.

EXAMPLE 11-12

Perform the following subtractions:

$$
\begin{array}{ll}
1000 & 1101 \\
-\,0100 & -\,0101
\end{array}
$$

Solution:

2's-complement the subtrahend by inverting each bit and adding 1 to the LSB.

$$
\begin{array}{ll}
(8_{10}) \quad 1000 & \text{(minuend)} \\
-\,(4_{10}) \quad \underline{0100} & \text{(subtrahend)} \rightarrow 1011 \quad \text{(1's-complement)} \\
(4_{10}) & \underline{+1} \\
& 1100 \quad \text{(2's-complement)}
\end{array}
$$

$$
\begin{array}{ll}
1000 & \text{(minuend)} \\
+\,\underline{1100} & \text{(2's complement of subtrahend)} \\
(1)\,0100 & (4_{10})
\end{array}
$$

↑
disregard carry

$$
\begin{array}{ll}
\quad\ 1101 & (13_{10}) \\
-\ \underline{0101} & (5_{10}) \rightarrow \quad 1010 \quad \text{(1's-complement)} \\
& (8_{10}) \quad + \underline{\quad 1} \\
& \qquad\qquad\quad 1011 \quad \text{(2's-complement)}
\end{array}
$$

$$
\begin{array}{l}
\ ^{1\ 1\ 1} \\
\ \ 1101 \\
+\ \underline{1011} \\
\textcircled{1}\ 1000 \quad (8_{10}) \\
\uparrow \\
\text{disregard carry}
\end{array}
$$

At some point it becomes necessary to deal with both positive and negative numbers. One method for dealing with this is to reserve the MSB position as a sign bit. A 0 in the MSB indicates a positive number and a 1 indicates a negative number. In digital systems that use the 2's-complement form for subtraction (as discussed in the previous paragraph), it is necessary to always represent negative numbers in their 2's-complement form. The number $+12_{10}$ would be

$$
\begin{array}{l}
01100 \\
\uparrow \\
\text{sign bit}
\end{array}
$$

and the number -12 would be

$$
\begin{array}{l}
1\ \ 0100 \\
\uparrow\ \ \smile \\
\text{sign bit} \quad \text{2's-complement of 1100}
\end{array}
$$

This format of always representing negative numbers in their 2's-complement form allows the use of the same circuitry for addition and subtraction of signed numbers. Whenever an answer to an arithmetic operation ends up being negative, it will show up in the 2's-complement form.

EXAMPLE 11-13

Perform the following arithmetic operations on signed numbers.

$$
\begin{array}{llllll}
\text{(a)} & (+8_{10}) & \text{(b)} & (+8_{10}) & \text{(c)} & (+4_{10}) \\
& +\ \underline{(+4_{10})} & & -\ \underline{(+4_{10})} & & +\ \underline{(-6_{10})}
\end{array}
$$

Solution:

(a)
```
     ⌐sign bit
     0 1000    (+8₁₀)
   + 0 0100    (+4₁₀)
   ─────────
     0 1100    (+12₁₀)
```

(b) To perform subtraction, 2's-complement the subtrahend and add to the minuend.

$$(+4_{10}) = 0\ 0100 \rightarrow 11011 \quad \text{(1's-complement)}$$

$$\uparrow \qquad\qquad \underline{+1}$$

$$\text{S.B.} \qquad\qquad 11100 \quad \text{(2's-complement)}$$

```
     0 1000     (+8₁₀)
   + 1 1100     −(+4₁₀)
   ─────────
   1 0 0100     (+4₁₀)
   ↑  S.B.
   ignore carry
```

(c) Make sure that negative numbers are always represented in 2's-complement.

$$+6_{10} = 0\ 0110 \rightarrow 1\ 1001 \quad \text{(1's-complement)}$$

$$\text{S.B.} \uparrow \qquad\qquad \underline{+1}$$

$$(-6_{10}) = 1\ 1010 \quad \text{(2's-complement)}$$

```
     0 0100
   + 1 1010
   ─────────
     1 1110     (2's-complement of 2 or −2)
     ↑
    S.B.
```

The sign bit indicates that the answer is negative; therefore, it is in the 2's-complement form. To see what the actual magnitude of the number is, 2's-complement it again.

```
   1 1110 →   0 0001    (1's-complement)
          + ────1
            ─────────
            0 0010 = 2₁₀    (2's-complement)
```

The same circuitry that is used for addition and subtraction can be used for multiplication and division. Multiplication is performed by a series of add and shift operations. Division is performed by a series of subtract and shift operations. As a matter of fact, all arithmetic operations performed by digital systems are broken down using an *algorithm* (breaking down into logical sequence of steps) and performed using just the basic operations of addition and subtraction.

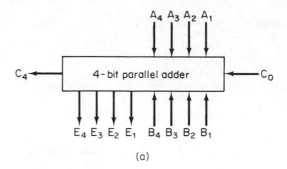

(a)

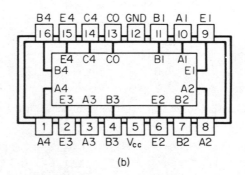

(b)

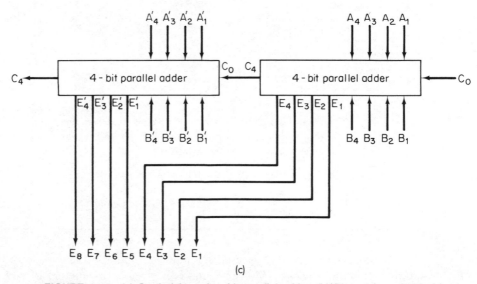

(c)

FIGURE 11-5 *(a) Symbol for a four-bit parallel adder; (b) Pin-out for a 7483; (c) Cascading adders to add larger numbers.*

336

In Section 11-2 the BCD code was introduced and some advantages and disadvantages were discussed. Another disadvantage of the BCD system shows up in addition. If two valid BCD digits were added, the sum could be one invalid code. To illustrate, add 6_{10} and 5_{10} in BCD.

$$\begin{array}{r} 0110_{BCD} \\ + \, 0101_{BCD} \\ \hline 1011 \end{array} \quad \leftarrow \text{invalid BCD code}$$

Obviously, this requires a two-digit answer, yet the addition did not provide it. This error occurs due to the fact that only 10 of the 16 possible combinations are used in BCD. Therefore, to correct the answer, $+6$ is added to any invalid BCD code group. For the illustration above, the addition of $+6$ to the invalid code group gives the correct answer to the addition.

$$\begin{array}{r} 1011 \quad \text{invalid BCD code} \\ + \; 0110 \\ \hline 0001 \mid 0001_{BCD} = 11_{10} \end{array}$$

The process of checking and adding $+6$ to any invalid code group during addition requires more circuitry than straight binary and is, therefore, a disadvantage to using BCD.

With the advances in integrated circuit technology, it is doubtful that one would ever build an adder circuit from individual logic gates. It is more likely that a manufactured IC adder circuit would be purchased. One of the most common versions available is a four-bit parallel adder that is fabricated on a single IC chip. Figure 11-5(a) shows the block diagram of this adder. The inputs are two four-bit numbers, A_4, A_3, A_2, A_1 and B_4, B_3, B_2, B_1, with the capability of a carry into the first position, C_0. Its outputs are E_4, E_3, E_2, E_1 and a carry out of the last position, C_4. Figure 11-5(b) is the actual pin-out for the SN7483, showing the pin assignment for the IC. The availability of carry-in, C_0, and carry-out, C_4, allows for cascading ICs to increase the number of bits capable of being added. Figure 11-5(c) shows cascading two of these chips, allowing the addition of two eight-bit numbers.

Subtraction can be performed using the same IC by feeding the inverted subtrahend (1's-complement) to one of the sets of inputs and making C_0 a logic 1 (effectively making the 2's-complement of the subtrahend). Therefore, the cost effectiveness of using the 2's-complement system for subtraction can now be easily seen, since the same arithmetic circuitry can be used for addition and subtraction.

11-4   COUNTERS AND REGISTERS

In Chapter 10 the flip-flop (FF) was introduced. The basic building block of counters is the FF. The simplest type of counter to build is called an *asynchronous counter*, as shown in Fig. 11-6(a). The output of FF X is tied to the clock input of FF Y

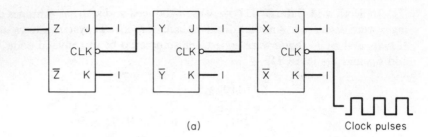

(a)

Clock pulses

Z	Y	X	Number of clock pulses
0	0	0	0
0	0	1	1
0	1	0	2
0	1	1	3
1	0	0	4
1	0	1	5
1	1	0	6
1	1	1	7
0	0	0	8 ← (causes
0	0	1	9 counter to recycle)

(b)

FIGURE 11-6 Three-bit asynchronous (ripple) counter.

and the output of FF Y is tied to the clock input of FF Z. All the FFs are the J-K type and with $J = 1$ and $K = 1$ they will toggle on the negative transition of their clock input. Figure 11-6(b) lists the sequence of binary states that the FFs will follow as clock pulses are applied to the clock input of FF X. Notice that the count is a straight binary sequence from 000 through 111 and then recycles. The fact that clock pulses are applied only to FF X and not to FF Y and FF Z is the reason this is called an asynchronous counter. In a *synchronous counter,* the clock signal would be fed to all the FFs and they would all change states at the same time. In the asynchronous counter, FF Y has to wait for FF X to change states before it could be triggered, and FF Z has to wait for the appropriate change in FF Y to occur before it can change states. In Fig. 11-6(b), when going from the third clock pulse to the fourth, the FF outputs go from 011 to 100. The way this output changes is that on the negative transition of the fourth clock pulse, FF X changes to a 0. There is a small time delay from when the negative transition of the clock occurs to when the FF output responds to this input. This delay is referred to as the FF's *propagation delay (t_{pd}).* The negative transition on the output of FF X toggles FF Y, making it go to the 0 state after a short delay. The negative transition on the output of FF Y toggles FF Z to the 1 state. Notice how the occurrence of the fourth clock pulse caused a sequence of events that, in effect, rippled through the counter. Asynchronous counters are often referred to as *ripple counters*. Also, since three FFs are involved and each FF has an inherent propagation delay, the counter will have a maximum frequency of operation equal to

$$f_{max} = \frac{1}{N \times t_{pd}}$$

where N is the number of FFs. By connecting the inputs of various gates to the outputs of the FFs in a counter, and the outputs of the gates to the FFs' SET or CLEAR inputs, the counter can be made to count through some other sequence than straight binary.

The counter in Fig. 11-6 counted through the sequence 000–111 for a total of eight different states. Generally, a counter consisting of N FFs can count from 0 up to $2^N - 1$, for a total of 2^N different states. The number of different states a counter can go through is referred to as its *MOD number*. The counter in Fig. 11-6 can count from 0 up to $2^3 - 1$, or from 0_{10} up to 7_{10}, for a total of $2^3 = 8$ different states; therefore, this is a MOD-8 counter.

In addition to counting input pulses, counters can be used for frequency division. In Fig. 11-6, FF X toggles on every input clock pulse; therefore, its output frequency would be one-half of the input clock signal. Since FF Y toggles on each negative transition on its input, FF Y will have an output frequency one-half that of FF X, or one-fourth of the original clock signal frequency. Similarly, FF Z will have an output frequency one-half that of FF Y, or one-eighth of the original clock signal frequency. In general, the MSB output of a counter will divide the input frequency by the MOD number of the counter.

A wide variety of counters are available as standard integrated-circuit packages, and therefore it is not usually necessary to construct a counter from individual FFs. Figure 11-7 shows a popular decade counter (10 states), the 7490. This IC contains four master-slave FFs interconnected to provide a divide-by-two and a divide-by-five capability. If the Q_A output is tied to the input B pin, the device will operate as a BCD decade counter. This counter can be reset to 0000 by making $R_{0(1)} = R_{0(2)} = 1$, or reset to 1001 by making $R_{9(1)} = R_{9(2)} = 1$. This reset function overrides the counting operation, and therefore to enable the counter to sequence through its various states, at least one of each of the reset inputs needs to be 0. The actual clock signal is applied to the input A pin.

Also available in IC form are counters that can count up or count down and are called *up/down counters*. There are also *presettable counters* which can be set to some desired starting number and counting continued from that point. Counters can come as *asynchronous*, which are simple but somewhat slow (propagation delay) or *synchronous* (all FFs receive the clock pulse at the same time), which are fast but more complicated in design. A compromise between the two are *asynchronous/synchronous counters*, which incorporate some of the features of both.

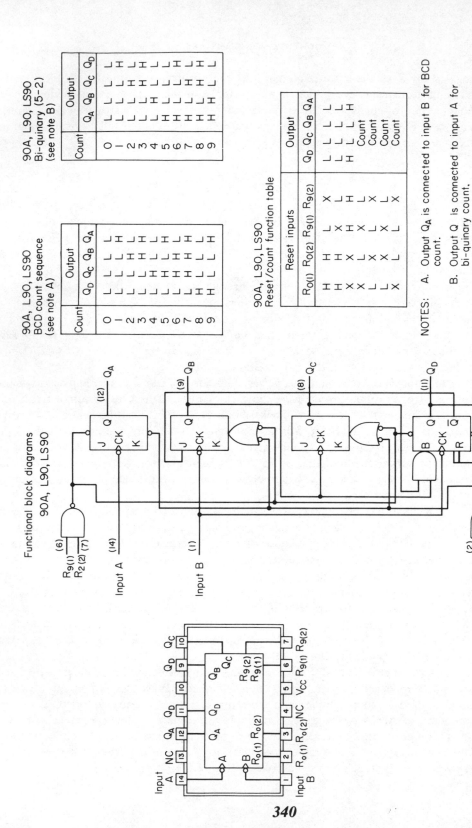

FIGURE 11-7 7490 decade counter.

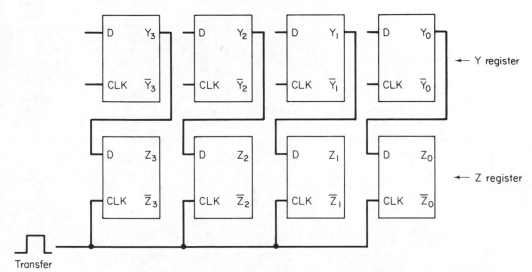

FIGURE 11-8   *Parallel transfer from register Y to register Z.*

Besides counting, FFs can be used as *registers*. A *register* is a group of FFs used to temporarily store information. One of the most frequent operations performed in digital systems is the transfer of information from one register to another. Figure 11-8 shows two four-bit registers set up for the parallel transfer of data from register Y to register Z. Each of the Y register FFs' outputs are tied to the corresponding Z register D inputs. The clock inputs of the Z register are all tied together and, upon application of a transfer pulse, the information stored in the Y register is transferred to the Z register in parallel. This arrangement for the transfer of information is very fast because only one pulse is necessary for the transfer to occur, but this method also requires the most interconnections between the two registers.

Serial transfer of information involves the use of shift registers, as shown in Fig. 11-9(a). Each clock pulse shifts one bit of information from the Y register to the Z register. Therefore, for a three-bit register, it takes three clock pulses to transfer the contents of the Y register to the Z register. Suppose that the Y register contained 011 before any clock pulses occurred and the transfer of this information to the Z register was to be undertaken. Figure 11-9(b) shows the progression of the data transfer after each clock pulse. Notice that for a three-bit register it takes three clock pulses for the transfer to be complete. Serial transfer requires more time, since it takes one clock pulse per bit, but it requires fewer interconnections between the two registers.

Most of today's registers are not made by interconnecting individual FFs, but are found in integrated-circuit packages. An example of one type of register, referred to as a *latching* register, is shown in Fig. 11-10. This register is a four-bit D-type latching register whose outputs will follow the D inputs while the clock is in the logic 1 state. When the clock goes through its negative transition, the information on the D inputs is latched by the FFs and becomes the new outputs. Information can not change again until the clock goes back to the 1 state. When the Clear input is made a logic 0 level simultaneously, all four D-type FF outputs are made 0.

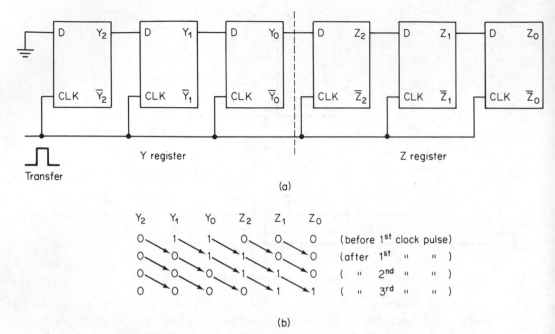

Y_2 Y_1 Y_0 Z_2 Z_1 Z_0

Transfer

Y register   Z register

(a)

Y_2 Y_1 Y_0 Z_2 Z_1 Z_0

0	1	1	0	0	0	(before 1st clock pulse)
0	0	1	1	0	0	(after 1st " ")
0	0	0	1	1	0	(" 2nd " ")
0	0	0	0	1	1	(" 3rd " ")

(b)

FIGURE 11-9   Contents of the Y register serially transferred to the Z register.

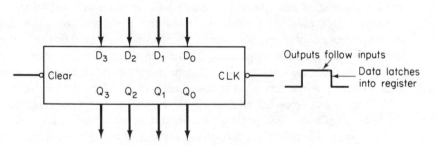

FIGURE 11-10   4-bit D-type latching register.

As opposed to latching registers, which allow the outputs to follow the inputs while the clock signal is high, *edge-triggered* registers allow the D inputs to affect the outputs only at the instant when the proper clock transition occurs.

A variety of shift registers can be found in integrated-circuit form. These devices differ in how data are entered and leave; for example, parallel-in, serial-out or serial-in, parallel-out, or shift left, shift right. Figure 11-11 shows an eight-bit *parallel-in, serial-out* shift register. Data can be entered into the register in two different ways. *Parallel* data can be entered via the D_0–D_7 inputs when the parallel-load input (P_L) goes to a logic 0 state. Once the data are entered, they can be serially shifted out by applying clock pulses to the CLK input. Data can be entered serially by applying

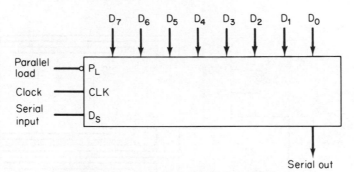

FIGURE 11-11   *Parallel-in, serial-out shift register.*

one bit at a time to the D_S input and generating a clock pulse. The serial output pin is tied internally to the output of the LSB of the shift register.

A relatively new entry into the register family is referred to as the *tristate* register. These types of devices are very useful in many digital systems, especially in digital computers, where a large majority of the computer operations involve the transfer of information from one register to another. By the use of tristate devices, many registers can be connected together utilizing a set of common connecting lines called a *bus*. Figure 11-12 represents a typical four-bit tristate register. With the input

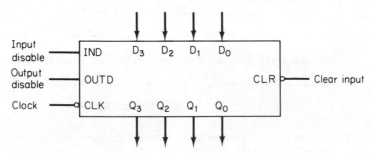

FIGURE 11-12   *Tristate register.*

disabled, $IND = 1$, the D_0–D_3 and clock inputs have no effect on information stored in the integrated circuit FFs. With the output disabled, $OUTD = 1$, the outputs of the FFs are essentially disconnected from the output pins by being placed in their high-impedance state. Although the outputs of the actual FFs can be changed, this will not show up at the output pins until $OUTD = 0$. The outputs of all the FFs can be cleared to 0 by a logic 0 level on the CLR input.

By connecting a number of these registers together with common interconnecting paths, as shown in Fig. 11-13, information can easily be transferred over this bus from one register to another by enabling the proper set of inputs and outputs while disabling all others.

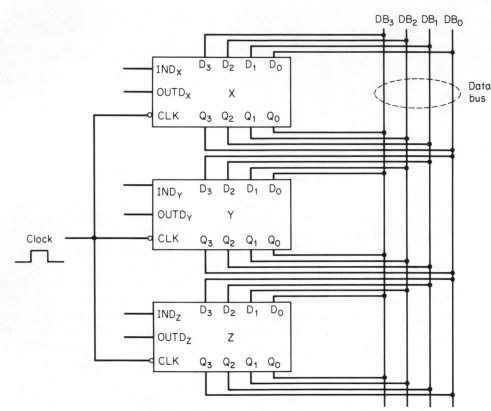

FIGURE 11-13 *Register-to-register transfer over data bus using tristate register.*

EXAMPLE 11-14

How can the contents of register X be transferred to register Z over the bus in Fig. 11-13?

Solution:

If $OUTD_X = IND_Z = 0$ while all other disable inputs $= 1$ and a clock pulse occurs, the data stored in register X will be transferred to register Z on the negative transition of the clock pulse.

11-5   DATA-MANIPULATING CIRCUITS

Decoders are a group of digital circuits that are designed to accept some input code group (binary, BCD, and so forth) and generate an output to indicate which of the possible input codes is present. When a particular input code being present is indicated by a logic 1, the device is said to have *active-high* outputs. If the presence of a particular code is indicated by a logic 0, it is said the device has *active-low* outputs. Figure 11-14(a) shows a *binary to octal* decoder with active-high outputs. It is described in this manner because it accepts a three-bit binary code as its input and indicates which code is present (000–111) by making one and only one of its eight outputs a logic 1. Eight outputs are necessary since $2^3 = 8$. This type of decoder is also sometimes referred to as a 3-line-to-8-line decoder. Figure 11-14(b) shows a *BCD-to-decimal* decoder with active-low outputs. This decoder accepts a four-bit BCD code as its inputs and indicates which code is present by making one of its outputs a logic 0

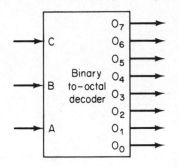

C	B	A	Activated Output = 1*
0	0	0	O_0
0	0	1	O_1
0	1	0	O_2
0	1	1	O_3
1	0	0	O_4
1	0	1	O_5
1	1	0	O_6
1	1	1	O_7

*All other outputs = 0

(a)

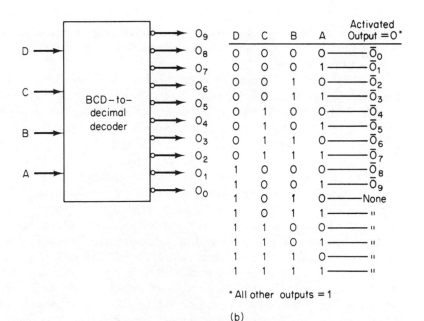

D	C	B	A	Activated Output = 0*
0	0	0	0	$\overline{O}_0$
0	0	0	1	$\overline{O}_1$
0	0	1	0	$\overline{O}_2$
0	0	1	1	$\overline{O}_3$
0	1	0	0	$\overline{O}_4$
0	1	0	1	$\overline{O}_5$
0	1	1	0	$\overline{O}_6$
0	1	1	1	$\overline{O}_7$
1	0	0	0	$\overline{O}_8$
1	0	0	1	$\overline{O}_9$
1	0	1	0	None
1	0	1	1	''
1	1	0	0	''
1	1	0	1	''
1	1	1	0	''
1	1	1	1	''

*All other outputs = 1

(b)

FIGURE 11-14 *Integrated circuit decoders.*

while all the other outputs remain a logic 1. Notice the small circles on the outputs of this decoder and how the outputs are labeled with an overbar ($\overline{0}_0$–$\overline{0}_9$). These circles are used to indicate that this is an active-low output device. This device is also sometimes referred to as a 4-line-to-10-line decoder. The circuit is designed so that if any illegal BCD code groups appear (1010–1111), all the outputs remain high, therefore not activating any output and not causing any erroneous decoding.

Encoders are devices that perform the opposite function of decoders. These circuits are designed to generate a specific code at their output (binary, BCD, or hexadecimal), corresponding to which input is activated. Inputs can be either active-low or active-high. Figure 11-15 represents an active-low input encoder which accepts one of eight

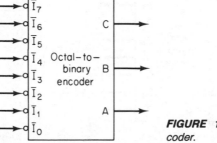

FIGURE 11-15 *Integrated circuit encoder.*

inputs and generates the binary output code corresponding to which input has been activated. For example, if $\overline{I}_4 = 0$, the outputs will be CBA = 100, respectively. This circuit is referred to as an octal-to-binary or 8-line-to-3-line encoder. Another example of an encoder could be a decimal-to-BCD encoder, which has 10 inputs. When one of these inputs is activated, the encoder will produce the BCD code corresponding to that input.

Multiplexers are logic circuits that have two or more input channels. Through the use of a *Select* control, only one of these input channels is allowed to reach the output. Each channel can be one or more bits. The simplest multiplexer circuit is shown in Fig. 11-16(a), where a two-channel single-bit multiplexer is shown. When the *Select* input is equal to a logic 1, the A input is passed to the output. If the *Select* input equals a logic 0, the B input is passed to the output. As can be seen from this example, multiplexers route the desired input channel, out of a number of input channels, to the output under the control of a *Select* input. To choose between two input channels, only a single-bit *Select* control is necessary. To choose between four-input channels, a two-bit *Select* control would be necessary. The number of *Select* control bits necessary is equal to the power that 2 must be raised to in order to equal the number of input channels. Most multiplexers are now found in integrated-circuit form and can come in a wide variety of configurations, some of which are two-channel three-bit, four-channel single-bit, two-channel four-bit, and eight-channel single bit. Figure 11-16(b) is a general diagram for a *multiplexer* which shows the input channels and *Select* input drawn as large arrows. These arrows indicate that one or more bits may be present.

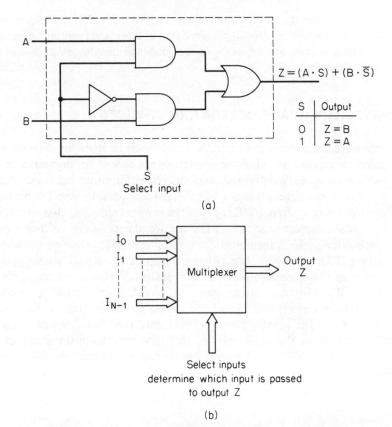

$$Z = (A \cdot S) + (B \cdot \overline{S})$$

S	Output
0	Z = B
1	Z = A

S
Select input

(a)

Output
Z

Multiplexer

I_0
I_1
I_{N-1}

Select inputs
determine which input is passed
to output Z

(b)

FIGURE 11-16 *Two-input and general diagram of multiplexers.*

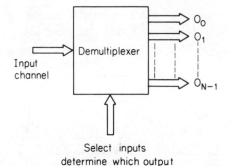

Demultiplexer

Input
channel

O_0
O_1
O_{N-1}

Select inputs
determine which output
will receive the input

FIGURE 11-17 *General diagram of a demultiplexer.*

Demultiplexers perform the reverse operation than that of a multiplexer. The demultiplexer takes a single input channel (one or more bits) and by use of a *Select* control determines which one of a number of output channels will receive this input. Figure 11-17 shows a general diagram for a *demultiplexer*. Again, the large arrows indicate the possibility of more than one bit per channel. The number of *Select* bits needed is controlled by how many output channels there are in the particular *demultiplexer*. Most *demultiplexers* come in integrated circuit form, and a wide variety of configurations are available concerning the number of channels and the number of bits per channel.

11-6   WAVESHAPING AND GENERATING CIRCUITS

Many logic gates require that the signals used as their inputs meet a minimum transition-time requirement. As these signals are passed throughout a system and various loads are applied, a deterioration of the transition times can occur. Another possibility for a problem occurs when a signal from the outside world is brought into a digital system, because often it will not have the proper transition characteristics. For example, if it was decided to use the 60-Hz line frequency as the reference clock for a digital clock circuit, the transition times on a sinusoidal waveform are too slow to reliably trigger TTL circuitry. The *Schmitt trigger* is a digital circuit quite often used to shape up the transition times of waveforms. The Schmitt trigger is a level-sensitive device that switches output states at two distinct triggering levels, one called the upper trigger point (UTP) and one called the lower trigger point (LTP). Figure 11-18(a) shows the symbol for a NAND gate that has Schmitt trigger inputs. Figure 11-18(b) shows the application of this gate to change the shape of the input signal for a digital clock circuit to one that is capable of reliably triggering TTL circuitry.

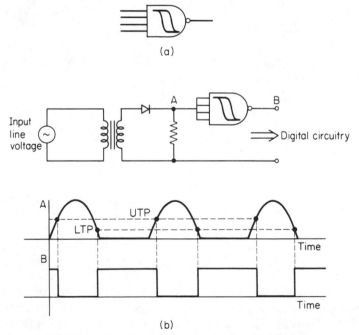

FIGURE 11-18   *Schmitt trigger symbol and application.*

One widely used waveform generator circuit is called the *one-shot* (OS), also referred to as the monostable multivibrator. The OS, like a FF, has two outputs, Q and $\overline{Q}$, which are the inverse of each other. However, unlike the FF, the OS has only one stable state (normally Q = 0, $\overline{Q}$ = 1), where it remains until it is triggered by an appropriate input signal. Once it is triggered, the OS outputs switch to the opposite states (Q = 1, $\overline{Q}$ = 0). The device will remain in this quasi-stable state for a fixed period of time, t_p, which is usually determined by an external *RC* circuit added to the IC. After the time t_p, the OS outputs return to their original states and remain there until triggered again. Figure 11-19 shows the logic symbol for an

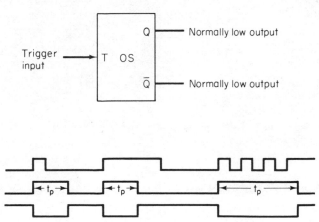

FIGURE 11-19 *O.S. symbol and response to input signal.*

OS and the output waveforms corresponding to the input signal shown. The value for t_p, can vary from nanoseconds to seconds and is controlled by the value of *R* and *C* chosen by the user of the circuit. Notice in the figure how the OS can be used to stretch out an input pulse, as with the first input signal, or to shorten an input pulse, as shown with the second input to the OS. The third input signal and output occur quite often in digital circuits and provide one of the major uses of the OS. This input condition can represent the waveform that could appear when a mechanical switch is closed. Upon closing, the actual switch contacts bounce and the waveform created by the bouncing could trigger digital logic circuitry many times for one switch closure. To eliminate this problem, the switch closure is used to trigger an OS whose t_p is longer than the bouncing is thought to last (usually ≤10 ms) and the output of the OS is used to feed the digital circuitry, therefore giving one output pulse per switch closure. Figure 11-20 shows the logic diagram of a commercial OS, the 74121. Notice that it has three inputs—if either one or both A inputs are held low, a positive transition on B will trigger the OS. Alternatively, if B is kept high, either A input going high (positive transition) will trigger the OS. The value of t_p is controlled by R_{ext} and C_{ext} and can vary from 40 ns to 40 s. If no external timing components are used and pin 9 is tied to the supply (pin 14), the OS will

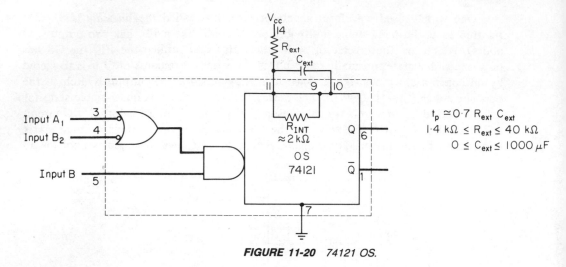

FIGURE 11-20   74121 OS.

produce a pulse width of approximately 30 ns, due to its own internal resistance and capacitance.

11-7 SYSTEM APPLICATION

A very popular system that most people have some exposure to is the digital clock system. This system will incorporate some of the circuits discussed in this chapter. This example will show how various circuits can be put together to build a system capable of performing a useful function.

   To have a useful digital clock it is necessary that an accurate and stable time base or clock signal be used in the system designed. For this example, 60-Hz power line frequency will be used, since utility companies regulate this to very stringent standards. In Fig. 11-21 a block diagram of the system is depicted. The first block will consist of waveshaping circuitry necessary to convert the 110-V 60-Hz power signal into a 60-Hz TTL compatible train used as the *clock signal* for the digital clock system. The 60-Hz pulse train is then fed to a Mod-6 counter, which divides the input frequency by 6, leaving a 10-Hz pulse waveform at its output. This 10-Hz pulse waveform is fed to a Mod-10 counter, which divides the input frequency by 10. At the output of this counter, a 1-Hz pulse waveform is available for the counters of the timing circuitry.

   The 1 pulse/s signal is fed to a set of counters and associated circuitry, which will count and display the seconds 0–59. This is accomplished by feeding the 1-Hz signal into a BCD counter (Mod-10), which will count 0–9 and recycle. This BCD counter is then keeping track of the units of seconds. The most significant bit (MSB) output from this BCD counter will provide a pulse every 10 s. This signal is fed to a Mod-6 counter, which counts 0–5 and then recycles back to 0. This counter is

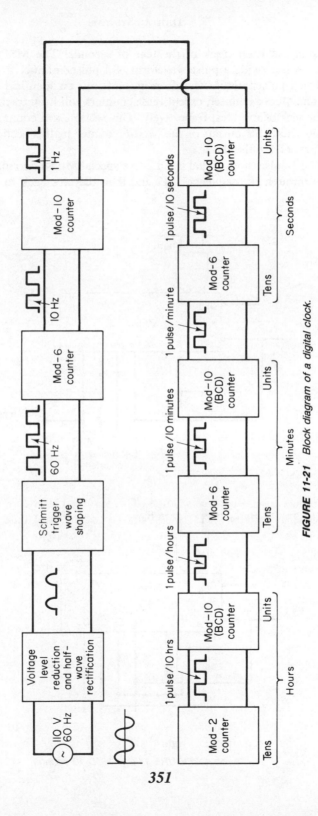

FIGURE 11-21 Block diagram of a digital clock.

351

then going to keep track of the tens of seconds. The MSB output of this Mod-6 counter will provide a pulse waveform of 1 pulse/minute.

The 1-pulse/minute waveform is sent into an identical section consisting of a BCD and Mod-6 counter, except these counters will keep track of the units of minutes and the tens of minutes, respectively. This section will count 0–59 minutes and then recycle. The MSB output of the Mod-6 counter in this section will provide a pulse waveform of 1 pulse/hour.

The 1-pulse/hour signal is fed into a special Mod-12 counter section. This section counts through the sequence 1–12 and then recycles back to 1, keeping track of the

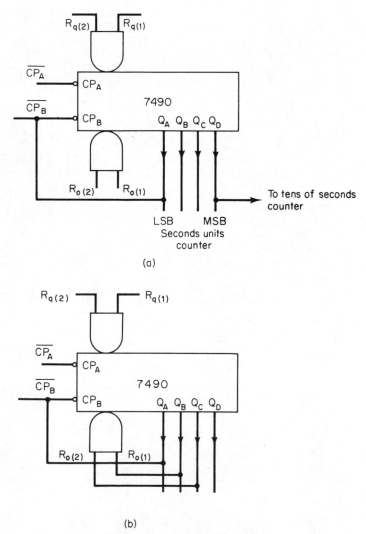

(a)

(b)

FIGURE 11-22 MOD-10 counter.

units of hours and the tens of hours. A more detailed description of this section will be covered when each block is gone over in more detail.

From Fig. 11-21 it can be seen that many counters are needed in the fundamental digital clock. These counters can be made using FFs and gates, as described in Section 11-4, or integrated-circuit counters can be made use of for this function. The 7490 decode counter chip might be a good choice for this application. Figure 11-7 shows the logic symbol and truth table for this counter. To see the application of this counter, let us take a closer look at the seconds section of the digital clock. In Fig. 11-22(a) the 7490 is set as a Mod-10 (BCD) counter. Notice that the LSB output, Q_A, is fed to the $\overline{CP}_B$ input. This, if you recall from the Section 11-4, is done because the 7490 can be used as two separate counters, a divide by two and a divide by five. Since the divide-by-ten or BCD count is going to be used, the Q_A output must be tied to the $\overline{CP}_B$ input. This counter will then count 0000–1001 and recycle back to 0000, keeping track of the units of seconds. When the counter is in the count of 1001 and the next clock pulse occurs, the recycling of the counter makes the MSB go from 1 to 0, which is a negative transition and can be used as a pulse (clock) input signal to the tens-of-seconds counter.

The 7490 is also used as the tens of seconds counter, except that the counting sequence is altered. The counter should count 0000–0101, then on the next clock pulse recycle back to 0000. This is accomplished as shown in Fig. 11-22(b), where the Q_B and Q_C outputs are fed to $R_{0(1)}$ and $R_{0(2)}$, respectively. When the counter is at the count of 0101, the next clock pulse would normally make it go to 0110, but as soon as Q_B and Q_C both equal 1, the counter is reset to 0000. Thus, the counter will count 0000–0101 and then recycle back to 0000. The Q_C output can be used as the clock signal input to the minutes section of the digital clock. In Fig. 11-23, the two 7490 chips are connected together to provide the seconds section.

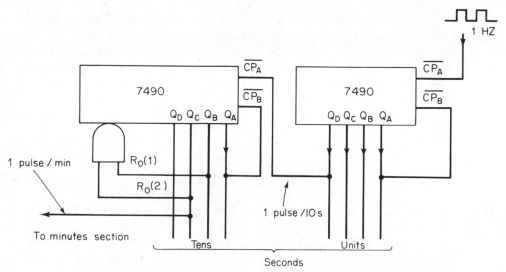

FIGURE 11-23 *Seconds section of digital clock using the 7490.*

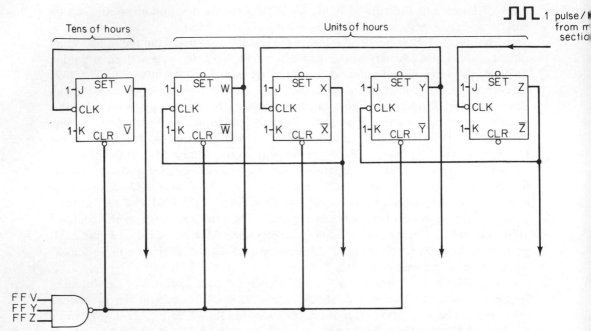

FIGURE 11-24 *Hours section of digital clock using J-K FFs.*

The minutes section is identical in construction to the seconds section. The output of the minutes section will be a 1-pulse/hour waveform. This signal will be fed to the hours section. This section consists of four J-K FFs set up as a BCD counter for the units of hours and a J-K FF for the tens of hours. Additional gates are utilized as shown in Fig. 11-24 so that the counter will sequence through the counting cycle 1–12 and then recycle back to 1. FFs W, X, Y, and Z are set up as a straight BCD counter. When the counter is in the 1001 state and the next clock pulse occurs, the BCD counter recycles back to 0000. The negative transition on FF W's output makes FF V go to the 1 state, indicating a count of 10. The next two clock pulses bring the counter to the count of 12, or 1 0010. The next clock pulse increments the counter to 13, or 1 0011, making FF V, Y, and Z all high. This condition makes the output of the NAND gate go low, clearing FFs V, W, X, and Y, and leaving FF Z = 1 and the hours counter section at the count of 1 for 1 o'clock.

The counter described so far has no means to display the value of time. One of the most popular display devices is the seven-segment display. This display device uses seven light-emitting-diode segments to produce the decimal characters 0–9. Figure 11-25 shows the seven-segment arrangement; by biasing certain segments, the various decimal numerals can be displayed. For example, the number 5 can be displayed by causing segments a, f, g, c, and d to be illuminated while segments b and e are dark. Each of these segments is a string of light-emitting diodes with, for example, all their anodes tied together. Whichever segment's cathode is grounded will light.

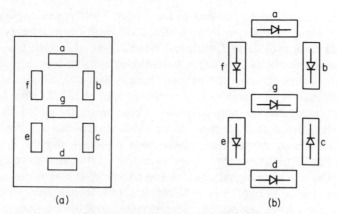

FIGURE 11-25 *Seven-segment display.*

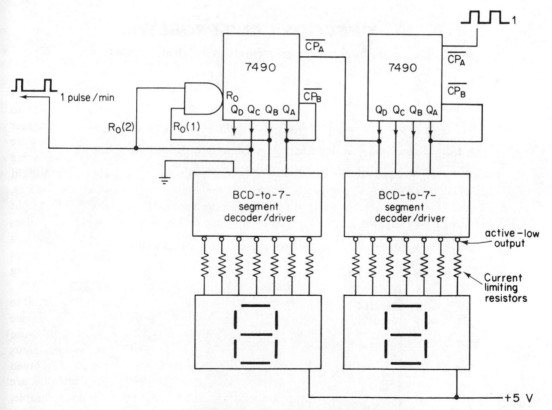

FIGURE 11-26 *Seconds section of digital clock.*

Those segments not grounded will not light. This type of display can be incorporated into the digital clock with the addition of some extra circuitry. The counters of the digital clock provide BCD outputs. What is needed is some type of decoding circuitry to convert the BCD code to the outputs necessary to bias the corresponding segments of the display to show the proper decimal numeral. This is accomplished by using integrated-circuit decoder chips which convert the BCD code to a proper seven-segment output for the applied input. These chips are called BCD to seven-segment decoders. Since each segment draws more current than a standard TTL output can handle, special output stages have been designed and incorporated into what are called BCD-to-seven-segment decoder/driver integrated-circuit chips.

The complete digital clock would consist of the basic waveshaping and counting section shown in Fig. 11-21. The outputs from the counters would become the inputs to BCD-to-seven-segment decoder/drivers, and the outputs of the decoder drivers would go to the seven-segment display. An example of the seconds section is shown in Fig. 11-26.

QUESTIONS AND PROBLEMS

11-1-1. Convert the following decimal numbers to their binary equivalent.
- (a) 15_{10}
- (b) 25_{10}
- (c) 127_{10}
- (d) 255_{10}
- (e) 9_{10}

11-1-2. Convert the following binary numbers to their decimal equivalent.
- (a) 1010_2
- (b) 100101_2
- (c) 0111_2
- (d) 1110_2
- (e) 10110101_2

11-1-3. Convert the following binary numbers to their octal equivalent.
- (a) 100110_2
- (b) 10101_2
- (c) 1110101_2
- (d) 10111101_2
- (e) 10110101_2

11-1-4. Convert the following octal numbers to their binary equivalent.
- (a) 35_8
- (b) 674_8
- (c) 310_8
- (d) 1547_8
- (e) 3_8

11–1–5. Convert the following decimal numbers to their hexadecimal equivalent.
   (a) 12_{10}
   (b) 196_{10}
   (c) 512_{10}

11–1–6. Convert the following hexadecimal numbers to their binary equivalent.
   (a) $67B_{16}$
   (b) $A5_{16}$
   (c) $6AF4_{16}$
   (d) $43C_{16}$
   (e) $10DE_{16}$

11–1–7. Convert the following binary numbers to their hexadecimal equivalents.
   (a) 10100011_2
   (b) 1011_2
   (c) 10010101_2
   (d) 1110101101_2
   (e) 11101011111_2

11–2–8. Convert the following decimal numbers into their BCD code representation.
   (a) 74_{10}
   (b) 126_{10}
   (c) 974_{10}
   (d) 5_{10}

11–2–9. Convert the following BCD encoded numbers to their decimal equivalent.
   (a) 10010111_{BCD}
   (b) 0100100110_{BCD}
   (c) 100101010110_{BCD}

11–2–10. Why is the BCD code inefficient compared to the binary number system?

11–2–11. What is the advantage of the Gray code over binary in representing incremental information?

11–2–12. What is the function of the parity bit in digital transmissions?

11–2–13. Add the necessary parity bit to the data below to conform to an odd-parity transmission scheme.
   (a) __ 10010101
   (b) __ 00010101
   (c) __ 11011010
   (d) __ 11110100

11–2–14. Repeat Problem 11-2-13 but set it up for an even-parity transmission scheme.

11–3–15. Perform the following binary additions.

(a)		(b)		(c)		(d)	
	10111		1010		1110110		1010111
+	10001	+	0001	+	0110	+	1111

11–3–16. Describe the operation of 2's-complementing a binary number.

11–3–17. Perform the following subtractions using 2's-complement addition.

    (a)   101101   (b)   1110100   (c)   11001   (d)   101010
           − 001011   − 11011   − 1111   − 10001

11–3–18. Convert each of the following decimal numbers to their binary equivalent, indicating both magnitude and polarity.
    (a) $+9_{10}$
    (b) -6_{10}
    (c) $+24_{10}$
    (d) -31_{10}

11–3–19. Perform the following BCD additions.

    (a)   0100   (b)   00010000   (c)   00010101   (d)   01100011
         + 0011   +   0111   +   1001   + 00101001

11–4–20. Draw the waveforms for the X, Y, and Z outputs for the asynchronous counter of Fig. 11-6 for 10 clock pulses.

11–4–21. Explain the difference between asynchronous and synchronous in conjunction with counters.

11–4–22. Explain what is meant by the term "propagation delay"?

11–4–23. Six FFs are connected as in Fig. 11-6. What is the maximum number this ripple counter can count to? How many possible states can the counter go through? What is the MOD-number of this ripple counter?

11–4–24. A 64-kHz clock signal is fed to the least significant FF in Problem 11-4-23. What is the frequency of the output signal at each FF output?

11–4–25. Why is serial transfer of data slower than parallel transfer?

11–4–26. What is the major disadvantage of parallel transfer of information over long distances?

11–4–27. Explain a major advantage of tristate registers over standard TTL registers, and show an example.

11–5–28. Explain the function of a decoder.

11–5–29. What is the function of an encoder?

11–5–30. Explain the function of a multiplexer.

11–5–31. How many Select control bits are necessary for a multiplexer to route four channels of information with each channel comprised of eight bits?

11–5–32. Explain the function of a demultiplexer.

11–6–33. Explain how a Schmitt trigger NAND gate can be used as a waveshaping circuit.

11–6–34. The one-shot shown in Fig. 11-27 has the waveform shown applied to its trigger input. Draw the output waveform assuming that the OS has a $t_p = 100$ ns.

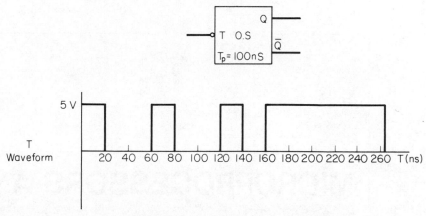

FIGURE 11-27 *Problem 11–6–34.*

11–7–35. Design the Mod-6 counter that comes after the waveshaping circuitry in Fig. 11-21. Use J-K FFs and gates.

11–7–36. How can the setting of the correct time be incorporated into the minutes and hours section of the digital clock of Fig. 11-21?

12

MICROPROCESSORS AND MICROCOMPUTERS

A recent advance in electronics has had great impact in both the electronics industry and our daily lives. With the arrival of the microprocessor, another evolutionary stage has taken place in the electronics field. With regard to its impact on the electronics industry, what the introduction of the transistor did a number of years back, the microprocessor is doing now. Every day, more and more applications of this device affect our lives in many ways. What are microprocessors? Why have they had such an impact on industry and our lives? The answer to these questions are given in this chapter. The topic of microprocessors and microcomputers encompasses many books; this chapter can only begin to introduce the reader to the basics.

12-1 COMPUTERS

Before an understanding and appreciation of microprocessors and microcomputers can be achieved, one must know how a computer works in general and what components are involved. A simplified description of how a computer works is to describe it as follows: *the computer executes a sequential set of instructions.* The instructions are in a binary-coded form and reside in the computer's memory. Each instruction has a unique code specifying a particular operation and has been placed in specific sequence by the computer programmer. The complete set of instructions is referred to as a *program,* and the program allows the computer to perform a useful function.

The computer can perform this function by taking *(fetching)* the first instruction from memory and performing *(executing)* the operation called for by the code. It then goes back to the memory unit and takes the next instruction in sequence (unless directed otherwise) and performs the operation called for by its code. This sequential *fetching* of an instruction and *execution* of that instruction continues until the final instruction is executed. The computer has then finished performing the function defined by the program and can either wait for a new set of instructions (program) or be directed to repeat the entire program over again.

Computers can be broken down into three main categories based on their size. The biggest types are those we see in large business corporations, banks, and scientific laboratories. An entire large room may be devoted to these *maxicomputers* and their associated peripheral equipment, such as magnetic tape units, card punchers, card readers, and line printers. The function of these units can range from scientific computation and engineering problem solving to large business-type operations, such as payroll, accounts keeping, inventory, and maintaining large files of data.

Minicomputers are much smaller in physical size and are used mostly for purposes such as industrial process control (see Chapter 13), scientific applications in research laboratories, and management of business records for small companies. These computers have great appeal because of their relatively inexpensive price tag compared to maxicomputers and their varied capabilities, making them very flexible and easy to package for a variety of applications.

The *microcomputer* is the least expensive and smallest of the three types of computers. Its greatest impact is in the realm of data acquisition and control in industrial process control, although many applications have been found in consumer electronics and the computer hobbyist market. The low cost of the microcomputer permits the average person to afford purchase of a personal computer. Microcomputers and the spin-offs of the technology involved in manufacturing them will change many aspects of our daily lives, such as how we pay our bills and educate and entertain ourselves.

12-2 BASIC COMPUTER SYSTEM ORGANIZATION

All computers, of whatever size, have four basic units:

1. An arithmetic/logic unit (ALU).

2. A control unit.

3. A memory.

4. An input/output (I/O) capability.

Figure 12-1 shows the interconnection of these units, and the arrows indicate the direction in which data, information, or control signals are normally flowing.

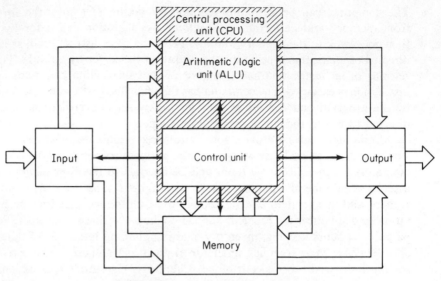

FIGURE 12-1 *Basic computer organization.*

Sometimes, the input and output units can be considered separately. The ALU and the control unit are often grouped together and referred to as the *central processing unit* (CPU). This is what the term "microprocessor" means. It is the ALU and control unit of a microcomputer. The microprocessor is usually one chip (IC) to which are added memory chips and I/O chips to make a complete microcomputer.

The *memory* unit serves as an area to store *instructions* and *data*. *Instructions* are binary-coded pieces of information that get decoded by the microprocessor and specify a particular operation to be performed. *Data* are coded pieces of binary information that get operated on by the CPU. The memory can also be used for storing intermediate and final results of arithmetic operations performed by the ALU. The control unit oversees the operation of the memory unit. By providing an appropriate address, information can be read from the memory unit and placed in the CPU or an output device. Information can also be written into memory from the CPU or an input device under the control of the control unit.

The CPU consists of the ALU and the control unit. The ALU is the area of the computer where arithmetic and logic operations are performed on data. The type of operation to be performed is determined by signals generated from the control unit. These signals are based on the decoding of an instruction that was read from memory. The data that get operated on can come from memory or an input device based on the instruction for that operation. The results of the operation can be placed in memory or in an output device, again depending upon the instruction specified.

The *control* unit has the function of controlling the functions performed by all the other components of the microcomputer. It must generate timing and control signals necessary to *read* (fetch) instructions from memory, decode these binary pieces of information utilizing the instruction decoder circuitry, and *execute* what is called for by the decoded instruction. It must also be able to communicate with I/O devices by performing *read* or *write* operations when called for by an instruction during the execution of a program.

The input/output unit consists of all the devices that allow the computer to communicate with the outside world. Some examples of *input* devices are keyboards, toggle switches, Teletypewriters, punch-card and paper-tape readers, magnetic-tape readers, and analog-to-digital converters. Some examples of *output* devices are indicator lights, LED readouts, Teletypewriters, printers, punched-paper-tape, CRTs, and digital-to-analog converters.

To get a better feeling for how a computer works, a simple program example will be discussed. This program will be set up to do the following:

1. *SUBTRACT* one number (X) from another number (Y).

2. *STORE* the result, Z, in memory location 0022_{16}.

3. *JUMP* to location 0050_{16} if the result equals zero.

4. *JUMP* to location 0060_{16} if the result is a positive number.

5. *JUMP* to location 0070_{16} if the result is a negative number.

The program to perform this task is listed in Table 12-1. The actual program stored in the computer memory would be the column listed "Contents of Memory Location." Since the computer is continually *fetching* and *executing* instructions, it must be operating on information in groups of bits. These groups of bits that the computer operates on, either transferring or manipulating, are referred to as the computer *word* size. The greatest majority of microcomputers use an eight-bit word size, indicating that the CPU deals with eight-bit transfers or manipulations of information. In Table 12-1, each memory location holds an eight-bit piece of information, or one computer word. The grouping of eight-bits into one piece of information for transfer or manipulation occurs so frequently that it is given the name *byte*. Therefore, eight bits is one byte and the computer used for the example program in Table 12-1 operates on one-byte words, so it can be described as an 8-bit microcomputer.

Larger computers, such as maxicomputers, generally operate with word sizes ranging from 16 to 64 bits, with 32 bits being the most common. These computers (32-bit) would then be described as having a word size of four bytes. Minicomputers operate with word sizes ranging from eight bits to 32 bits, with 16 bits being the most common.

The information stored in memory shown Table 12-1 can be broken down into two broad categories: data words and instruction words. *Instruction words* are listed in Table 12-1 as being stored in memory locations 0000_{16}–0011_{16}. The instruction format contains two basic pieces of information: the code that tells the CPU what operation is to be performed *(op code)*, and the operand address *(address)*, which represents the location in memory where the data to be operated on are stored. Since most microcomputers operate with an eight-bit word size, the instruction format

TABLE 12-1 Machine language program

| Memory Address (Hex) | Contents of Memory Location (Word) | | Symbolic Code | | Description |
	Binary	Hex			
0000	00111010	3A	LDA	20	Load Y into the accumulator from
0001	00000000	00			memory location 0020_{16}.
0002	00100000	20			
0003	10010111	97	SUB	21	Subtract X (located in memory loca-
0004	00000000	00			tion 0021_{16}) from the accumulator.
0005	00100001	21			
0006	00110010	32	STA	22	Store result Z in memory location
0007	00000000	00			0022_{16}.
0008	00100010	22			
0009	11001010	CA	JZ	50	If Q = 0, Jump to memory location
000A	00000000	00			50_{16}.
000B	01010000	50			
000C	11110010	F2	JP	60	If Q > 0, Jump to memory location
000D	00000000	00			60_{16}.
000E	01100000	60			
000F	11000011	C3	JMP	70	If Q < 0, Jump to memory location
0010	00000000	00			70_{16}.
0011	01110000	70			
0012					
⋮					
0020	$Y_7 Y_6 Y_5 Y_4 Y_3 Y_2 Y_1 Y_0$				
0021	$X_7 X_6 X_5 X_4 X_3 X_2 X_1 X_0$				
0022	$Q_7 Q_6 Q_5 Q_4 Q_3 Q_2 Q_1 Q_0$				
0023					
⋮					

most commonly used with these eight-bit computers will be described. Figure 12-2 shows the three common instruction formats for these 8-bit computers: single-byte, two-byte, and three-byte. The single-byte instruction only specifies an operation code (op code), which defines a specific function to be performed. No address is given in this type of format. An example might be to clear the accumulator (CLA). Since the accumulator is internal to the CPU, no address is necessary, so all that is needed is the code for the operation to be performed. Some other examples would be: rotate the accumulator right (RAR), rotate the accumulator left (RAL), halt (HLT), and complement the accumulator (CMA).

Most microprocessors specify a 16-bit memory address and, therefore, two bytes are necessary to define a specific memory location. Figure 12-2(c) shows a three-byte instruction where the first byte specifies the op code. The second and third bytes define the operand address. Some microprocessors allow for storing data in specific areas of memory, in which case a special op code for the operation to be performed is used and only eight bits of address are needed for the complete instruction,

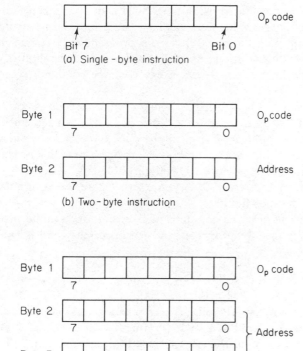

Bit 7   Bit 0

(a) Single - byte instruction

Byte 1 ... O_p code

7   0

Byte 2 ... Address

7   0

(b) Two - byte instruction

Byte 1 ... O_p code

7   0

Byte 2

7   0   } Address

Byte 3

7   0

(c) Three - byte instruction

FIGURE 12-2 *Instruction formats used in 8-bit microcomputers.*

as shown in Fig. 12-2(b) for a two-byte instruction. This feature allows the programmer to write programs that take up less memory space, since the address is only one byte and the program runs faster because not as many read operations have to be performed to fetch the complete instruction. Instructions are stored in memory locations 0000_{16}–0011_{16} in the example listed in Table 12-1. The first instruction occupies memory locations 0000_{16}–0002_{16}, with the op code stored in location 0000_{16} and the operand address specified by the next two sequential memory locations. Some microprocessors specify that the eight most significant bits of the 16-bit operand address be the second byte of a three-byte instruction. Other μPs state that the second byte must be the eight least significant bits of the 16-bit operand address. The user of the particular μP would have to check the specification for the particular device for which the program is being written.

Besides the instructions in the program of Table 12-1, there are *data* words stored at memory locations 0020_{16} and 0021_{16}. These two memory locations hold the data that will be operated on by the ALU during the execution of the program. These data words can be representing numerical quantities using the binary number system, or they may be signed numbers, in which case the 2's-complement representation would be used. The numerical information could also be stored using the BCD code or alphanumeric information stored using the ASCII code (American Standard Code for Information Interchange). The point is that a given location in memory is going to be storing a string of 1's and 0's. It is the programmer's responsibility, when writing the program, to define where and in what format information is going to be stored, so that when the program is being executed the computer interprets the data properly.

There is a column in Table 12-1 labeled *symbolic code*. This code is comprised of a few letters used as an aid in programming since it is easier to remember the symbolic code than the binary op code. For example, once the user knows that the symbolic code LDA stands for "load the accumulator," the next time this instruction has to be specified it is easier to remember LDA than 00111010_2. Also, it is easier to remember that STA, rather than 00110010_2, represents "store the accumulator." The actual op codes can always be found at a later time.

Let us next take a look at how the computer will execute the sample program listed in Table 12-1. Assume that the program has been in memory and is ready to be executed by the computer.

1. The user initiates program execution by entering the four-hex-digit address of the first instruction in this program (0000_{16}) and pushing a RUN or START or GO control key. This places 0000_{16} in a special internal register called the *program counter* (P.C.). It is the function of the P.C. to always be pointing to the address of the next instruction the computer will fetch after finishing the execution of the present instruction. This causes the control unit to perform a read operation from memory location 0000_{16} and place the contents of this location into a special internal register called the *instruction decoder register* (I.D.).* It is important, therefore, that the programmer has a valid op code stored in memory location 0000_{16}; otherwise, whatever is stored in that memory location will be handled as an op code and result in the computer being directed to perform an operation not called for in the program. Once the instruction at location 0000_{16} is placed in the I.D. register, the control unit increments the P.C. so that now it contains the address 0001_{16}. The computer now decodes the op code in its I.D. and determines that a "load the accumulator (LDA)" operation is to be performed and that the next two sequential memory locations hold the address where the data to be placed in the accumulator are stored. The computer therefore performs two more *Read* operations, placing the contents of memory locations 0001_{16} and 0002_{16} into a special 16-bit register called the *memory address register (MAR)*.

    This finishes what is called the *instruction cycle* of computer operation, where the computer obtains an instruction. Next the computer goes into an *execution cycle*. In completing the instruction cycle, the P.C. is left containing 0003_{16}, which points to the next instruction, and the computer contains

---

* Some microcomputers automatically go to a specific location, such as 0000_{16}, when the GO key is depressed.

the op code and operand address of the first instruction. Therefore, the execution cycle of this first instruction involves the control unit performing a read operation at the address specified by the MAR, which, in this example, would be 0020_{16} and loads the contents of this location into the accumulator. This completes the execution cycle and the computer continues by beginning another instruction cycle. As a matter of fact, the entire operation of the computer revolves around fetching an instruction and then executing that instruction and then repeating the cycle over and over, progressing through the entire program.

2. The control unit goes into the next instruction cycle by placing the contents of the memory location specified by the P.C. (0003_{16}) into the I.D. and incrementing the P.C. Here the op code is decoded and found to be a subtract command with the data stored at an address specified by the next two memory locations. The instruction cycle then continues with the control unit performing two more read operations to get the operand address into the MAR (0021_{16}) and updating the P.C. (0006_{16}). The execution cycle involves a *read* of the contents of the memory location specified by the MAR (0021_{16}) and subtracting these data from those already in the accumulator. The result is left in the accumulator for now.

3. The next instruction cycle takes the contents of the memory location specified by the P.C. (0006_{16}) and places it in the I.D. and increments the P.C. (0007_{16}). The op code is decoded and found to call for a "store the contents of the accumulator (STA)" at the address specified by the next two memory locations. Therefore, the instruction cycle continues with two more read operations to get the address of where to store the contents of the accumulator into the MAR (0022_{16}) and updating the P.C. (0009_{16}). The execution cycle consists of a write operation being performed and placing the contents of the accumulator into the address specified by the MAR (0022_{16}). The contents of the accumulator is not changed by this write operation.

    This sequence of instruction and execution cycles, with the P.C. always indicating where to go for the next instruction during the instruction cycle and the MAR storing the address of where to get or put information during the execution cycle, continues until the program is complete. The following description of the continuation of the program will be abbreviated since the idea of the steps involved in the execution of a program were covered in detail in steps 1–3.

4. The next instruction located at 0009_{16} is referred to as a *conditional jump* instruction. When decoded, this instruction tells the control unit to check the accumulator. If the previous operation resulted in the accumulator being exactly zero, the contents of the next two memory locations are placed into the program counter. This forces the control unit to get the next instruction during its instruction cycle from the address specified (0050_{16}), rather than

the next sequential location in memory ($000C_{16}$). In other words, this instruction *can* alter the flow of program execution. If the condition is true, the control unit jumps to a new location (0050_{16}) in memory to continue its instruction and execution cycles. If the accumulator is not exactly zero, the P.C. is updated and points to the next sequential instruction location ($000C_{16}$).

5. If the accumulator was not exactly zero, the control unit would fetch its next instruction from memory location $000C_{16}$. This is another conditional jump instruction. This time, if the accumulator had a number greater than zero as a result of the subtraction performed earlier, the contents of the next two memory locations would be placed in the P.C. The control unit, when it enters the next instruction cycle, would then obtain its next instruction from memory location 0060_{16}. If the accumulator is not positive (greater than 0), the P.C. is updated and the next instruction would come from memory location $000F_{16}$.

6. This last instruction is referred to as an *unconditional jump*. This instruction unconditionally alters the normal sequencing through memory of the control unit. The contents of the next two memory locations are placed into the P.C. (0070_{16}) and the control unit gets its next instruction from 0070_{16} during the next instruction cycle.

In this particular program example, two numbers were subtracted. The result, which is in the accumulator, is then stored in memory location 0022_{16}. A number of tests are then performed on the result of this subtraction. First, if the result was zero, the program directs the control unit to new instructions starting at memory location 0050_{16}. If the result was not zero, a test is performed to check if the number is positive. If positive, the program directs the control unit to a new set of instructions starting at memory location 0060_{16}. If the result was neither zero or positive, it must be negative, so the program unconditionally directs the control unit to a set of instructions at memory location 0070_{16}.

This concludes the somewhat simplified explanation of the operation of a computer. Some important points to remember are:

1. The computer is continually going through *instruction* and then *execution* cycles.

2. The instructions that are being fetched and executed are in memory and were placed there by a programmer. It is the programmer who is in reality controlling the overall operation of the computer. The computer *cannot* think; it can only fetch and execute instructions in a program.

3. The real power of the computer is that it only takes microseconds to fetch and execute an instruction. Therefore, many thousands of instructions can be performed within a second.

All the operations performed by the control unit in response to program instruction must be synchronized. This is accomplished by logic circuitry inside the control unit and a master clock signal. This clock signal acts as a time reference frame for all operations. An instruction cycle might take three clock cycles to complete and an execution cycle two clock cycles or more, depending upon the instruction. Various operations are synchronized to the occurrence of these clock cycles. These clock signals may be fed to various parts of the computer system, such as memory and I/O, so that transfer of information can be synchronized.

12-3   WHAT ARE MICROCOMPUTERS? WHY ARE THEY SO POPULAR?

Microcomputers are microprocessors to which memory and I/O capability have been added, allowing for the same basic functioning as the computer discussed in Section 12-2. The microprocessor (μP) is equivalent to the CPU section of a computer and is usually contained on a single LSI (large-scale integrated circuit) chip. The microcomputer contains the four basic units of any computer, except that they are usually implemented in integrated circuit form. Therefore, one μP chip, a few memory chips, and some I/O chips are basically all that is needed to put together a microcomputer. All these devices can be placed on one circuit board and cost less than $200. In fact, there are complete computers on one LSI chip, although they have limited memory and I/O capability. Figure 12-3 shows the basic elements of a typical micro-

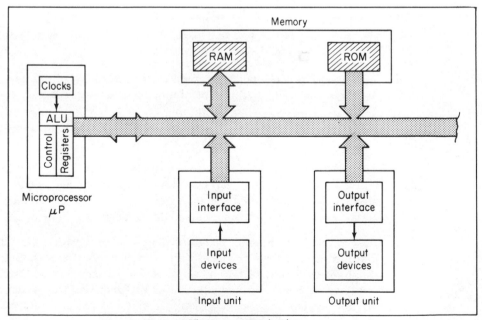

FIGURE 12-3   Basic elements of a μC.

computer (μC). The μP is the central processing unit (CPU) portion of the μC. It contains the ALU and control unit together with special-purpose and general-purpose internal registers for the temporary storage of information during program execution. The memory unit may consist of a *ROM* (read-only memory) section, which is usually one or more LSI chips used to store instructions and data that are not going to change. Or the memory section may have a *RAM* (random-access memory) section. Again, this is implemented using one or more LSI chips and is designed to store programs and data that will change throughout the use of the computer. The I/O section consists of special interface circuits, usually LSI chips, which allow the I/O devices to properly communicate with the rest of the computer. More details on each of these sections will be presented later in the chapter.

Why have μCs become so popular in the recent years? There are several reasons for the superiority of μP-based designs over conventional logic. The term "conventional logic" means implementing functions with networks of gates, flip-flops, counters, and registers. The microcomputer implements its functions based on a set of instructions stored in memory elements. Data and certain types of programs are stored in RAM memory, and the basic operating program of the system is stored in ROM memory elements. The μP performs the specified functions by fetching instructions from memory and then executing these instructions. The results are usually communicated to the outside world via the μC's I/O ports. For example, an 8080 μP executing instructions stored in a single 2048-byte ROM memory element can perform the same logical functions that previously may have required up to 1000 logic gates. Some of the benefits of designing a system using a microcomputer are:

1. Since one microcomputer chip set replaced many standard random logic elements, the overall size and parts count of the total system is reduced. This reduces assembly cost and total cost.

2. As the number of components decreases, the probability of a failure occurring decreases. Since all the logic control functions that were controlled by various hardware components are now replaced by a few ROM chips which are nonvolatile (contents not lost even after power is removed), the probability of failure decreases, so product reliability increases.

3. Since the operation of the system is under program control, any system redesign that is necessary can be accomplished by simply changing the program. This requires simply replacing or reprogramming a ROM chip. This is much easier than rewiring a standard random logic system.

4. Because of the computing power of the μP, manufacturers can offer many more features in their products with little, if any, extra cost. Oscilloscopes can have write-outs of the values of measured parameters. Cash registers can have automatic tax computation by programming the particular state's tax information into the ROM chip. If the tax structure changes, only the ROM chip would have to be reprogrammed.

The applications of microprocessors and microcomputers are increasing every day. They seem limited only by the imagination of the designers involved in this field. A typical application might be an automatic computing weight scale for a supermarket. Figure 12-4 shows the components involved in such a system. There are two input devices: one for a keyboard entry so that the operator can select specific functions and enter information concerning price per unit weight. Two output devices

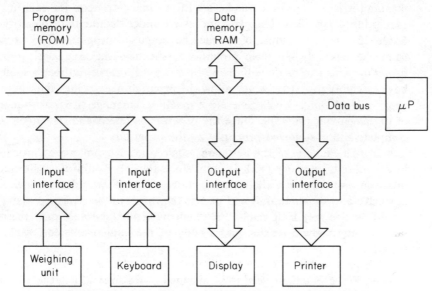

FIGURE 12-4 *Application of μC in automatic computing scale.*

are shown: one is a readout consisting of several seven-segment displays used to display the total price, and the other is a printer for hard copy so that the cost can be tagged onto the package.

The μP is going to accept data inputs and function information from the keyboard, weight information from the weight unit and, based on these inputs, generate a total price and send it to the display and printer. The ALU's only function is to perform the necessary operations to arrive at the total price based on the weight and the price per unit weight. This is a multiplication process that can be accomplished by an algorithm stored in the μC's memory. The designer of this system starts out by generating a *flowchart,* indicating which signals must be read, what functions are going to be performed (generating the multiplication algorithm), and what output signals are going to be needed. The term "flowchart" refers to a diagram showing all the logical steps the program must follow, including all jumps and branches. The designer then writes an *assembly language* program. Assembly language (symbolic) programs use mnemonics for each instruction and labels for all addresses plus comments to explain the purpose of various instructions. "Mnemonic" refers to a technique in aiding the memory. An instruction such as **LDA** would mean "load the accumulator"; LSR would mean "logic shift right." The mnemonics of the example program in Table 12-1 are listed under the column "Symbolic Code."

Op codes and addresses are not used in the program yet. This assembly (symbolic) program can be scrutinized for correctness of logical thought and corrected if necessary. Once it has been determined that the logic of the program is correct, the *machine language* program can be generated from the assembly language program. A machine language program is one that is written so that the computer can use it directly, meaning it is written in 1's and 0's with a binary op code for each instruction and each address specified. The program tested under "contents of memory location" in Table 12-1 is an example of a machine language program. The process of going from the assembly language program to the machine language program is called *assembly*. This can be done by hand by the programmer, although some μC systems have assembly programs which allow the programmer to enter the assembly language program and the μC will generate a machine language program under the control of the assembler program. Once the program is tested and debugged, the system is complete and customized primarily by the program.

In this example of the automatic scale, the program would probably be placed in a read-only memory (ROM) chip, since the μP would always execute the same program, implementing the scale functions. The μP would continually monitor the keyboard and weighing unit and update the display whenever necessary. This system would require very little read-write memory (RAM), because only the price per unit weight, intermediate results, and a copy of the final results for display need to be stored.

This entire system could be designed and implemented using standard TTL logic devices. When functions need to be changed or features added, the TTL-based system may require extensive rewiring and circuit-board layout change and possibly even refabrication. However, with the μC system, since the control is under the program stored in memory, the memory chip can be replaced or reprogrammed with the new information, making this a relatively simple change. The only custom circuit design in the μC system would be the interface circuitry, and this would be necessary even in the standard TTL logic design. Other applications of μCs include intelligent terminals, gaming machines, cash registers, accounting and billing machines, telephone switching control, numerically controlled machines, process control, and last but not least, the home hobbyist computer market.

12-4   INSIDE THE MICROPROCESSOR

The typical microprocessor consists of three fundamental interconnected units:

1. Registers.
2. Control and timing.
3. Arithmetic/logic unit (ALU).

These units must perform the functions of:

1. Fetching instructions and data from memory.

2. Decoding instructions.

3. Executing instructions involving arithmetic and logic operations.

4. Transferring data to and from I/O devices.

5. Responding to signals generated by I/O devices such as Reset and Interrupt.

6. Providing the necessary timing and control signals for all the various elements comprising the μC system.

Figure 12-5 shows these three sections and the busses that are of interest. While the scope of this book does not allow for a detailed look into each of these sections, a brief exposure should be of great help for future dealings and comparisons of μPs.

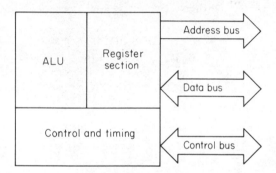

FIGURE 12-5 *Fundamental units of a μP.*

Control and Timing

The control and timing circuitry has the task of fetching and decoding instructions from a program stored in memory and then generating all the necessary control and timing signals required by the ALU and register section to execute this instruction. Since the μP is continually fetching, decoding, and executing instructions, a clock signal is necessary for an orderly sequence of events. This clock signal provides a reference frame for all μP actions. During the instruction cycle, the control unit generates the necessary signals to fetch and decode an instruction from memory. The control unit then proceeds to generate the necessary control and timing signals to execute that instruction. The control unit must also be capable of responding to signals from the outside world, such as *Interrupt.* When an interrupt signal is presented to the control unit, it proceeds to finish executing the instruction it is working on and then temporarily stops the execution of the main program and jumps to a special program *(interrupt service routine)* to satisfy the interrupting device. The control unit, after servicing the device, automatically returns to the main program, where it is left off and continues the execution of that program.

The control unit must generate control and timing signals and pass them to many components of the μC. This is accomplished by sending these signals over a bus called the *control bus*. Some typical signals comprising a control bus are those in the following list.

R/W: This is a μP output signal used to inform the rest of the μC that the μP is in either a *READ* (obtaining information from an input device or memory) or a *WRITE* (sending information to an output device or memory). The terms **READ** and **WRITE** are always referenced with respect to the μP; therefore, when a **READ** is being performed, the μP is obtaining information. When a **WRITE** operation is being performed, the μP is sending information.

CLOCK: Some μPs require two clock signals whereas others do not. Whichever the case, the clock signal is the basic timing element of the system. It is the clock signal that generates the time intervals during which system operations take place.

RESET: A μP input used to restore the μP's internal registers to zero. It is also used many times to regain control of the μC by forcing the μP to start executing a special program stored in ROM memory dedicated to servicing the keyboard. This allows the user to enter new data or programs, thereby taking over control of the system.

MREQ (Memory Request): A μP output used to indicate that the μP is in the process of accessing memory.

IORQ (I/O Request): A μP output used to indicate that the μP is accessing an I/O device as opposed to memory.

READY: A μP input utilized by slow memory or I/O devices used to force the μP into a type of idle state until the device is ready to transfer the data.

IRQ (Interrupt Request): A μP input that, when activated by I/O devices, causes interruption of the current program by forcing the μP to finish executing the instruction it is dealing with and jump to a special service routine in memory designed to take care of the interrupting device. After servicing the device, the μP automatically returns to the program it had been working on when the interrupt occurred. The return to the main program is at the instruction following the one where the interrupt occurred, since the μP always finishes executing an instruction before going to the service routine.

INTE (Interrupt Enable): A μP output used to indicate whether or not the μP can be interrupted at this point. If INTE = 1, the μP can be interrupted by applying the appropriate signal to the IRQ input. If INTE = 0, the internal μP interrupt circuitry is disabled and therefore the μP cannot be interrupted. The programmer has control over the setting (INTE = 1) or clearing (INTE = 0) of the INTE control by software (program) instructions.

NMI (Nonmarkable Interrupt): Many μPs have two interrupt inputs: one as previously described under IRQ and INTE, and the other which cannot be disabled. If the proper signal is presented to the NMI input, the μP will be interrupted no matter what state INTE is in at that time.

Arithmetic/Logic (ALU)

Every μP contains an ALU which contains circuitry for performing arithmetic and logic operations on binary data. Some μPs allow for operations to be performed on BCD data as well as binary. The ALU contains basic logic circuitry to provide the capabilities of LOGIC ANDing, ORing, EXCLUSIVE ORing, SHIFT, ROTATE, COMPLEMENT, and CLEAR. In addition, the ALU must contain circuitry to provide ADDITION and SUBTRACTION capabilities between the contents of two registers in binary arithmetic. Some μPs allow for BCD arithmetic operations to be performed. The programmer can then write programs using standard algorithms for providing the capability of multiplication and division. This gives the μP fairly large arithmetic capabilities.

Register Section

Registers are temporary storage units inside the μP. Some of these registers have dedicated functions used solely by the μP. Examples of such registers include the program counter (PC), instruction register (IR), stack pointer register (SP), and status register (SR). Other registers are for more general-purpose use, and these include the accumulator (A) and general-purpose registers (GP). The advantage of having GP registers inside the μP is that information can temporarily be stored and when needed, the μP does not have to perform a memory READ operation, thereby speeding up program execution. Also, since the information stored in GP registers is internal to the μP, no address bytes are necessary when specifying an instruction, thereby cutting down on program size and memory space used to store the program. Some of the most common registers found in μP are described next.

Program counter (PC): This register has been discussed previously. It stores the address in memory of the next instruction (or portion of instruction-operand address bytes) that the μP is to fetch. The μP automatically increments the PC after each use and therefore executes the stored program sequentially unless some instruction execution forces a new address to be placed in the PC. This occurs during the execution of a JUMP or BRANCH instruction.

Instruction register (IR):   When the μP fetches an instruction, it is temporarily stored in this register until decoding is complete.

Memory address register (MAR):   This register stores the operand address portion of an instruction. In other words, this register contains the address of data the μP will read from or write into memory during the execution cycle. The outputs of this register and the PC are multiplexed to the address bus, since both specify an address during different cycles of operation. The PC specifies the address of an instruction and puts it out on the address bus during the instruction cycle; the MAR specifies an address and places it on the address bus during the execution cycle.

Status register (SR):   This register is a group of individual bits used to indicate specific conditions inside the μP at any time. These bits are usually called *flags,* because they indicate specific conditions occurring inside the μP at some point in time. Different μPs will have different flags, but some typical flags are:

Carry flag (C)—The carry flag is automatically set (C = 1) whenever the μP performs an addition or subtraction that results with a carry out of the most significant bit (MSB) position. The C flag is automatically cleared (C = 0) when no carry is generated. The programmer can, under software control, set or clear the C flag by using appropriate instructions in the program.

Zero flag (Z)—This flag is automatically set (Z = 1) whenever a data transfer or data manipulation instruction results in all zeros. The flag is automatically cleared (Z = 0) under any other conditions.

Decimal mode flag (D)—Some μPs allow arithmetic operations to be performed in BCD. The programmer can specify BCD arithmetic operations by placing an instruction in the program to set the D flag (D = 1). Subsequent operations will be performed in BCD arithmetic. If the D flag is cleared (D = 0), arithmetic operations will be handled in straight binary.

Interrupt disable flag (I)—This flag indicates whether or not the μP can be interrupted. If the programmer uses a set the interrupt disable flag (I = 1), instruction from that point on cannot be interrupted unless it has an NMI input. Once the instruction to clear the I flag (I = 0) is encountered, the μP can be interrupted by the appropriate signal to the IRQ input.

Overflow flag (V)—This flag is automatically set (V = 1) whenever an arithmetic operation involving two *signed* numbers results in a carry into the sign-bit position. This flag is automatically cleared under all other conditions. The programmer under software control can set or clear this flag.

Negative flag (N)—Whenever the result of a data transfer or data manipulation results in the MSB being a 1, the N flag is set (N = 1). If signed numbers are being used, this indicates a negative number. If the result of the data transfer or manipulation leaves a 0 in the MSB, the N flag is cleared (N = 0).

Besides indicating the state of the μP at any particular time, the flags are used in conditional branch instructions. For example, if a jump on result equal zero instruction (JZ) is encountered, the control unit will check the zero flag (Z). If Z = 1, the previous operation resulted in all zeros and the control unit will take the operand address portion of the JZ instruction, place it in the PC, and thereby direct operation to a different area in memory. If, however, Z = 0, the control unit will go straight into the next instruction cycle and the PC will direct operation to the next sequential instruction in the program.

Stack pointer register (SP):   Often, the programmer finds it necessary and advantageous to use *subroutines.* Subroutines are programs within programs. When a sequence of instructions has to be repeated many times throughout a program, it is easier and more efficient to write this sequence of steps once and direct the μP to these steps whenever necessary rather than to rewrite the sequence of instructions each time they are needed. The μP handles subroutines in a special way to make sure that the μP returns to the main program and can continue on. When an instruction to "jump to subroutine" (JSR) is encountered, the control unit increments the PC, which will now contain the address of the next instruction in the main program. The contents of the PC are then stored in a reserved area of RAM memory called the *STACK.* The STACK contains the address of the next instruction in the main program. The control unit now takes the operand address portion of the JSR instruction and places it into the PC. The next instruction fetched during the instruction cycle will therefore come from the subroutine. The last instruction of any subroutine is a *RETURN* from subroutine instruction. When this instruction is decoded, the control unit goes to the STACK and replaces the contents of the PC with what was stored on the top of the STACK. Since the stack contained the address of the next instruction to be fetched in the main program, the μP will return to the main program where it had left off prior to the JSR.

How does the control unit know where the STACK is located in RAM memory? The μP contains an internal register called the *stack pointer* (SP) register, which always holds the address of the next available location on the stack. The control unit automatically takes care of decrementing and incrementing the SP whenever information is placed (pushed) on the stack or removed (pulled) from the stack. The programmer has to initialize the SP when the μC is first powered up and does this by writing an instruction to load the SP with a specified address in RAM memory.

In addition to the contents of the PC, the contents of various internal μP registers may be placed on the stack by writing appropriate instructions. This allows use of the registers during the subroutine. When returning from the subroutine, the contents can be read off the stack and placed in the appropriate registers, thereby leaving the μP in the same state that it was in prior to the JSR.

The same sequence of events can take place when an interrupt occurs, since the μP needs to know where to come back to in the main program after servicing the interrupting device. The contents of various internal registers also need to be saved if the interrupt service routine will be using these registers.

Accumulator: The accumulator takes part in most ALU operations. A usual instruction might call for the addition of the contents of some memory location to the contents of the accumulator and storing the results of this addition in the accumulator. Therefore, the accumulator often ends up being the source of one of the operands and the destination of the result of the operation called for by the instruction.

The accumulator will also be used many times as a temporary storage element when communicating with I/O devices. Information will be READ from an input device and placed in the accumulator before being directed to a specific place in memory for final storage. Information being transferred to an output device will be placed into the accumulator and then written to the output device.

General-purpose registers (GP):   The number of GP registers available varies from one μP to another. The function of the GP register is temporary storage of information so that there is no need to send intermediate results back and forth between memory and the accumulator. Since the GP register is internal, the control unit does not have to perform a memory READ operation, thereby increasing the operating speed. Also, since the register is internal to the μP, no operand address is specified in the instruction using the GP register. This results in faster programs because the control unit does not have to perform two memory READ operations during the instruction cycle to get the operand address bytes. Also, the programs take up less memory space, because only one byte is necessary to specify the instruction since no operand address is required.

Index register (IR):   The index register is like a GP register except that it has the added special feature of taking part in determining the address of data the CPU is trying to get. This is a special addressing mode called *indexed addressing.* The basic idea of indexed addressing is that the actual *(effective)* operand address specified by an instruction is the sum of the operand address portion of the instruction and the contents of a specified index register.

effective address = operand address portion of instruction
+ contents of specified index register

This form of addressing allows for the handling of tables or arrays of data with relative ease.

Table 12-2 lists some of the features of four of the most common μPs in use today. From this table you will get an idea of the flexibility and power of one μP over another.

TABLE 12-2 *μP comparisons.*

	μP Type			
	8080A	*Z-80*	*6800*	*6502*
Program counter	16-bit	16-bit	16-bit	16-bit
Accumulator	8-bit	Two independent 8-bit	Two independent 8-bit	8-bit
Status register	8-bit	Two 8-bit	8-bit	8-bit
General-purpose register	Six 8-bit	Twelve 8-bit	*	*
Index register	0	Two 16-bit	16-bit	Two 8-bit
Stack pointer register	16-bit	16-bit	16-bit	8-bit
ALU	8-bit	8-bit	8-bit	8-bit
Interrupt vector register		8-bit		
Memory refresh counter		8-bit		
Speed†	3.5 μs	2.8 μs	4 μs	4 μs

*   Index registers can be used as general-purpose registers.

†   Time to perform addition of two binary numbers with one in the accumulator.

12-5 MEMORY

Most μPs in use today utilize semiconductor memory chips as the main memory element in the system. We will therefore concentrate our description of memory to semiconductor memories. In general, memory can be thought of as a large information storage area where binary information is stored in groups of bits called *words*. A word is the basic unit of information that the computer deals with. Most μPs deal with eight-bit words, so each memory word would be eight bits. Integrated-circuit memory chips generally have common values for the number of words per chip, such as 64, 256, 512, 1024, 2048 or 4096, with the number of bits per word being one, four, or eight. By combining memory chips of a certain number of words and bits per word in various configurations, a total memory can be made to obtain specific word sizes and number of words in total memory. Each memory location is assigned a specific address and internal circuitry decodes these binary addresses to enable the μP to communicate with one and only one specific location in memory at some given point in time. Most μPs have a 16-bit address bus, which means that they are capable of communicating with $65,536 = 2^{16}$ different memory locations. In reality, only a small portion of this total address capability contains actual memory. Figure 12-6 shows a typical computer system in block-diagram form. Notice that the address bus, data bus, and control bus are all connected to the memory section. There are two possible operations the μP performs in dealing with memory: the READ operation, during which information is acquired from a specific memory location, and the WRITE operation during which the μP places information into a specific memory location.

READ operation—the μP places the binary code of the specific location in memory onto the address bus. The control unit then generates the READ signal,

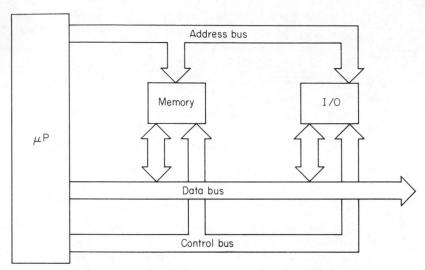

FIGURE 12-6 *Typical computer system in block diagram form.*

which is sent out over the control bus to memory. By the decoding of the address on the address bus, a specific memory location is selected and when the READ signal is generated, the contents of this memory location are placed on the data bus. At the appropriate time, determined by a signal on the control bus, the information on the data bus is latched into the μP.

WRITE operation—the μP places the binary code of the specific location in memory onto the address bus. The control unit generates the WRITE signal, which is sent out over the control bus to memory. By the decoding of the address on the address bus, a specific memory location is selected. The μP places the word to be written into memory onto the data bus, and at the appropriate time, determined by a signal on the control bus, the information is stored into this specific memory location.

Since the memory is either receiving information during a WRITE operation or sending information during a READ operation, IC manufacturers often combine the data input and data output functions using common I/O pins, thereby conserving on IC pins. During a WRITE operation, the I/O pins act as inputs and during a READ operation, the I/O pins act as outputs. Most IC memory chips include one or more chip select (CS) inputs, which are used to enable or disable the entire chip. This feature allows for the combining of memory chips in various configurations to increase memory size. Semiconductor memory ICs can be broken down into two broad classifications with regard to the frequency of read operations versus write operations to be performed.

Read-only memories (ROM): This type of memory device is designed primarily for having information read from it during normal computer operation. As a matter of fact, once placed into the system, the μP can only perform memory READ operations. ROMs are used to store information that is not going to change. ROMs are nonvolatile memory elements, meaning that when power is turned off, the information stored is *not* lost but remains intact. Figure 12-7 shows the pin configuration for a typical ROM which stores 256 words, each word consisting of eight bits. Notice that there are no data input pins or READ/WRITE control pin, since the ROM does not have a WRITE operation as part of its normal operating sequence.

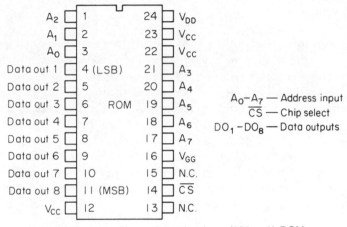

FIGURE 12-7 *Pin configuration for a (256 × 8) ROM.*

ROMs can be subdivided into several categories, depending on how information is initially programmed into the various memory locations.

Mask-programmed ROM—This type of ROM has the information programmed into its memory locations by the manufacturer at the time the IC is fabricated. A photographic negative called a *mask* is used to control the electrical interconnections. This type of ROM is only economical when a large quantity of ROMs, all having the same information programmed into them, is going to be required.

Programmable ROM (PROM)—This type of ROM comes to the user with all 1's or 0's stored in every memory location. The user can then program each memory location individually by supplying specified amounts of current to particular pins on the chip, thereby changing the content of that memory location. Once programmed, the contents of these chips cannot be changed and therefore have to be replaced if any changes need to be made. Any mistake in programming results in lost time and the cost of the chip, since these devices cannot be reprogrammed. This type of ROM has been widely used by the computer hobbyists because the PROM chip is relatively inexpensive and the circuitry necessary for programming not very complex.

Erasable Programmable ROM (EPROM)—One of the latest to come out, this ROM can be programmed by the user as above but can also be erased and reused. Once the EPROM is programmed, it is like an ordinary ROM, in that when placed in the μC system, it will only be used in the READ operation. The contents of this device are nonvolatile, meaning that they will not be lost when power is turned off. If, at a future time, the stored program needs to be changed, the EPROM can be removed from the μC system and placed under ultraviolet light. The UV light passes through a quartz window on the chip. This procedure erases the entire contents of the EPROM. The new information can now be programmed into the EPROM and the chip returned to the μC system.

Read-write memories (RAM): This type of memory is designed so that it can easily be read from or written into under normal computer program execution. RAM stands for *random-access* memory, which means that the memory is capable of having any one of its address locations accessed without having to sequence through other locations. Although ROMs are also random access in nature, popular usage of terms has evolved such that RAM stands for a read/write random-access memory. When a user talks about having 4K of RAM memory, he or she is indicating that the system has 4K of read/write memory.

RAMs are used for temporary storage of programs and data that will change throughout the use of the computer. RAMs are volatile, meaning that once power is lost, the contents of this memory element is lost. When power is returned, the memory must be reprogrammed. Figure 12-8 shows the pin configuration for a typical 256×4 RAM memory IC chip. Notice the sharing of I/O pins, which is controlled by the level presented to the R/W input. Also notice that there are two chip select inputs, which can be used for easy memory expansion.

There are two types of RAM memory with different operating modes. *Static* RAM consists of storing binary information in FFs. Therefore, a memory location can be thought of as a FF register. This type of memory is fairly complex and can take as many as 10 transistors per bit of storage, and therefore the memory capacities

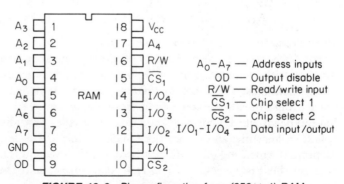

FIGURE 12–8 *Pin configuration for a (256 × 4) RAM.*

per chip are fairly small. A maximum of approximately 4096 bits per chip is typical. *Dynamic* RAM memory involves storing information as charge on the gate capacitor of a MOS transistor. This requires only a few transistors per bit of storage, therefore more memory capacity per chip. The disadvantage of dynamic RAM memory is that this charge leaks off and must be periodically *refreshed*. Refreshing consists of performing a read cycle at each address every 2 ms.

12-6 INPUT/OUTPUT

The μP must be able to communicate with devices that exist in the outside world. Somehow, information must be able to be placed inside the μP and the results of any μP operations must reach the outside. Devices such as keyboards, paper tape, floppy disks, Teletypewriters, and displays are typical examples of devices used to input information to the μP and store the results of the various μP operations.

The first distinction in interfacing the μP to I/O is how the μP will deal with I/O devices. There are two basic approaches to this: the μP communicates with I/O just as it does with any memory location (memory-mapped I/O), or the μP treats I/O separately and distinguishes whether it is dealing with memory or I/O by separate control signals (isolated I/O).

Most μPs use the *memory-mapped I/O* configuration. This allows the user to treat I/O just as another memory location. The instructions used in dealing with operations performed with memory can be used for I/O. This is possible since one or more of the most significant address lines would be fed to decoding circuitry, which would generate a chip select (CS) signal to the I/O interface circuitry for the address specified. As far as the μP is concerned, it is simply either performing a READ or WRITE operation at a specified address—it neither knows or cares whether it is memory or I/O. Figure 12-9(a) shows how memory address allocations need to be reserved for use by I/O when organizing the total μC system. One of the disadvantages of memory-mapped I/O is that it takes away address locations that may have been used for memory. Most μC systems never use all of the 65,536 addressable locations for memory, so memory space is not a major problem.

Isolated I/O, on the other hand, treats I/O separately from memory. The user specifies special input and output instructions in the program when communicating with I/O. These instructions cause separate control signals to be generated on the control bus to be fed to the I/O interfaces. Figure 12-9(b) shows the isolated I/O configuration with its separate memory and I/O allocations, allowing for all 65,536 addressable locations to be used for memory. One of the advantages of isolated I/O is that by using the special instructions in dealing with I/O, an eight-bit address may only be needed to specify a device. This allows for two byte instructions, which *could* result in smaller programs and faster running times, since fewer read operations have to be performed during the instruction cycle. The Intel 8080A μP allows for either memory-mapped I/O or isolated I/O configuration.

The I/O elements can range from something as simple as a set of switches as

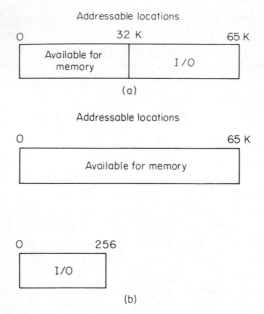

FIGURE 12-9 Memory-mapped I/O, isolated I/O.

input and a set of LEDs as output, to something as sophisticated as a floppy disk system for storage and retrieval of information. Let us take a look at the simplest I/O system to see what is involved. If the system is designed for memory-mapped I/O, the interfacing can be fairly simple. The input could be a set of eight switches, which would allow the entry of logic 1's or 0's. The outputs of these switches would go to tristate buffers, whose outputs could be tied directly to the data bus. Only at the appropriate time would the information from the switches be placed on the data bus. An instruction such as load the accumulator (LDA) from a specified address would be used to generate a read during the execution cycle. The address portion of the instruction would specify the switches as the input device. When the address was placed on the address bus, decoding circuitry would decode this information and generate a signal that could be gated with the READ signal to enable the tristate buffers. This allows the information from the switches to get onto the data bus. Once this information is on the data bus, the μP would latch it into the accumulator and the input operation would be complete. If a different address is used, or the same address when a write operation is occurring, the tristate buffers would be disabled, thus disconnecting the switches from the data bus.

Writing information to an output device can be handled in a similar fashion. An instruction such as store the contents of the accumulator (STA) at a specified address can be used to output information. During the execution cycle, when the specified address is on the address bus, decoder circuitry would generate a signal that can be gated with the WRITE control signal and be used to clock a latch whose inputs are connected to the data bus, and therefore store the information present on the data bus that the μP is trying to transmit to the output device. The outputs of the latch can be fed to a buffer chip and used to drive LEDs, which would then give an output that was a visual indication of the contents of the accumulator which the μP was transmitting to the outside world.

The scope of this book does not allow for a detailed look at interfacing of I/O to the μP, but it can be assumed that the more sophisticated the I/O, the more complex the interfacing. Levels of sophistication go from switches as inputs to hexadecimal keyboard, full ASCII keyboard, audio tape, and floppy disk. For outputs, the level of sophistication goes from LEDs to seven-segment displays, teleprinter and CRT, audio tape and floppy disk.

D/A and A/D conversion: Many applications of μPs involve some type of process control (see Chapter 13). Under these circumstances, it may be necessary for the μP to acquire data from some transducer, manipulate these data in some fashion, and then generate some type of process control signal to regulate the variable being measured or controlled. This generally involves both analog and digital signals. The μP deals only with digital information, although most signals in the real world are analog in nature. It is therefore necessary to convert the analog signals of the real world to digital equivalents (analog-to-digital conversion, A/D) which the μP can deal with, and convert the digital process control signal generated by the μP to an analog equivalent (digital-to-analog conversion, D/A) which can be used to control the process variable. Since many A/D conversion methods utilize the D/A conversion process, let us examine D/A conversion first.

Since the binary number system is a weighted number system, it is necessary to generate a voltage for each bit position. The value of the voltage should be proportioned to the binary weight of the bit. In Fig. 12-10(a) a simple D/A converter circuit is shown utilizing an op amp as a summing amplifier and another op amp as an inverting amplifier. The circuit is set up to accept a four-bit binary number and produce an output voltage proportioned to the binary input. In a four-bit binary

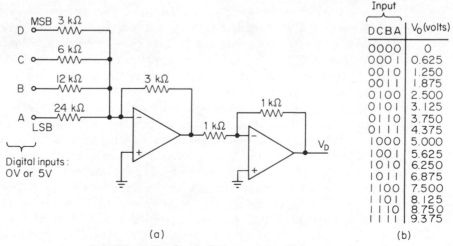

Input	
DCBA	V_0 (volts)
0000	0
0001	0.625
0010	1.250
0011	1.875
0100	2.500
0101	3.125
0110	3.750
0111	4.375
1000	5.000
1001	5.625
1010	6.250
1011	6.875
1100	7.500
1101	8.125
1110	8.750
1111	9.375

(a)   (b)

FIGURE 12-10 *Simple D/A converter utilizing op amps.*

number, the weight of the MSB is eight times greater than the LSB. As can be seen from the circuit, the gain of bit A is $-\frac{1}{8}$, B is $-\frac{1}{4}$, C is $-\frac{1}{2}$, and D is -1, so that the MSB has a gain eight times greater than the LSB. If binary inputs have values of 0 V for a logic 0 and 5 V for a logic 1, outputs would be produced as listed in Fig. 12-10(b) for the various input combinations. The output changes in 0.625-V steps as the binary input is incremented. D/A converters have a specification referred to as *resolution,* which is equal to the smallest change that can occur in the analog output. For the D/A converter in Fig. 12-10, the *resolution* would be 0.625 V; however, often the specification % resolution is utilized, which is found by the equation

$$\% \text{ resolution} = \frac{\text{step size}}{\text{full-scale output (F.S.)}} \times 100\% \qquad \textbf{(12-1)}$$

For the D/A converter of Fig. 12-10, the % resolution would be

$$\% \text{ resolution} = \frac{0.625 \text{ V}}{9.375 \text{ V}} \times 100\% = 6.67\%$$

Since the step size is dependent upon the number of input bits for a given full-scale output, the % resolution can be made smaller by increasing the number of input bits. The smaller the resolution, the more closely a binary number on the input can come to generate a specific analog voltage. The accuracy of a D/A converter depends on how accurate the supply voltage and the binary input voltages are, since any variation of input voltage will change the output voltage. This is why standard single-chip D/A converters will usually include on the chip some type of stable reference voltage supply and input-level amplifiers, which will produce fixed output voltages for varying binary input voltage levels.

The process of analog-to-digital (A/D) conversion can involve different techniques. We will look at the simplest conceptual technique, which is the digital-ramp A/D converter. The block diagram in Fig. 12-11 shows the basic components involved in this technique of A/D conversion. Essentially, clock pulses are fed to a binary counter. The output of the counter is fed to a register, whose outputs are fed to a D/A converter. As the binary count is incremented, the output of the D/A converter starts from 0 V and increases accordingly. Each increment in the binary count changes the output voltage of the D/A converter by one step size. The output of the D/A converter is fed to one input of a comparator; the other input is the unknown analog voltage. When the output of the D/A converter just exceeds the actual analog voltage, the output of the comparator changes state abruptly, and this signal is used to stop the clock pulses from incrementing the converter. The counter is then left with a binary number stored in the register which represents the binary equivalent of the actual analog voltage. When the next conversion is to take place, the counter and register are cleared and the process repeats. How closely one analog voltage input can be identified from another analog voltage input depends upon the A/D converter's

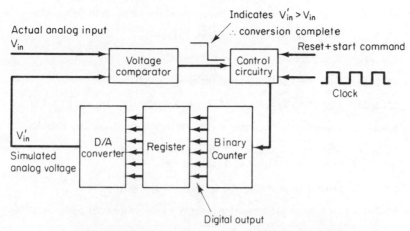

Indicates $V'_{in} > V_{in}$
∴ conversion complete

Actual analog input
V_{in}

Voltage comparator

Control circuitry

Reset + start command

Clock

V'_{in}
Simulated analog voltage

D/A converter

Register

Binary Counter

Digital output

FIGURE 12-11 *Digital-ramp A/D converter block diagram.*

resolution, which, in turn, depends upon the internal D/A converter's resolution. As mentioned previously, the resolution can be increased by increasing the number of bits.

D/A and A/D converters can be purchased as single-package ICs at a cost of $5 to $100, depending upon the type of conversion process used, the speed of conversion, and the accuracy (number of bits). The A/D converters are more expensive, because they incorporate the D/A conversion circuitry also. Recently, A/D converters have been developed with tristate outputs so that it is relatively easy to interface the outputs of the A/D converter to the µP data bus, making the acquisition of analog information a relatively simple process.

QUESTIONS AND PROBLEMS

12–1–1. What are the three main general classifications of computers according to size?

12–2–2. List the four basic units that comprise any computer.

12–2–3. What does the term "CPU" stand for, and what elements are involved in it?

12–2–4. What is the function of the memory unit?

12–2–5. Define a computer instruction.

12–2–6. What is the function of the ALU?

12–2–7. State the function of the control unit.

12–2–8. If a computer is said to have a four-byte word size, how many bits does it operate on at one time?

12–2–9. List the three formats that instruction words can utilize.

12-2-10. Why is it necessary to have three-byte instruction formats?

12-2-11. What does the term "op code" mean?

12-2-12. What does "symbolic code" mean?

12-2-13. Describe the function of the program counter (P.C.).

12-2-14. Describe the function of the instruction decoder register (I.D.).

12-2-15. Describe the function of the memory address register (MAR).

12-2-16. The entire operation of the computer can be thought of as a repetitive process of two cycles. What are these cycles?

12-2-17. How does a conditional branch instruction operate?

12-2-18. How does the computer process an unconditional jump instruction?

12-2-19. Why is a clock signal necessary in a computer system?

12-3-20. Explain the difference between μP and μC.

12-3-21. Describe the four benefits of designing around a microcomputer-based system as opposed to conventional logic design.

12-3-22. Define what is meant by a flowchart.

12-3-23. What is an assembly language program?

12-4-24. What three units comprise a typical μP?

12-4-25. List six functions the μP performs in a μC system.

12-4-26. What is the purpose of the read/write (R/W) signal on the control bus of a μP?

12-4-27. What is the purpose of ϕ_1 and ϕ_2 signals on the control bus of a μP?

12-4-28. Describe two types of interrupt inputs and how they differ.

12-4-29. What is the function of the status register of a μP?

12-4-30. List five typical flags of a status register and their purpose.

12-4-31. What is a stack? How is a stack pointer register utilized?

12-4-32. Define what a subroutine is and how the μP deals with it.

12-4-33. Why are internal general-purpose registers useful for a μP?

12-4-34. What is the advantage of an index register in a μP?

12-4-35. How is an effective address arrived at?

12-5-36. READ and WRITE operations are performed with respect to what reference?

12-5-37. What is a ROM?

12-5-38. List three types of ROMs and explain their differences.

12-5-39. What are RAMs?

12–5–40. Explain the difference between static and dynamic RAM.

12–6–41. Explain memory-mapped I/O.

12–6–42. Explain isolated I/O.

12–6–43. What is the function of a D/A converter?

12–6–44. How is the resolution of a D/A converter determined?

12–6–45. What is the function of an A/D converter?

12–6–46. How is the resolution of an A/D or D/A converter improved?

12–6–47. What is an advantage of a tristate A/D converter?

13

CONTROL SYSTEMS

13-1 INTRODUCTION

A *control system* is, quite simply, a means to control something. Some examples of a control system include the following:

1. A system that automatically controls the filling of cereal boxes to the proper weight on a production line.

2. A system that senses the proper sound output of a radio and controls the receiver's gain to maintain that level. This is called automatic gain control (AGC) and is incorporated in all receivers.

3. A system that controls traffic lights based on the existing traffic flow at an intersection.

The preceding examples show that control systems are a part of everyday life. The basic operations of a control system are shown in the block diagram of Fig. 13-1. A *sensor* looks at the system to be controlled and determines its condition. The sensor is a device that normally converts from one form of energy into another. It is then, by definition, a *transducer*. As an example, the cereal box measurement system may utilize a pressure transducer as its sensor. It may provide a variable capacitance, based upon the force of the cereal box upon it.

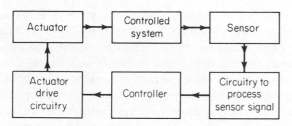

FIGURE 13-1 *Control system block diagram.*

The sensor's output signal is applied to the signal conditioning circuitry, which provides a standard range of voltage (or current) at its output. The controller receives this as its input. The *controller* calculates the correct response and issues a signal to the servo in Fig. 13-1. In a complicated system the controller may well be a computer. The *actuator* accepts the controller output signal and accomplishes the necessary change. In the cereal example, the actuator may be a solenoid which opens and closes a chute that delivers the cereal to the box. The actuator in this case and in most cases is also a transducer, since it converts from electrical energy to mechanical motion. Another name sometimes used for an actuator is a *servo*.

The control system just described has a closed loop of activity from the controlled system around and back to itself. For this reason it is termed a *closed-loop* system. The controlling process is performed automatically, and these systems are therefore called *automatic control systems*. The first electronic automatic control system was the AGC process previously described for radio receivers. It was developed in the 1920s.

13-2 MOTION AND FORCE SENSORS

The sensor block shown in Fig. 13-1 is obviously a key element in an automatic control system. This section provides a brief look at a few of the many types of sensors that are available.

Linear-Motion Potentiometer

Many industrial processes require sensors that detect the linear position of a moving device. The *linear-motion potentiometer* shown in Fig. 13-2 is able to sense linear motion by presenting a resistance that is proportional to linear position. It is a variable

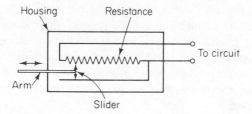

FIGURE 13-2 *Linear-motion potentiometer symbol.*

resistor whose resistance is varied by the movement of a slider over its resistance element. The slider is moved over the resistance element by means of an arm connected to the moving component. This slider shorts out a portion of the resistance. When the arm in Fig. 13-2 moves to the left, the resistance decreases, whereas a movement to the right causes increased resistance. Thus, the resistance change in the potentiometer is an indication of the amount of linear movement, and movement direction is indicated by increasing or decreasing resistance. Other commonly used sensors of linear position include the linear-motion variable inductor, the linear-motion variable capacitor, and the linear variable differential transformer.

Bonded Wire Strain Gage

The *bonded wire strain gage* is shown in Fig. 13-3(a). It is used to sense pressure or force. It consists of a length of fine wire arranged in a grid and bonded to paper or plastic. If the strain gage were cemented to the side of a steel column and the

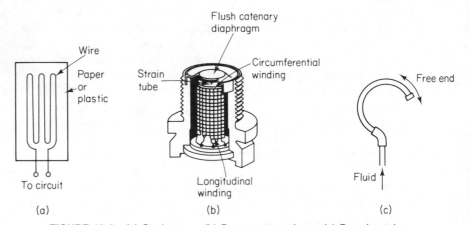

FIGURE 13-3   *(a) Strain gage; (b) Pressure transducer; (c) Bourdon tube.*

steel were deflected, the strain gage would present a changed resistance. For example, as the strain gage is compressed, the length of its wire is shortened and its cross-sectional area is increased. This causes a reduction in its resistance. The bonded wire strain gage can also be used to sense tension and twisting force (torque).

A variation of the strain gage is shown in Fig. 13-3(b). The *pressure transducer* is a sensor that is used to measure variable air or gas pressures. The pressure to be measured is imposed upon the diaphragm. As the pressure increases, the cylindrical strain tube decreases in length while increasing in diameter. These dimensional changes are detected by the wire strain gages bonded directly to the strain tube in such a way that circumferential winding resistance increases while the longitudinal winding resistance decreases. These resistance changes are made as nearly proportional to the applied pressure as possible.

Bourdon Tube

The pressure exerted by a fluid (liquid or gas) can be sensed by a *Bourdon tube,* shown in Fig. 13-3(c). The tube is usually made of thin, springy metal in the form of an incomplete circle. The bottom end is fixed and the free end is sealed. As pressure in the tube is increased, its circular shape tends to straighten out. This causes a movement of the free end that, for example, could be used to drive the arm in a linear-position potentiometer. The free end could also be connected to a pointer that indicates pressure on a scale. The simple single-curve Bourdon tube illustrated in Fig. 13-3(c) can be modified into a multiple-curve type that provides more free-end movement for a given pressure change.

13-3 TEMPERATURE AND LIGHT SENSORS

Thermocouple

If two dissimilar metal wires are joined, a low voltage dependent on temperature is generated across the open ends of the two wires. This is known as the *Seebeck effect.* Such a device is called a *thermocouple* and is used as a temperature sensor. Although any two different metals can be used to create a thermocouple, the most common variety uses iron and constantan (an alloy of copper and nickel). A graph of the voltage generated versus junction temperature is shown in Fig. 13-4. As can be seen, the voltage–temperature relationship is a linear one. The voltage developed

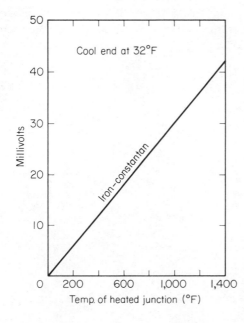

FIGURE 13-4 *Voltage versus temperature for an iron-constantan thermocouple.*

393

across the cool ends is measured so that the temperature of the hot end can be determined. If necessary, the millivolts generated can be amplified (usually by an operational amplifier) so that it is able to actuate some control device.

Thermistor

The *thermistor* is a semiconductor of ceramic material made by sintering mixtures of various metallic oxides. Unlike metals, their resistance drops with temperature increases. The resistance drop is quite large, going from thousands of ohms to several ohms for a change from 0°C to 100°C in a typical unit. The thermistor is obviously a very sensitive device. They come in many sizes and shapes, ranging from tiny beads a few thousandths of an inch in diameter, to washers about an inch in diameter.

Thermostat

A *thermostat* is a device that actuates electrical contacts or produces physical movement as a result of temperature changes. It is based upon the principle that metallic materials change their dimensions when temperature is changed. Since different metals respond differently to temperature change, a bending action is caused when two dissimilar metals are joined together and experience a temperature change. This is known as the *bimetallic effect.*

Consider the thermostat shown in Fig. 13-5. It is the type used in many home-heating systems to maintain the desired temperature. The curved portion is made up of two different metals. A temperature increase causes the two metals to expand at a different rate and thus its free end (at *A*) to move to the right. If point *A* moves away from the temperature adjust control, an electrical circuit is opened that causes the furnace to turn off.

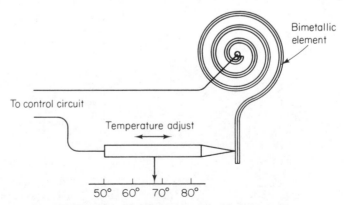

FIGURE 13-5 *Thermostat for home-heating system.*

Photovoltaic Cells

Semiconductors under certain conditions are found to generate a voltage that is proportional to the amount of light striking their surface. These devices are called *photovoltaic cells,* or sometimes *solar cells.* They can be used in control systems that are to be a function of illumination or ambient light. There is also considerable interest in these devices as a source of electrical energy. The symbol for a photovoltaic cell is shown in Fig. 13-6(a). An example of an automatic control system using a photovoltaic

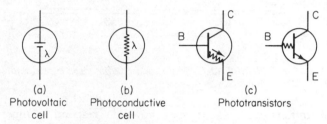

(a)
Photovoltaic
cell

(b)
Photoconductive
cell

(c)
Phototransistors

FIGURE 13-6 *Light-sensor symbols.*

cell is in a camera that has its shutter aperture size controlled by the amount of light from the scene to be photographed.

Photoconductive Semiconductor Devices

Certain semiconductor materials exhibit a reduced resistance proportional to increased light intensity striking them. These devices are termed *photoconductive cells* and are very popular (and inexpensive) light sensors. The symbol for a photoconductive cell is shown in Fig. 13-6(b). An example of a simple control system using a photoconductive cell is a TV set that has its picture contrast automatically adjusted based upon the amount of light in a room.

The photoconductive effect is also used in a special class of transistors called *phototransistors.* These transistors are packaged with a small lens focusing the light on either the base or emitter lead. The transistor conduction, and thus current flow, are in direct proportion to the amount of light. The phototransistor can be made quite small and is more sensitive than the photoconductive cell. They are frequently employed in computer punched-card machines, where the light shining through a hole in the card (or tape) may actuate a counter or some other device. The symbols for a phototransistor are shown in Fig. 13-6(c).

This same effect is used to good advantage in the *photodiode.* Recall that the reverse current flow of a diode is quite low. If, however, the junction is exposed to light, the current rises in almost direct proportion to the light intensity. This effect is shown by the current versus reverse voltage curves for various levels (footcandles) of light in Fig. 13-7(a). The symbol for a photodiode is shown in Fig. 13-7(b). The photodiode does not offer increased sensitivity over the photoconductive cell that

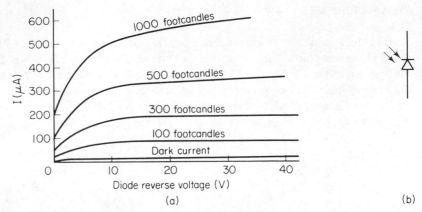

FIGURE 13-7 *Photodiode characteristic curves and schematic symbol.*

the phototransistor does. It does, however, respond to much faster light changes than the photoconductive cell does—a feature it shares with the phototransistor. Thus, the diode and transistor devices can typically respond to light fluctuations of 1 MHz, whereas the photoconductive cell is limited to perhaps several hundred hertz.

13-4 OPEN-LOOP CONTROL SYSTEMS

The simple control system shown in Fig. 13-8 is monitoring two important aspects on a production line—the number of objects passing by and the rate (frequency) of passage. This is an example of an *open-loop control system,* in that no means for the results of the count or frequency operations is made to feed back and control the production line. Rather, this *process control system* (it is involved with controlling a production *process*) relies upon human monitoring to ensure that the process is properly controlled. For instance, a production worker might periodically monitor the number of objects passing by and when a prescribed number had been reached,

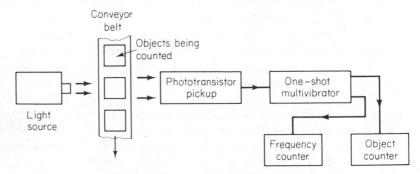

FIGURE 13-8 *Open-loop control system.*

shut the line down. This might be done to facilitate a modification to the product at an earlier stage of production for a different customer. Thus, human intervention "closes the loop" so to speak, in this open-loop system.

The production worker would also monitor the rate of objects passing by. This rate is simply the frequency of objects passing by per unit time. If the frequency counter were reading objects per minute, there might be an allowable range for suitable operation of this production line. For example, a nominal speed of 20 per minute with a range of from 19 to 21 may be acceptable. If the allowable limits were not met, the production might be shut down to determine the problem and make necessary repairs.

The light source in Fig. 13-8 is blocked from reaching the phototransistor whenever an object is passing by. This creates a pulse output that is "cleaned up" by the one-shot multivibrator shown. The one-shot's output is then used to drive a counter and frequency meter to provide the readouts. Commercially available frequency counters often include a count mode so that one unit switching back and forth between count and frequency may suffice.

13-5   CLOSED-LOOP CONTROL SYSTEM

The open-loop process control system described in Section 13-4 requires human monitoring for proper operation. In many process control systems automatic control is desirable. In this case a controller senses a condition and automatically causes the necessary change to occur. For instance, if the production line were going too slowly, it might send out a command signal to speed up the motor driving the line. This would then be an example of a *closed-loop system*, as illustrated in Fig. 13-1.

An automatic closed-loop control system of medium complexity exists in most homes. It is the home-heating system and a block diagram for a typical installation is shown in Fig. 13-9. It is a natural gas, forced-air unit, but the basic concepts to be described certainly are easily applied to oil and/or hot-water variations.

The system shown in Fig. 13-9 includes a source of natural gas. The gas comes in past a manual on–off control. A small gas line feeds to the pilot light, which must be constantly burning. The main gas line comes in to an electric gas valve which is controlled by the controller. The thermostat in the house is a sensor, often an adjustable bimetallic thermostat such as the one shown in Fig. 13-5. It provides a contact closure to the controller when the house temperature has fallen below the set point. When that occurs, the controller actuates the electric gas valve, which opens and causes the burner to ignite (from the pilot-light flame). If for some reason the burner does not ignite after a short period of time (perhaps due to a blown-out pilot light or a failed gas valve), a burner sensor detects that and the controller causes the whole system to shut down.

Assuming that the burner does ignite, another sensor determines when the metal plenum, being heated by the burner, has reached a suitable temperature. Once the controller has received that signal, it turns the blower on. The blower receives air

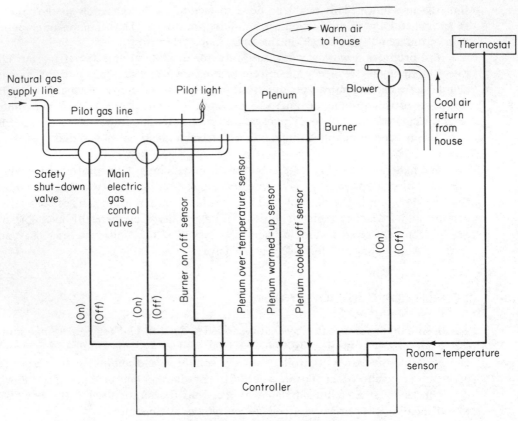

FIGURE 13-9 *Home-heating system.*

from the house cold-air return ducts, blows it over the hot metal plenum, and sends it into the house as warm air. Once the house thermostat senses that the home has reached the proper temperature, it sends a signal (open contacts) to the controller which causes the main electric gas valve to shut and thereby turns the burner off. The blower keeps operating, however, until another sensor determines that the plenum is cooled off. This extracts the last possible heat from each cycle of operation.

If extremely cold weather or a failure cause the burner to stay on long enough to heat up beyond a safe level, the plenum overtemperature sensor sends a signal to the controller which shuts off the main gas line safety valve. This extinguishes the burner and allows the plenum to cool down.

13-6 COMPUTER CONTROL SYSTEMS

The home-heating control system just described is moderately complex. In some cases, the complexity or special requirements of a system makes the use of a computer (full size, mini, or micro) desirable. This is especially true now that microcomputers with high capability are available at very low cost. When using computer control, the sensors make up the computer's input, and the actuators (motors, electric valves, relays, etc.) are connected to the computer's output. The computer is able to make intelligent decisions based upon the condition of a number of sensors, as prescribed by its programmed instructions. The ease with which programs can be changed provides an added advantage to computer control.

An example of computer control is the use of computerized thermostats, which can cut home heating and cooling cost by close to 30%. The extra features afforded compared to the simple bimetallic unit used in the previous section include:

1. Automatic control of two, three, or more set-back and set-up temperature periods a day.

2. Provides a different schedule of times and temperatures for weekdays and weekends.

3. Digitally shows time, day, and actual and desired temperatures.

4. Allows you to override the set-back/set-up program by flipping a switch.

The set-back/set-up schedule for a typical weekday might go like this. At 6:30 A.M., the thermostat actuates the furnace to raise the house temperature from 58 to 68°F so that you will be comfortable when you awaken at 7:00. At 8:30, just before the last member of the family leaves for the day, the thermostat sets back to 58°F. At 4:30 P.M., it goes up again to 68°F so that the house is comfortable by 5:00. At 10:30 P.M., when you retire, it sets back again to 58°F. In this example, your furnace is set back (to 58°F) for 16 hours per day, and this can mean a 30% fuel savings as compared to keeping it set to 68°F constantly. The schedule of heat is "programmable" to a different timetable for the weekend. A typical computerized thermostat is priced at about $100 and is programmed via slide switches.

The computer control system just described is a relatively simple one. The process control of industrial production lends itself to extremely complex systems. Direct computer control of petroleum refineries and sheet-steel production are examples of these highly automated process controls. Another example of a rather complex closed-loop computer control is the automotive system introduced in the next paragraphs.

Automotive Computer Control

The computerized thermostat is an example of a rather simple microcomputer control application. A somewhat more demanding example is the computerized control of automobiles. The trend to microcomputer usage was forced by government controls regarding exhaust emissions and fuel mileage requirements. Auto makers can most easily meet these requirements by using a microcomputer air/fuel and ignition regulator system. These *electronic engine control* (EEC) systems use sensors to monitor various parameters—crankshaft, throttle, and exhaust-gas-valve positions, coolant and

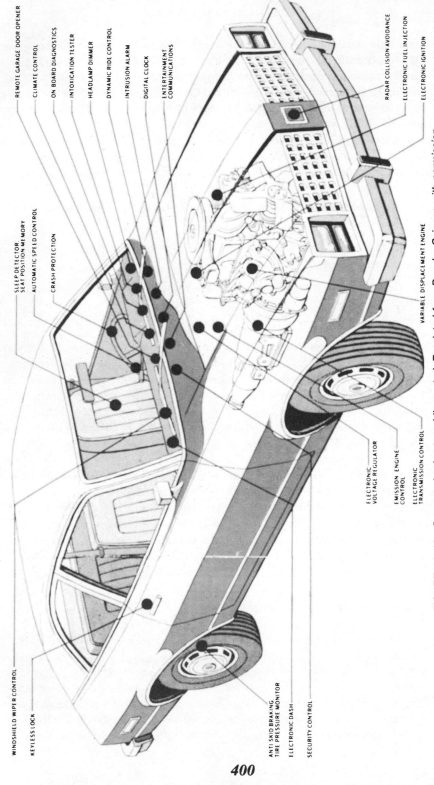

WINDSHIELD WIPER CONTROL

KEYLESS LOCK

ANTI SKID BRAKING
TIRE PRESSURE MONITOR

ELECTRONIC DASH

SECURITY CONTROL

REMOTE GARAGE DOOR OPENER

CLIMATE CONTROL

ON BOARD DIAGNOSTICS

INTOXICATION TESTER

HEADLAMP DIMMER

DYNAMIC RIDE CONTROL

INTRUSION ALARM

DIGITAL CLOCK

ENTERTAINMENT
COMMUNICATIONS

SLEEP DETECTOR
SEAT POSITION MEMORY

AUTOMATIC SPEED CONTROL

CRASH PROTECTION

RADAR COLLISION AVOIDANCE

ELECTRONIC FUEL INJECTION

ELECTRONIC IGNITION

VARIABLE DISPLACEMENT ENGINE

ELECTRONIC
VOLTAGE REGULATOR

EMISSION ENGINE
CONTROL

ELECTRONIC
TRANSMISSION CONTROL

FIGURE 13–10 Computerized automobile control. Reprinted from Popular Science with permission
© 1979, Times Mirror Magazines, Inc.

inlet air temperatures, manifold absolute pressure, barometric pressure, exhaust oxygen, and more. The computer analyzes all this information and sends out signals to actuators that control engine performance. These include controls on the carburetor air/fuel mixture and the ignition timing and dwell. As you drive, this closed-loop control system made up of the sensors, computer controllers, and actuators, may be fine-tuning your car 30 times or more a second. In this way, optimum exhaust pollution control is provided while offering 20% fuel mileage improvement over non-computer-controlled cars.

The computer is not being used to capacity when just performing its EEC function. So once the computer was in the car, manufacturers began giving it additional tasks to perform. Some of those currently incorporated, as well as possible future applications, are shown in Fig. 13-10. One of the more interesting computer control functions shown is the variable-displacement engine concept. In this system, the microcomputer senses when the vehicle does not require much power, for instance when traveling at a constant speed on level or downhill terrain. Under those conditions, some of the cylinders are disabled, so that a four-cylinder car may actually be running on only two cylinders. When the need for greater power is sensed, the other two cylinders are reactivated until they are no longer needed. This approach provides an appreciable increase in fuel economy.

QUESTIONS AND PROBLEMS

13–1–1. Define a control system. Briefly describe two different control systems not mentioned in this chapter.

13–1–2. Provide a block diagram of a closed-loop control system and give a brief description of each block's function. Your descriptions should include definitions of sensor, controller, and actuator.

13–2–3. A linear-motion potentiometer exhibits 50 Ω resistance at position 1 and 125 Ω at position 2, 3 cm to the right. Assuming perfect linearity, what resistance is expected at position 3, 5 cm to the right of position 1?

13–2–4. Explain the principle whereby a bonded wire strain gage is able to sense pressure or a force.

13–2–5. Explain the operation of the pressure transducer strain gage shown in Fig. 13-3(b).

13–2–6. Provide a sketch of a Bourdon tube and explain its ability to detect the pressure of a liquid or gas.

13–3–7. Define the Seebeck effect and explain its usefulness as a sensor. What potential is expected when an iron–constantan thermocouple is 500°F above its open-wire temperature?

13–3–8.   The resistance of copper wire rises very slightly with a temperature increase. Describe how a thermistor's performance contrasts with copper wire as a function of temperature.

13–3–9.   Describe the *bimetallic effect* and explain how it is used in a thermostat.

13–3–10.   Explain the difference between photovoltaic cells and photoconductive cells. Provide the symbol for each one.

13–3–11.   Describe the advantage that a photodiode and phototransistor have as compared to a photoconductive cell.

13–4–12.   Define an open-loop control system. List three commonly encountered open-loop control systems.

13–4–13.   What is a process control system? Using a block diagram, show how the process control system in Fig. 13-8 could be converted from open-loop to closed-loop operation.

13–5–14.   Describe the difference between an open-loop and a closed-loop control system.

13–5–15.   Show, via a block diagram, an improvement of your own choice to the home-heating system shown in Fig. 13-9.

13–6–16.   Explain the advantages that a computer offers as compared to non-computer-controlled systems.

13–6–17.   Describe some of the major benefits that an EEC computer control system provides to automobile operation.

Appendix A

SCIENTIFIC NOTATION

The standard notation for numbers is called *ordinary* notation. It uses *arabic* numerals, and this system has been taught to you since your early childhood days. Examples of this system are the numbers 55 and 7. Calculations using these numbers are easily accomplished. For instance, if you travel at 55 miles per hour (mph) for 7 hours, you have traveled a total distance of

$$55 \text{ miles/hour} \times 7 \text{ hours} = 385 \text{ miles}$$

This computation was easily accomplished. However, in many science and technology calculations, the use of very small or very large numbers makes ordinary notation cumbersome. As an example, consider a radio transmission from a distant space probe. It travels at about 300,000,000 meters per second and is a distance of 470,-500,000 meters from earth. To calculate the time of transmission, we would have

$$470,500,000 \text{ meters} \div 300,000,000 \text{ meters/second}$$

We will not proceed any further with this calculation. Suffice it to say that mathematical operations using ordinary notation with very large (or small) values are difficult.

A better method for dealing with this situation is offered by *scientific notation*. In scientific notation we overcome the inconvenience of ordinary notation by using *powers of 10* to represent large and small values. An example of powers-of-10 usage

403

is the number 2×10^3. That means that the number is 2 times 10 raised to the third power, or

$$2 \times 10^3 = 2 \times 10 \times 10 \times 10 = 2 \times 1000 = 2000$$

A table of powers of 10 and their decimal equivalents is provided in Table A-1. Also included in this table are the prefixes used to represent multipliers commonly used in electronics.

TABLE A-1 *Powers of 10 and prefix.*

Power of 10	Decimal Equivalent	Prefix	Prefix Symbol
10^{-12}	0.000000000001	pico	p
10^{-9}	0.000000001	nano	n
10^{-6}	0.000001	micro	μ
10^{-3}	0.001	milli	m
10^0	1.0		
10^1	10.0		
10^2	100.0		
10^3	1,000.0	kilo	k
10^6	1,000,000.0	mega	M
10^9	1,000,000,000.0	giga	G

Scientific notation involves expressing a number as a product of two numbers. The first is always a number between 1 and 9.999 . . . and the second is a power of 10. The power of 10 is used to "keep track" of the decimal point.

Scientific Notation—Conversion from Ordinary Notation for Numbers above 1.0

1. Shift the decimal point to the left until just one digit remains to the decimal point's left: e.g.,

    377.5   (two places to left)

2. The number that remains is called the *coefficient:* e.g.,

    3.775

3. The number of places the decimal point was moved becomes the *positive* exponent for the power of 10: e.g.,

    10^2 (decimal point was moved *two* places to the left in step 1)

4. The result in scientific notation is the product of the coefficient and power of 10: e.g.,

    3.775×10^2 (scientific notation expression for 377.5)

Scientific Notation—Conversion from Ordinary Notation for Numbers below 1.0

1. Shift the decimal point to the right until just one digit remains to the decimal point's left: e.g.,

$$0.0000173 \quad \text{(five places to right)}$$

2. The number remaining is the coefficient: e.g.,

$$1.73$$

3. The number of places the decimal point was moved becomes the *negative* exponent for the power of 10: e.g.,

$$10^{-5} \quad \text{(decimal point was moved five places to the right in step 1)}$$

4. The result in scientific notation is the product of the coefficient and power of 10: e.g.,

$$1.73 \times 10^{-5} \quad \text{(scientific expression for 0.0000173)}$$

Conversion from scientific to ordinary notation is simply the reverse of the process just shown. To illustrate,

$$3.775 \times 10^2 \quad \text{is} \quad 3.775 \quad \text{or} \quad 377.5$$

and

$$1.73 \times 10^{-5} \quad \text{is} \quad 00001.73 \quad \text{or} \quad 0.0000173$$

Addition and Subtraction Using Scientific Notation

In order to add and subtract using scientific notation, the powers of 10 must be identical. For example, the following operation could not be accomplished immediately because the powers of 10 are not the same.

$$6.22 \times 10^3 + 5.1 \times 10^2 = ?$$

In order to perform this operation, one of the two numbers must be changed. To keep its value unchanged, increasing the power of 10 by one requires that the decimal point in the coefficient be moved one place to the left. Similarly, a power-of-10 decrease by one means that the decimal point must be moved one place to the right. Thus,

$$6.22 \times 10^3 = 62.2 \times 10^2$$

and

$$5.1 \times 10^2 = 0.51 \times 10^3$$

Therefore, this addition can be accomplished as

$$
\begin{array}{cc}
6.22 \times 10^3 & 62.2 \times 10^2 \\
+ \underline{0.51 \times 10^3} \quad \text{or} \quad & + \underline{5.1 \times 10^2} \\
6.73 \times 10^3 & 67.3 \times 10^2 = 6.73 \times 10^3
\end{array}
$$

The subtraction process $3.65 \times 10^{-3} - 7.2 \times 10^{-4}$ can be accomplished as

$$
\begin{array}{r}
3.65 \times 10^{-3} \\
- \underline{0.72 \times 10^{-3}} \\
2.93 \times 10^{-3}
\end{array}
$$

since $7.2 \times 10^{-4} = 0.72 \times 10^{-3}$.

Multiplication and Division Using Scientific Notation

Multiplication in scientific notation is accomplished by multiplying the coefficients and *adding* the exponents in the powers of 10. For example:

$$
\begin{aligned}
3.9 \times 10^5 &\times 6.4 \times 10^2 \\
&= (3.9 \times 6.4) \times 10^{5+2} \\
&= 24.96 \times 10^7 \\
&= \underline{2.496 \times 10^8}
\end{aligned}
$$

and

$$
\begin{aligned}
1.95 \times 10^6 &\times 9.7 \times 10^{-3} \\
&= (1.95 \times 9.7) \times 10^{6+(-3)} \\
&= 18.915 \times 10^3 \\
&= \underline{1.8915 \times 10^4}
\end{aligned}
$$

Division in scientific notation is accomplished by dividing the coefficients and subtracting the exponents in the powers of 10. For example:

$$
\begin{aligned}
5.1 \times 10^{12} &\div 1.7 \times 10^7 \\
&= (5.1 \div 1.7) \times 10^{(12-7)} \\
&= \underline{3 \times 10^5}
\end{aligned}
$$

and

$$
\begin{aligned}
8.4 \times 10^{-5} &\div 9.2 \times 10^{+3} \\
&= (8.4 \div 9.2) \times 10^{(-5-3)} \\
&= 0.913 \times 10^{-8} \\
&= \underline{9.13 \times 10^{-9}}
\end{aligned}
$$

The convenience of scientific notation is made apparent by now completing the example mentioned at the beginning of the appendix.

EXAMPLE A-1

A radio transmission from a space probe is picked up by an earth station. The space probe is 470,500,000 m from earth and the radio waves travel at about the speed of light, 300,000,000 m/s. Calculate the time of transmission.

Solution:

$$470{,}500{,}000 \text{ m} = 4.705 \times 10^8 \text{ m}$$

$$300{,}000{,}000 \text{ m/s} = 3.0 \times 10^8 \text{ m/s}$$

$$\text{time} = \frac{\text{distance}}{\text{velocity}} = \frac{4.705 \times 10^8}{3.0 \times 10^8}$$

$$\simeq 1.57 \times 10^0 \text{ s}$$

$$= \underline{1.57 \text{ s}}$$

INDEX

P

Parallax error, 44
Parallel circuits, 15–20
Parallel data transfer, 341
Parallel LC circuits, 81–83
Paraphrase amplifier, 204
Parity, 331
Peak reverse voltage (PRV), 124
Peak-to-peak voltage, 38
Period, 38
Permanent Magnet, 87
Permeability, 90
Permittivity, 55–56
Phase, 64
Phase inverting, 170, 204
Phase modulation (PM), 267
Phase shift, 65, 72
Phase shift oscillator, 275–77
Photoconductive devices, 395–96
Photovoltaic cells, 395
Pierce oscillator, 275
Pinch-off, 176
PNP, 155
Poles, 88
Positional-value system, 321
Positive feedback, 269–70
Positive logic, 316
Potential difference, 4
Potential energy, 268
Potentiometer, 30–31
Power, 32
Power amplifier, 194–207
Power supplies, 131–48, 190–94
Powers of ten, 404
Presettable counter, 339
Pressure transducer, 392
Primary winding, 107
Program, 360
Program counter (P.C.), 366, 375
Propagation delay, 296
Pulsating DC, 37
Pulse width, 297
Push-pull amplifier, 201–07

Q

Q, 78–79
Q Point, 158, 164
Quality factor (Q), 78–79
Quiescent current, 157–58

R

Radio frequency amplifier, 208
Random access memory (RAM), 370, 382–83
RC circuits, 60–65
RC filters, 66–69, 208–11
RC phase shift oscillator, 275–77
Reactance, 59
Read, 362
Read-only memory (ROM), 370, 381–82
Reciprocity, 287
Rectification, 122–30
Rectifier, 122
Register, 341–43
Regulation, 134–37, 190–94
Relative permittivity, 55–56
Relay, 113
Resistance, 6
Resistive/capacitive (RC) circuits, 60–65
Resistor, 7
Resistor color code, 8–9
Resistor failure, 14
Resistor measurement, 42–43
Resolution, 386–87
Resonance, 74–78
Resonant frequency, 74, 269
Revolutions slip, 104, 105
Rheostat, 31
Ripple, 96, 131
Ripple counter, 339
Rise time, 297
ROM, 370
Root-mean-square (RMS), 40, 123
Rotor, 94
RS flip-flop, 309

S

Satellite sky wave, 288, 290
Sawtooth, 38
Schmitt trigger, 348
Scientific Notation, 403–07
Secondary winding, 107
Seebeck effect, 393
Selective, 79
Selectivity, 280
Semiconductor, 2
Sensitivity, 280
Sensor, 390–97
Serial data transfer, 341
Series circuit, 10
Series motor, 103–04